FLUID ANALYSIS FOR MOBILE EQUIPMENT

Condition Monitoring and Maintenance

DIEGO NAVARRO & BLAINE BALLENTINE

Edited by Michael D. Holloway

INDUSTRIAL PRESS

Industrial Press, Inc.

1 Chestnut Street
South Norwalk, Connecticut 06854
Phone: 203-956-5593
Toll-Free in USA: 888-528-7852
Email: info@industrialpress.com

Authors: Diego Navarro and Blaine Ballentine
Title: Fluid Analysis for Mobile Equipment: Condition Monitoring and Maintenance
Library of Congress Control Number: 2023942927

ISBN (print)   978-0-8311-3691-8
ISBN (ePUB)   978-0-8311-9638-7
ISBN (eMOBI)   978-0-8311-9639-4
ISBN (ePDF)   978-0-8311-9637-0

Publisher/Editorial Director: Judy Bass
Copy Editor: James Madru
Compositor: Patricia Wallenburg, TypeWriting
Proofreader: David Johnstone
Indexer: ARC Indexing

books.industrialpress.com
ebooks.industrialpress.com

CONTENTS

Foreword by Mike Vorster...xxi

Introduction xxiii

Acknowledgments...xxv

CHAPTER 1 **Condition-Based Maintenance.................................1**

Wear No Matter What! 2

Wear: Does Size Matter? 3

Wear: Does Technology Matter?..3

Do We Really Control Wear? 4

Maintenance Paradigms..4

    None of These Creates Equipment-Saving Opportunities 4

Types of Maintenance 5

    Flaws of Scheduled Maintenance and RAF..........................5

    Basic Principles of CBM 6

How Can We Listen Better? 7

How Can We See Better? 7

    Oil Analysis 7

    Coolant Analysis 8

    Fuel Analysis 9

The Unseen World..10

    Maintenance Is a Matter of Visibility 12

    When Do the Diagnostics Take Place?.............................12

    Looking at Wear with CBM Eyes 13

The Performance–Failure Curve 14

CBM versus Scheduled Maintenance: Cultural Differences........... 15

Impact of CBM on Operating Costs 15

Inspections.. 16

Operator 17

Telematics... 17

Fluid Sensors.. 19

Managing the Data.. 20

Machine Health Correlations 20

Root-Cause Analysis and Failure Scene Investigation 22

Most Popular RCA Methodologies................................ 23

Conclusion 25

CHAPTER 2 **Lubrication** 27

Lubrication Fundamentals 27

Tribology 27

Lubricant Functions 27

Petroleum... 28

API Base Oil Groups... 29

Solvency 31

Viscosity 32

Viscosity Index 32

SAE J300 Specification and the Introduction of Additives............ 32

SAE J300 Standard for Temperature and Viscosity.................. 35

Gear Lube Viscosity.. 35

ISO Viscosity Grades... 36

API Diesel Engine Categories................................... 37

Gasoline Engine Oil Classifications 39

Oil Breakdown 40

Contaminants and Filtration.................................... 41

Types of Lubrication 41

Hydrodynamic Lubrication Example 42

Boundary and Dynamic Lubrication 43

Herztian Forces.. 44

Elastohydrodynamic Lubrication................................ 44

Additives and Their Functions.................................... 44

Foam Inhibitors.. 46

Additive Antioxidant Synergies 46

Zinc Dialkyl Dithiophosphate and Tricresyl Phosphate 46

Dispersants and Detergents..................................... 47

Additives versus Load 47

Additive Synergies versus Temperature 48

Additives Fight for Surface. 49

Water and Rust Inhibitors 49

Dispersants for Soot. 50

Copper Passivators. 50

Etching versus Physical Erosion 51

Friction Modifiers 51

How Additives Show Up in Oil Analysis 52

Lubricants and Additives Used in Mobile Equipment. 52

Typical Signatures of Various Lubricants 53

Hydraulic Fluids 53

Multi-Viscosity Fluids 55

Tractor Fluid and Automatic Transmission Fluid Signatures. 55

Gear Oils Signatures 56

Engine Oil Signatures 57

Viscosity Decline at 40°C. 58

Physical Properties. 59

Lubricant Compatibility. 61

Where Lubrication Happens. 63

Lubricant Optimization 63

Lubricant Storage 67

Greases. 70

Grease Classification 71

Grease Temperature Performance 71

Grease Colors 72

Performance Specifications for Automotive Use
(Mobile Equipment Included) 73

NLGI High-Performance Grease Specifications for
Mobile Equipment (2021) 74

HPM Core 75

HPM + WR (Water Resistance). 77

HPM+CR (Saltwater Corrosion Resistance). 77

HPM+LT (Low Temperature). 78

HPM+HL (High Load) 78

Grease Selection 79

Grease Compatibility 80

Choosing the Best Option for a Fleet. 81

Example of a Greasing Lubrication Decision 81

Greases for High-Load and High-Shock Applications. 82

Automatic Lubricators 83
Conclusion 83

CHAPTER 3 **Contamination** **85**
What Is Contamination? 85
What Are Those Contaminants? 85
Particle Sizes and Visibility 86
Filtration 92
Heat and Air. 96
Things to Look for during a Visual Inspection of a Machine 98
Metals as Catalysts 98
Static Current 99
Water Content: Parts per Million and Percentage of Saturation 99
Visibility of Water. 100
Visibility of Particles 102
Wear Metals and Contaminant Sources 102
Measuring and Counting 105
Particle Counts Can Hide Something Else 106
Cleanliness versus Life of Components: British Hydrodynamics. 107
Particle Counts for Engines 109
Soot in Engines 109
Cleaning Soot in Engines 110
Micropatch and Microscope: Creating the Patch 110
ISO Code Visibility: Quantitative Code Catalog 111
Oil Pump or Dirt Pump? 112
Component Tolerances 112
Internally Generated Contamination 113
Conclusion 114

CHAPTER 4 **Hydraulic Cleaning Procedures** **115**
Where Does the Debris Collect After a Failure? 115
Intelligent Filter Caddies. 116
How Much of the Old Fluid Remains in the System after Flushing?. 119
Identifying Type of Contamination and Cleaning Procedures
in Hydraulic Systems: Case Discussion 120
The Machine Shows Dirty Fluid and High Particle Counts 120
The Hydraulic System Is Contaminated with Water. 121
The Fluid Is Oxidized or Has a High TAN 122
The Hydraulic System Shows Mixed Fluid 123

A Component Is Wearing Out Slowly, and the Fluid Is
Showing Growing Metal Readings. 123
A Component Has Failed and Has Contaminated the Whole
System (Machine Is Not Operable). 125
The System Is Contaminated with Varnish 126
The System Shows Blackened Fluid 127
Conclusion 128

CHAPTER 5   **Oil Sampling** **129**
Sampling Methods 130
Sampling Valve Method. 131
Vacuum Pump Method 132
Bottle Types 133
Submitting Samples 133
The Laboratory. 134
Sampling Points 134
Sampling Valve Installation. 135
Sampling Valve Dead Ends 135
Conclusion 136

CHAPTER 6   **Lubricant Testing**. **137**
Oil and Fluid Testing 137
Wear Metals (ICP AES Spectrometer ASTM D5185,
D6595 [RDE]) 138
Viscosity Tests (ASTM D445/D727) 140
Water Content Tests (Karl Fischer ASTM D6304/E203) 143
Alternative Methods of Measuring Water Content (ASTM D2412). . 144
Soot: Shimadzu Attenuated Total Reflection (ATR) Method
(IR/ASTM D77844). 145
Glycol by Sodium and Potassium Readings 146
Engine Oil Fuel Dilution 146
Particle Counts (ISO 11500) 148
Total Base Number (TBN) Test (ASTM D2896 and D4739) 153
Oxidation (ASTM 943, D5846, D8048 FTIR [Fast Fourier
Transform Infrared] Spectroscopy). 155
Nitration (ASTM D943) 157
Sulfation (ASTM D7415) 157
PQ Index (ASTM D8184). 158
Grease Testing 159
ASTM Grease Testing Methods. 160

Grease Thief Tests.. 163
Conclusion 164

CHAPTER 7 **Oil Analysis Basics................................... 165**
Oil Analysis Interpretation Knowledge Fundamentals.................. 165
Why We Need Oil Analysis..................................... 166
What Can We Measure with Oil Analysis? 166
Oil Analysis Formats 167
Horizontal versus Vertical Displays.............................. 168
Required Information 169
What Should We Test in Engines? 169
1. Dirt 171
2. Glycol 173
3. Viscosity.. 175
4. Soot.. 175
5. Fuel 176
6. TBN and TAN 176
7. Metals.. 176
8. Additives 178
9. Oxidation... 179
10. Sulfation and Nitration................................... 179
11. Lab Comments... 179
12. Oil Type, Brand, and Hours of Operation 179
Oil Analysis for Hydraulics.. 179
1. Particle Counts... 181
2. Dirt 181
3. Moisture.. 181
4. Wrong Fluid/Mixes 181
5. Metals.. 181
6. Acidity 182
7. Degradation, Aging.. 182
8. Temperature Records...................................... 182
9. Fluid Changed 182
10. Total Hours and Fluid Hours............................... 182
Oil Analysis for Power Shift Transmissions and Gearboxes.............. 183
1. Dirt 184
2. Moisture.. 184
3. TAN.. 184
4. Fluid Signature... 184
5. Oxidation... 184

6. Viscosity Changes 184

7. Metals 184

8. PQ Index 185

9. Particle Counts 185

10. Comment Box 185

11. Fluid Changed 185

12. Information on Type of Fluid and Hours 186

Oil Analysis for Axles and Final Drives 186

1. Dirt 187

2. Water 187

3. Friction Modifier/TAN 187

4. Additive Signature 187

5. Oxidation 188

6. Viscosity 188

7. Metal Generation 188

8. PQ Index 190

9. Oil Hours Information 190

10. Comments from the Lab 190

Lab Website and Results 190

Some Rules Can Help 191

Elements for a Good Oil Analysis Interpretation 192

Logic Flow Path for High Viscosity on a Diesel Engine 192

Logic Flow Path for High Copper in Hydraulics Example 193

Maintenance Applications 194

Conclusion 195

CHAPTER 8   **Wear Tables and Standard Deviations** **197**

What Are Wear Tables? 197

Why Do We Need Wear Tables? 197

How Do We Calculate Wear Tables? 197

What Is Standard Deviation? 197

Types of Standard Deviations 198

Which Standard Deviation Is More Appropriate for an Equipment Fleet? . . 198

Population Standard Deviation 198

Sample Standard Deviation 198

Parametric Bell Curve 200

Example Applied to a Fleet 201

Parametric Curve and Cumulative Curve 201

Nonparametric Curve for High Reference Values 202

Nonparametric Curve for Zero Reference Values 202

Measurements Handled without Standard Deviations 203

Mean or Median Values? 203

Why Do We Need So Many Wear Tables? 204

Geography and Utilization. 204

Can We Use Wear Tables from One Country in Another Country? . . 204

Collecting Information 204

Using Excel to Calculate Standard Deviations. 205

Cleaning the Data. 205

Normalizing Time-Dependent Elements. 207

Establishing Abnormal and Critical Values. 208

Wear Tables for Harsh Applications 208

Minor Metals and Contaminants. 208

Metals Not Typically Used by Mobile Equipment 208

Handling of Potassium (K) and Sodium (Na) 209

Example of a Wear Table 209

Example of a Wear Table for Elements That Do Not Need

Standard Deviations. 209

Table for Particle Counts, PQ Index, TBN, TAN, and

Viscosity Changes. 210

Conclusion 210

CHAPTER 9   **Hydraulic Fluid Analysis** **211**

Keeping the Balance 212

Basic Cleanliness Levels 212

Hydraulic Contamination Sources 212

Testing. 213

Wear Thresholds 213

Case Discussion. 213

1. Dirt and High Particle Counts in a Large Production Bulldozer

with Combined Hydraulics/Hydrostatics 213

2. Silt Accumulation in a Production Hydrostatic Bulldozer with

Combined Hydraulics and Hydrostatics. 215

3. Water Contamination in a Construction-Sized Excavator. 216

4. High Particle Counts: Gelling and Mixing in a

Construction Excavator 218

5. High Copper Production (Etching) in a Hydrostatic Crawler 219

6. High Copper Generation (Progressive Component Failure)

in a Large Production Excavator. 221

7. High Iron Particle Production in a Medium-Sized Excavator. 222

8. Fluid Oxidation and Thermal Records in a Medium-Sized
   Excavator 223

9. Operation with Bypass Filtration in a Large Production
   Excavator 225

10. Hydraulic Fluid Foaming in a Construction-Sized Excavator. 226

11. Atypical Particle Counts in a Four-Wheel-Drive Loader 228

12. Cheese Curding in Biodegradable Hydraulic Fluid in a
    Production Excavator. 229

13. Hydraulic Hammer and Dirt in a Medium-Sized Excavator 230

14. Varnish in a Construction-Sized Excavator 232

15. Hydraulic Microdieseling in a Large Excavator. 234

16. Aluminum and Titanium Readings in a Backhoe 235

17. Sluggish Hydraulics in Winter in a Conversion Excavator. 236

18. Sudden Failures in a Production Excavator: Can Oil
    Analysis Foresee Them?. 237

19. Life of Components versus Lubricant Choices. 239

20. Contamination Transfer from Scraper to Tractor
    in a Large Hydrostatic Crawler 241

21. Increased Water Readings after Fluid Change
    in a Production Excavator. 242

Conclusion 244

CHAPTER 10 **Engine Oil Analysis** **245**

Case Discussions. 246

1. Four-Wheel-Drive (4WD) Loader with Coolant Leak
   Through the Oil Cooler. 246

2. Backhoe Loader Coolant Leak Through Liner Pitting 247

3. Excavator, Coolant Leak through the Exhaust Gas
   Recirculator (EGR) 248

4. 4WD Loader with Water Contamination 250

5. 4WD Loader with High Soot. 250

6. 4WD Loader with Fuel Dilution. 253

7. Backhoe with Oil Thickening Due to Biodiesel 255

8. Bulldozer with Low TBN and High TAN. 256

9. 4WD Loader with Copper Passivation 257

10. Articulated Dump Truck (ADT) with Aluminum Readings
    from Accessory Compressor 258

11. 4WD Loader with High Iron and Copper and Time on Oil 259

12. 4WD Loader with Sulfation and Oxidation. 259

13. Bulldozer with High-Sulfur Fuel 260

14. Backhoe Loader with Fuel Dilution and Soot Combined 262

15. Cummins Engine: Weak Acids. 262

16. Use of Oil Analysis as a Forensic Tool 263

17. Choosing Lubricants for Extended Service Intervals
    (Power Stroke 6.4 and 6.7 Engines with 5W-40 Synthetic Oil) ... 265

Conclusion 270

CHAPTER 11 **Powertrain Oil Analysis** **271**

Components Covered 271

Brief Descriptions of Components 272

Case Discussions 278

1. High Aluminum and Silicon in an ADT Allison HD 4500
   Powershift Transmission 278

2. High Silicon and High TAN in a Backhoe ZF Powershift
   Transmission. 279

3. High Copper Generation in a Motor Grader with a
   Funk Transmission 281

4. High Aluminum in an ADT ZF Powershift Transmission 282

5. High Iron and Copper Readings in a 4WD Loader with
   ZF Powershift Transmission with Converter Locking Clutch..... 284

6. Fluid Mixing in an ADT ZF Ecomat 2 Powershift Transmission .. 285

7. Water and Dirt in an ADT with an Allison HD 4500R
   Transmission. 287

8. Low Viscosity in an ADT ZF Transmission with Integral
   Transfer Case 288

9. Low Viscosity in a Backhoe ZF Transmission with High
   Silicon and Aluminum Readings.............................. 290

10. High Iron Content in an Unfiltered 4WD Loader Axle
    with Multidisc Brakes...................................... 291

11. Water in a Filtered 4WD Loader Axle with Multidisc Brakes 293

12. Lead and Copper in an Axle with Single-Disc Brakes 294

13. Iron Reading/Type of Oil in a Small 4WD Loader ZF
    Common-Sump Multidisc-Brake Outboard Planetary Axle 296

14. Wrong Oil for a Liebherr-Spicer Outboard Final Drive
    with an Inboard Multidisc-Brake Axle....................... 298

15. High Iron in a ZF Backhoe Inboard Final Drive Axle 299

16. Perfect Readings in a ZF Outboard Final Drive Axle............. 301

17. High Iron Readings and Low TAN in a ZF 240 Large
    Loader Axle 302

18. Abnormal Additive Readings in a 4WD Loader ZF 240 Axle 304

19. Wrong Oil and Low TAN in an ADT Dana Axle 305

20. Inconsistent Gear Oil in a Construction Crawler Final Drive 306

21. Internal Leak in a Liebherr Large Construction
    Crawler Final Drive 308

22. High Iron Readings in a Large Excavator Final Drive 309

23. Dirt Entry in Liebherr Hydrostatic Bulldozer
    Splitter Drives/Pump Drives 311

24. High Iron in an Excavator Swing Gearbox Drive 313

25. Water in a Motor Grader Tandem Drive 314

26. High Iron and Chromium in a Forestry Forwarder Bogie 316

27. High Metals in an ADT Nonfiltered Transfer Gearbox 317

28. Oil Analysis for an ADT Filtered Transfer Gearbox 318

Conclusion 320

CHAPTER 12 **Engine Coolant Analysis** **321**

Coolant Development History 321

Types of Coolants 322

How Are Coolants Produced Today? 322

Coolant Requirements 322

How Much Heat Can Coolants Remove? 323

Marine Coolant Operation 323

Specific Heat of Coolants 324

Requirements for New Engine Technologies 324

Coolant Standards 324

Engine Coolant Composition 325

Conventional Coolants and Silicates 325

Coolant Sampling 325

Heavy-Duty Coolants and Extended-Life Coolants 327

Organic Acid Technology (OAT) Coolants 327

Nitrite–Organic Acid Technology (NOAT) Coolants 327

Hybrid–Organic Acid Technology (HOAT) Coolants 327

Heavy-Duty ELC Inorganic Signatures 327

Inorganic Additive Functions 328

Organic Additives in Heavy-Duty Coolants 328

Main Organic Additive Functions 328

How Corrosion Inhibitors Work: Conventional Coolants versus ELCs 329

Anodic Inhibitors 329

Cathodic Inhibitors 329

Adsorption Inhibitors 329

Coolant Freeze Protection 330

Boiling Protection 330

What Are the Issues with Using Only Water as a Coolant? 331

Cavitation 331

Coolant Specifications 331

Conventional Coolants 332

Basic Coolant (High Ph) in ELCs 333

Supplemental Coolant Additives 333

Degradation of Glycol 333

Lab and Field Tests for Coolants 333

    1. pH Value 334

    2. Reserve Alkalinity 334

    3. Glycol Concentration 335

    4. Nitrites 335

    5. Total Dissolved Solids 336

    6. Cumulative Organic Acids (Ultrapressure Liquid
       Chromatography) 336

    7. Ion Chromatography 337

    8. ICP Spectroscopy for Metals 337

    9. Water Quality (Hardness, Chlorides, and Sulfates) 337

On-Site Coolant Testing 338

    1. HOAT and OAT, pH, Concentration, and Organic Acid Content . . . 338

    2. pH with Pocket Tester 338

    3. NOAT, pH, Concentration, Nitrites, and Molybdates 339

    4. All Coolants: TDSs with a Pocket Tester 339

    5. Water Quality Test (Hardness, Chlorides, and pH) with
       Three-Way Strips 340

Case Discussions 342

    1. Low Freeze Protection in an Excavator with Low Glycol Content 342

    2. Overconcentration in a CAT Off-Road Truck 343

    3. Low pH, High Acidity in a CAT SR4B Generator Set 344

    4. High pH in an Excavator with OAT Coolant 344

    6. High Lead Readings in a Motor Grader 346

    7. High Glycol/High SCA Concentrations in a Production
       Bulldozer 347

    8. Low SCA Concentration in a European Tree Harvester 348

    9. High Copper Readings in a Construction 4WD Loader 349

    10. High Iron Readings and Cavitation in a Production Crawler 351

    11. Hazy Brown and High TDSs in a Navistar Truck 352

12. Corrosion in a Detroit Diesel V16 149 Two-Stroke
   Diesel Engine 353
Conclusion 355

CHAPTER 13 **Fuels and Fuel Analysis**. **357**
   Fuels 357
      Refining Process 359
   Fuels from Coal 360
      Fischer–Tropsch Process 360
      Types of Diesel Fuels. 361
      Diesel Fuel Viscosity 361
   Biodiesel. 361
      Biodiesel Production. 362
      Energetic Values of Biodiesel Blends. 363
   Gasoline 364
      Detonation (Knocking) in Gasoline Engines. 366
      Preignition in Gasoline Engines. 367
      Octane Rating 368
      Engine Compression Ratios 368
   Ethanol. 369
      Ethanol Blends and Disclaimers. 370
      Propane 370
      Natural Gas. 370
      Fuel Energy Values. 371
      Emission Control and Stoichiometric Engines 371
   High-Pressure Injection Systems. 372
      Fuel Filtration 373
      Sulfur in Fuels Around the World 374
   Diesel Fuel Additives 374
      Cetane Improver. 375
      Diesel Winter Additives 376
   Additives for Gasoline 377
      Detergents 377
      Friction Modifiers 377
      Corrosion Inhibitors, Demuslsifiers, and Solvents 378
      Octane Boosters 378
   Fuel Sampling. 378
      Fuel Farm Sampling. 378
      Machine Fuel Sampling 379

Fuel Farm Tank Maintenance 380

Water Condensation 381

Long-Term Fuel Storage Package................................ 382

Field Test: Water-Finding Paste 382

Machine Fuel Tank Maintenance................................. 383

Sample Information Form (SIF) 383

Fuel Sampling Bottles 384

Diesel Fuel Testing 384

1. Flash Point (ASTM D93)................................... 386

2. Water and Sediment (ASTM D2709) 386

3. Water (ASTM 6304) 387

4. Fuel Cleanliness, Particulate Matter (ASTM D2276 or D5452) .. 388

5. Distillation (ASTM D86)................................. 388

6. Cetane Number (ASTM D613-18a) 389

7. Cetane Index (ASTM D976/D4737) 390

8. API Gravity (ASTM287/D1298) 390

9. Biodiesel Content by Fourier Transform Infrared (FTIR)
   Spectroscopy (ASTM D7371) 391

10. Sulfur (ASTM 2622/5453) 392

11. Acid Number (ASTM D644) 392

12. Cold Filter Plugging Point (ASTM D6371) 393

13. Cloud Point (ASTM D2500)................................ 393

14. Pour Point (ASTM D97 and D5949) 394

15. Viscosity (ASTM 445 at 40°C) 394

16. Ash Content (ASTM D482) 395

17. Copper Strip Corrosion (ASTM D130)..................... 395

18. Carbon Residue (ASTM 524)............................... 396

19. Lubricity (ASTM D6079) 396

20. Long-Term Storage (ASTM D4625) 397

21. High-Temperature Stability (ASTM D6468) 397

22. Conductivity (ASTM D2624/4308).......................... 398

23. Bacteria (ASTM D6469)................................... 399

Fuel Analysis 399

Case Discussions... 400

1. Water-Excavator with Fuel Pump Failures 400

2. Hard Cold Starting in a Production 4WD Loader 401

3. Injector Failures in a Motor Grader 402

4. Heavy Knocking and Smoke in a Backhoe 403

5. Fuel Starvation During Winter in a 4WD Loader............ 404

6. Turbocharger Failures in a Production Bulldozer.......... 405

7. Filter Plugging in a Production Bulldozer. 406

8. Engine Smoking in a Tracked Tree Harvester 407

9. Injection Pump Failure in a Grapple Skidder. 408

10. Quality Check—Fleet Operation 409

Conclusion 410

CHAPTER 14 Diesel Exhaust Fluid Analysis. 411

Why the 32.5% DEF-to-Water Ratios? 412

Field Testing. 413

Lab Testing. 414

Why Those Tests?. 415

Biuret 415

Aldehydes 415

DEF Analysis 416

Case Discussions 416

1. Engine Deration in a Production 4WD Loader. 416

2. Engine Derating Intermittently in a Utility Bulldozer 417

3. SCR Failure in a Production Excavator 418

4. Fault Codes in a Dump Truck 419

5. Quality Check: Fleet Inventory. 420

Conclusion 421

APPENDIX A Acronyms 423

APPENDIX B Machine Profiles and Application 427

Index 433

FOREWORD

I grew up around construction equipment. Early school days were spent tagging along with the mechanics. Covered in oil and grease up to their elbows, they were my heroes. They did real work—they fixed things. And there were a lot of things to fix.

High school summers were spent setting up preventive maintenance checklists. We had learned that it was better to change and adjust things before they broke and that if we did this properly, we could save a lot of downtime. We also, I suppose, started with condition-based maintenance because when we wiped the oil from the dip stick and noticed that it was darker than usual, we made a mental note to drain and fill the crankcase sooner rather than later.

Things certainly have changed. Best-in-class companies now spend most of their repair and maintenance dollars on preventive and condition-based maintenance. Reliability has changed beyond recognition, and we have eliminated most of the waste associated with unplanned failures and emergency repairs. Repair-before-failure decisions are no longer an act of faith. We have the knowledge, data, and insights needed to listen to our machines and hear what they are telling us.

This book is a major contribution to both the science and art of fluid analysis. It is destined to become the cornerstone of every successful condition-based maintenance program and belongs on the bookshelves of everyone involved in the subject.

In fact, it is three books in one:

1. Chapters 1 to 7 are for *maintenance managers*. The chapters provide all the information needed to implement a world-class condition-based maintenance program and understand the complexities of fluid analysis.
2. Chapters 8 to 11 are for *technical experts in oil analysis*. These chapters provide an exceptionally deep look at the subject and present insights gained from a lifetime of experience. The work here is, in my opinion, without peer.

3. Chapters 12 to 14 are for the *technical experts who want to expand their knowledge* of fluid analysis to include the critically important areas of coolants, fuel, and diesel exhaust fluid.

Many of the insights provided are simple, straightforward, and practical:

- "Oil analysis is a powerful tool if there is knowledge to interpret the data."
- "The tests chosen determine whether the user will be able to use oil analysis as a proactive maintenance tool. This is true whether a human or a computer does the analysis."
- "The lab is usually happy to provide training on the essentials, but it may steer the user toward the lab's capabilities rather than the user's needs."
- "Oil analysis is not about the read metals, which many people interpret as the ultimate job, but rather about looking for the root causes of those abnormal readings."

The technical work presented herein is detailed and in depth, and a wide range of case studies is presented to show how theory translates into practice.

We all benefit greatly when individuals with a lifetime of experience and undoubted mastery of their field take the time to commit their knowledge to paper so that it may be handed on and used as a foundation for future work. Diego and Blaine have done this, and I am absolutely certain that many generations will benefit.

Use this book to build your expertise. Use this book to learn how to listen to your machines and hear what they are telling you.

Mike Vorster, CE, MBA, PhD

Mike Vorster is the David H. Burrows Professor Emeritus at Virginia Tech, where he has taught in the Construction Engineering and Management Program since 1986. Mike is a member of the National Academy of Construction and the Virginia Tech Academy of Teaching Excellence. He holds a BS in civil engineering, an MBA from the University of Cape Town, and a PhD from the University of Stellenbosch. He is the recipient of numerous awards, including the State Council of Higher Education for Virginia Outstanding Faculty Award, the American Society of Civil Engineering Peurifoy Award for contributions to construction research, and the South African Institute of Civil Engineers Basil Reid Gold Medal for contributions to construction. He is a fellow of the South African Institution of Civil Engineers and is registered as a professional engineer in South Africa.

INTRODUCTION

Welcome to the practical world of fluid analysis utilization!

Real-life practices have inspired this book and should suit equipment managers very well. There are plenty of laboratories around the world processing millions of oil, coolant, and fuel samples every year. Most of them do their work in a professional manner, but the information provided by them could fall into two main categories: the information is incomplete for a real machine health assessment or, at the user's end, nobody acts on the information at a level that would allow good proactive maintenance activity.

Fluid analysis as a science is a complex subject, and unless the basic knowledge is present at the user's end, data interpretation becomes difficult and useless. There is plenty of capital involved in making this technology prevail but much less in making it work. This situation diminishes communication with machines to a minimal level, converting equipment maintenance into a reactive activity that does not take advantage of the savings potential. The advantages of fluid utilization sometimes are hard to realize.

In addition, current equipment technology is at a level that requires a much closer look at mechanical behavior. The room for mistakes in maintenance has narrowed as machines have become more efficient and complex. Machines also are becoming more compliant with emissions laws, and this demands a much more specialized and wiser maintenance.

In the field of lubricants, refrigerants, and fuels, these fluids also have evolved to complement the efforts of the industry. Coolants are more resistant to new power regimes and temperatures, engine oils have been adapted to the demands of exhaust systems without causing damage to exhaust filters and catalysts, hydraulic fluids have become more resistant to high temperatures and longer working hours, and fuels with low sulfur content support the demands of high-pressure injection systems.

It is a fact that very few people make use of the valuable information contained in fluids. This book attempts to support the activities around fluid analysis in a document that is easy to use and that fleet managers can use in laying a more solid foundation for maintenance.

This guide takes you down an organized path with plenty of theory. The examples and recommendations will have direct application to a real *predictive maintenance* program. Patience and devotion are essential to digest the information described here. It is difficult to absorb this material in a short time because it takes a long time to become an expert on this subject.

We hope that the examples used for the case discussions add value for users because they come from real-life situations and should reflect what many fleet managers encounter in their daily challenges.

ACKNOWLEDGMENTS

Author Diego Navarro offers his thanks to John Deere Construction and Forestry, who welcomed him in 1976 and allowed him to move up the corporate ladder until his retirement in 2014. My thanks also go to my coauthor Blaine Ballentine, who has been an inexhaustible source of theoretical resources and an invaluable eye in helping to perfect this book.

I offer a special mention to my friend and collaborator Augusto Almeida for his valuable opinions in the composition of this book. And I send many thanks to Professor Mike Vorster at the Construction Equipment Management Program for allowing me to be his co-presenter at his equipment symposiums.

My thanks also go to Italo Lui, who provided valuable information to perfect this work.

I must also acknowledge the valuable training provided by Noria Corporation and the Society of Tribologists and Lubrication Engineers as well as the support from PALL Filtration, Stauff, Hydac, and Puradyn. In the field of oil analysis, my thanks to ALS and, in the field of lubricants and additives, to Lubrizol, Shell, and Chevron.

CHAPTER 1

CONDITION-BASED MAINTENANCE

Condition-based maintenance (CBM) helps users of equipment plan repairs ahead of time when it is the right time to do it. With CBM, unexpected failures decrease and machine longevity improves.

- **How is CBM different from preventive (scheduled) maintenance?** Most preventive maintenance (PM) activities rely on scheduled maintenance based on hours or miles or even fuel consumed. The recommended service intervals determine when to service the machine. The intervals are for "average applications" and do not consider the application or environment in which the individual machine works in adjusting maintenance activities accordingly.
- **Does PM disappear?** CBM takes preventive maintenance to the next level. It uses machine-specific information, such as fluid analysis and telematics, to determine if services need to be done sooner, or if the readings suggest that a mechanical problem is developing, or even if an abnormal contamination is present. In doing so, it takes into consideration the application and environment in which the machine works. Normal PM activities still exist, but we do them at a higher level. In summary, scheduled maintenance activities do not disappear, but we do them better.
- **Thought process.** CBM involves a different thought process and a culture in itself. In a world where productivity is so important, saving time on maintenance by being proactive is a natural proposition. However, changing from a PM-only culture to a CBM-capable industry does not happen in one day.

The CBM portfolio is composed of many disciplines that enable practitioners to analyze machines from different perspectives. The CBM-capable individual is able to communicate with machines on a different level and understands machine needs much sooner than traditional technicians.

This chapter introduces a new line of thought that helps individuals understand the tasks so as to become assertive by communicating with machines with knowledge of the signals the machine provides through the different methods of communication. This guide takes the reader on a trip through machine health issues and the microscopic world that is involved. It also takes the reader through a discussion of the types of maintenance involved and the benefits of CBM.

The complete CBM subject is complex and can take several years to master. CBM is the driving force for telematics, but telematics cannot go forever without a more powerful and defined purpose, and interaction with machine health/operation signals is necessary.

WEAR NO MATTER WHAT!

Dr. Ernest Rabinowicz, a professor at the Massachusetts Institute of Technology (MIT), created the graphic in Figure 1-1 to represent the results of his analysis of wear. This graphic is a classic worldwide didactic from Dr. Rabinowicz's era, and the information is still valid because wear still happens as he had concluded.

Wear is an opportunity to excel in service through proper identification of its source and causes, which, in turn, provides a window to improve uptime and machine longevity. If 70% of loss of usefulness comes from surface degradation, this means that there is a real opportunity for improvement not only for lubricated components but also for components that are in direct contact with the elements, such as tires and bucket teeth.

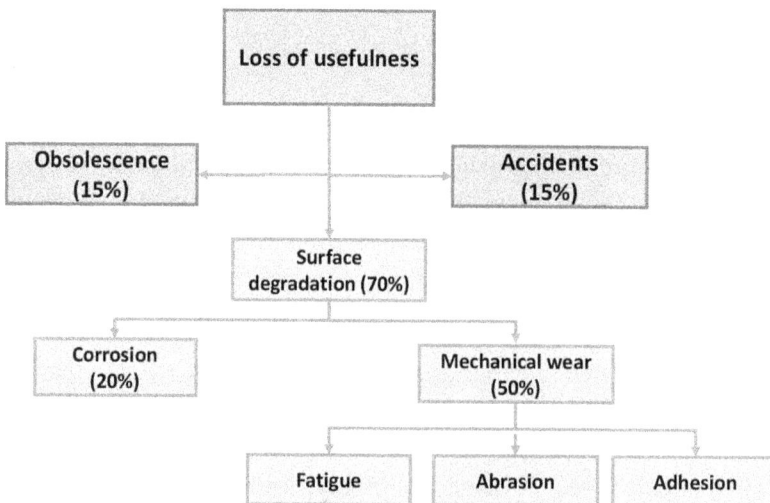

FIGURE 1-1 Sources of wear. (Courtesy of Dr. Ernest Rabinowicz, MIT.)

WEAR: DOES SIZE MATTER?

The industry tends to believe that wear is only important in large, expensive machines and often disregards very small machines (Figure 1-2). However, wear is going to happen anyway, regardless of the size of the components. The commercial implications of this attitude, of course, play a role in maintenance decisions, especially when the component is expensive and critical for the operation of a whole fleet.

FIGURE 1-2   Giant versus small turbocharger.

It is understandable that a giant turbocharger is going to require more scheduled maintenance than a small one that, when it fails, is simply replaced. However, wear will be present in both.

WEAR: DOES TECHNOLOGY MATTER?

When it comes to technology, there is a tendency to believe that old technology did not require maintenance and that it could run forever with little or no maintenance. This is not true to any extent—there are no maintenance-free components. Providing maintenance at the right time will make all the difference.

FIGURE 1-3   Reeves steam tractor versus Case Magnum 400 tractor.

DO WE REALLY CONTROL WEAR?

When parts fail, there are always questions about the root causes for the failure. The failed parts usually tell the story of what caused the failure. Bearings, for example, are not the exception with regard of failure signatures. Bearings can fail for multiple reasons, but a high percentage of them fail just from contamination or due to bad handling during installation.

Figures 1-4 and 1-5 are just two examples of failure modes. Knowing how to identify failures is essential in practicing CBM to avoid failure repetition.

FIGURE 1-4 Overload.

FIGURE 1-5 Loose fit.

MAINTENANCE PARADIGMS

When it comes to maintaining fleets, all kinds of approaches exist. There are users who perform strict scheduled maintenance, others who buy the cheapest consumables, and others who repair exclusively after failure. There are also users who base their maintenance activities on expense control, comparison studies, or correct cost assignments and do not use individual machine needs in their decisions.

The individual needs of machines in mixed fleets are critical. If the individual needs of a given machine are always lagging because some of the required maintenance does not match the service logistics of more important pieces of equipment, these unique machines will be at a disadvantage. Take the case of machines that require special lubricants, but the lube truck has no room for additional oil tanks. In such cases, the machine requiring a special lubricant will potentially operate with the wrong fluid.

None of These Creates Equipment-Saving Opportunities

Follows the book:

- Based on hours, days, or fuel consumed
- Uses the cheapest supplies
- Repair after failure

Accounting focused:

- Expenses control
- Comparison studies
- Correct cost assignments
- Few machine health considerations

TYPES OF MAINTENANCE

Figure 1-6 summarizes the impact of more advanced types of maintenance activities used in optimizing uptime and performance.

- **Reactive or repair-after-failure (RAF) maintenance** is the most common approach in maintenance. Its actions provide a measure to service response time, but it does not promote uptime before failure.
- **Scheduled or preventive (PM) maintenance** tries to minimize downtime by providing the machine with required services based on the manufacturer's recommendations. There are undeniable benefits to this maintenance approach, but it falls short in predicting abnormal conditions.
- **Condition-based maintenance (CBM)** or **predictive maintenance** maximizes uptime through a much closer look at critical indicators and by reaching the machine sooner for service depending on the signal(s) the machine is sending.

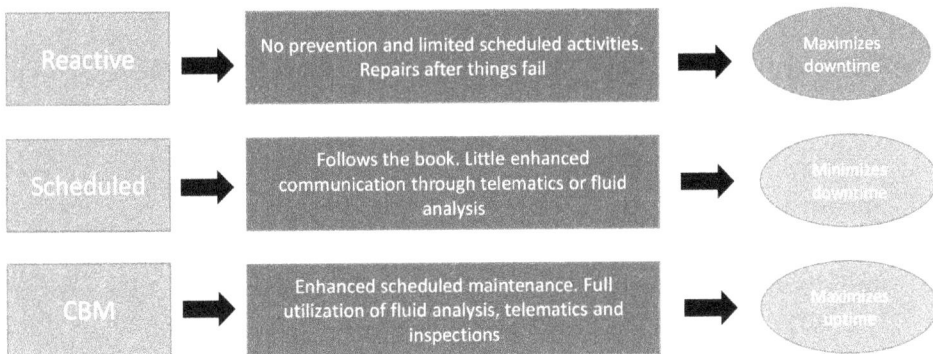

FIGURE 1-6 Types of maintenance.

Flaws of Scheduled Maintenance and RAF

The activities behind RAF are destined to fail because they run backwards without visibility to conditions, there is no breakdown prevention, and they try to get the machine back in operation as soon as possible (see Figure 1-7).

1. Most equipment owners service machines at intervals suggested by the maintenance manual. This gives the owner some peace of mind but in reality a false sense of secu-

rity. Unfortunately, this approach is blind because it cannot prevent failures and does not practice root-cause analysis. It is in fact the most expensive approach.

2. Under this type of maintenance, there is no knowledge of the machine's conditions, just assumptions that everything is going well. Failure waits around the corner, and a critical condition could occur in the middle of the work duty that wears the machine little by little without anyone noticing it.

3. When the machine finally breaks, the task is to put the machine back in operation in the fastest possible way, which is usually at the highest possible cost, and eventually, the machine is back in a cycle to fail again. During these events, the relationships among the manufacturer, seller, and service personnel could take a bad turn because everything happens inadvertently and always as a surprise to the user.

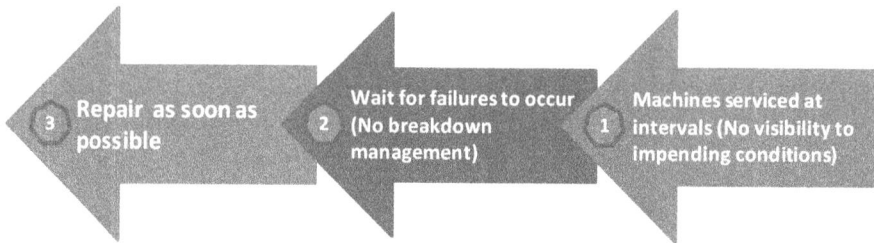

FIGURE 1-7   Repair after failure.

Basic Principles of CBM

The new trilogy supporting CBM (Figure 1-8) involves prefailure analysis, prognosis, and uptime enhancement (compare Figure 1-7 with this new approach):

1. Assessing the condition of the machine via modern tools, such as oil analysis or telematics, is power.

2. This leads to the practice of advanced prognosis, which is management of the potential failure, by planning the repair ahead of the events. How well we manage this information determines the degree of success in the application of CBM.

3. Doing root-cause analyses of the failures leads to avoidance of event/failure repetition, thus improving uptime. The cycle then is repeated, resulting in a fleet that seldom breaks down and where repairs happen when it is more convenient for the maintenance crew to address them.

The goal is to diminish surprise failures and replace them with planned services.

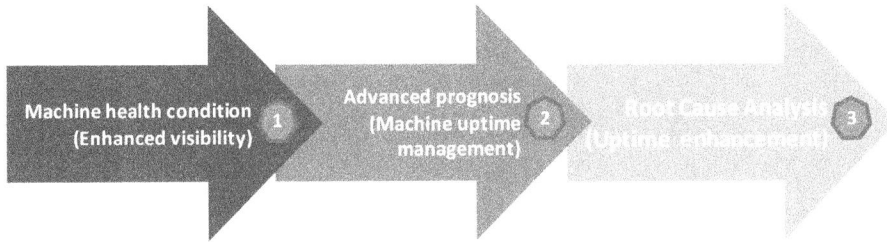

FIGURE 1-8 Basic principles of CBM.

CBM is about enhanced communications between service personnel and the machine. Elephants help us understand an interesting analogy with CBM. Elephants use different levels of communication than humans, including ultrasound waves, which are undetectable to our senses. In the same way, machines communicate their issues to us. Elephants also can communicate complex messages at ultrasonic levels, and machines do the same. The trick is to find the tools that can collect these subtle messages from machines, interpret them, and plan the actions when the messages are telling us something important.

Typically, we humans tend to ignore what we cannot hear, see, feel, or smell. These are our human limitations, and they result in our ignoring everything that does not stand up in a visible or audible way. These messages could include contamination, abnormal sounds at ultrasonic levels, and/or abnormal temperatures that human touch cannot measure.

Oil Analysis

Why do humans depend on blood testing? The answer is obvious: through blood tests, doctors learn whether a person's health is within given parameters and that the most critical of those parameters are in check. Today, doctors pay attention to levels of glucose in the blood of patients because excess glucose is an indication of poor sugar metabolism that impairs the cardiovascular system, and this could result in diabetes and, ultimately, vision issues. What is the advantage? Doctors can make healthy recommendations and improve bad indicators by adjusting the patient's diet, prescribing medicines, and/or advocating a healthier lifestyle.

Hydraulic systems and engines are very much comparable to humans, and many correlations exist between them. Oil analysis for hydraulics is the equivalent of a blood test. However, understanding what each number means requires training because many areas of knowledge play a role—the machine itself, the lubricant, the environment, and the application.

In the example in Figure 1-9, which is a hydraulic report, there are several observations that are unknown to most maintenance crews. Is copper reading within acceptable limits? Where do the wear tables to compare these results come from, who creates the tables, and how? With regard to the silicon readings, are those the result of dirt, gasket maker sealant, or fluid foam inhibitor? Here, knowledge of whether the fluid is to blame for the abnormal readings is essential. Is the water level within normal limits? Finally, why is the report showing high particle counts? There is a whole story behind high particle counts and their causes.

Oil analysis is a powerful tool—if there is knowledge to interpret the data. (Note that Chapters 7–11 describe oil analysis in detail.)

METALS							ADDITIVES						
Iron (Fe)	Chromium Cr	Lead (Pb)	Copper (Cu)	Tin (Sn)	Aluminum (Al)	Nickel (Ni)	Magnesium (Mg)	Calcium (Ca)	Barium Ba	Phosphorus (P)	Zinc (Zn)	Molybdenum (Mo)	Boron (B)
3	<1	<1	27	<1	<1	<1	10	**1831**	1	1131	1049	**<1**	<5

CONTAMINANTS					WEAR		PHYSICAL PROPERTIES			INFORMATION			
Aluminum (Al)	Silicon (Si)	Sodium (Na)	Potassium (K)	Water (%) KF	PQ Index	Particle Counts	Viscosity (cSt @ 40C)	Oxidation	TAN	Fluid Changed	Fluid Type	Total Hours	Fluid Hours
<1	9	2	<5	0.12	NA	23/19/11	34.4	NA	1.03	Yes	Donax TD	3602	3602

HYDRAULIC OIL ANALYSIS	Excavator with atypical particle counts and high copper

FIGURE 1-9 Oil analysis hydraulics.

Coolant Analysis

When it comes to laboratory results from an engine's coolant, the additive concentrations and physical properties are the primary factors to evaluate. The example shown in Figures 1-10 and 1-11, which is from a real case, shows that the organic acid additive concentration is almost nonexistent. The lack of silicates, combined with a high pH, indicates that this engine is very vulnerable to aluminum corrosion. In addition, chlorides are high for this sample. Inspection of the engine in question confirmed that it had been suffering from continuous corrosion issues. (Note that Chapter 13 describes coolants in detail.)

FIGURE 1-10 Detroit Diesel 16V 149.

PHYSICAL/CHEMICAL										INFORMATION		
Freeze Point D3321	Glycol Content	pH	Reserve Alkalinity	Nitrites D5827	Nitrates	Molybdate	Silicates	OA UPLC	Sodium	Coolant Type	Engine Hours	Coolant Hours
°F	%	D1287	HCl	ppm	ppm	ppm	ppm	%	mg/Kg			
32	0	9.3	1.3	708	<10	<2	<2	0.13	758	Nitrite	45,700	3000

CORROSION METALS				WATER HARDNESS					VISUAL APPEARANCE			
Pb	Fe	Al	Cu	Ca	Chlorides	Sulfates	Total Hardness	TDS	Color	Clarity	Oil Layer	Sediment
ppm	ppm	ppm	ppm	ppm	ppm	ppm	ppm	ppm				
<2	<2	56	<2	<5	205	11	100	1050	Clear	Clear	No	Trace

COOLANT ANALYSIS	Detroit V16 149. High aluminum corrosion

FIGURE 1-11 Coolant analysis.

Fuel Analysis

Diesel fuel analysis does not happen enough for mobile equipment, whether because of the high cost of testing or because few people can interpret the results. The lack of fuel testing leaves customers running dirty fuel without knowing it. Fuel analysis is as good as oil or coolant analysis and can tell a lot about fuel quality and cleanliness. (Chapter 13 describes fuel analysis in detail.)

A good fuel analysis needs to test for cleanliness (e.g., water, particulates, and bacteria) in addition to sulfur content, distillation point, cetane index, and biodiesel content (see Figure 1-12):

- **Distillation point.** This test checks that the diesel fuel was properly refined and not mixed with other fuels.
- **Cetane index.** Although not a direct measure of cetane, this index tells us whether easy starting in cold weather and at high altitudes will be an issue.
- **Biodiesel content.** This test determines whether the fuel is within the allowed percentage range provided by the manufacturer and whether it will create an issue in cold weather.
- **Other tests.** These are driven by the season, such as cold filter plugging point and cloud point.

PHYSICAL/CHEMICAL											
API Gravity	Cetane Number	Flash Point Deg. F	Water by Karl-Fischer	Water by Distillation	Viscosity	Sulfur %	Biodiesel	Acid Number	Cloud Point	Cold Filter Plugging	Total Particulate
D287	D4737	D93/D7094	D6304	ASTM D95	cSt	D4294	Volume %	mgkKOH/g	D2500	D6371	D6217
43	45	110	150	0.01	1.1	3000	<0.1	0.03	NA	NA	<10

DISLTILLATION D86						OTHER	MICROBIOLOGICAL		APPEARANCE D4176		
Initial Boiling Point	10% Recovered	50% Recovered	90% Recovered	End Point	% Recovered	HFRR	Microbial Growth	Organisms per ml	Clarity	Free Water	Particulate
°F	°F	°F	°F	°F	Volume %	ASTMD6079	Culture	Culture			
360	425	502	576	603	98	625	NEG	NO	CLEAR	NO	NO

FUEL ANALYSIS

FIGURE 1-12 Fuel analysis.

THE UNSEEN WORLD

We humans tend to judge everything based on the capabilities of our senses. If we can see, feel, or smell something, then we consider it to exist. However, we are very limited in our senses. If compared with an eagle, we are practically blind; if compared with a fox, we are deaf; and if compared with a snake, we are insensitive to heat, just to name a few traits. Because of our highly developed intelligence, we are also highly incredulous and do not easily acknowledge the existence of things we cannot see, feel, or hear.

Contamination happens in the micro world, which, unfortunately, plays a negative role in developing good maintenance strategies. It is hard for us to visualize how microscopic particles cause accelerated wear in machine components. For example, without the use of a microscope, it is not evident that most American pennies have a figure of President Abraham Lincoln sitting in an engraving of the Lincoln Memorial in Washington, DC (see Figure 1-13).

FIGURE 1-13 Figure of President Lincoln inside the Lincoln Memorial on a penny.

Extending the power of human senses requires sophisticated tools, such as infrared imaging and airborne ultrasound, that other industries are already using. Use of these tools is in its infancy in the mobile equipment industry, but they are very promising. Figure 1-14 shows how infrared technology can display the different temperatures in the concealed area of a machine. By providing the thermal charts showing the normal temperatures for various components and areas, this technology could quickly identify when parts are operating outside their normal range and do wonders in the prevention of wear and failure.

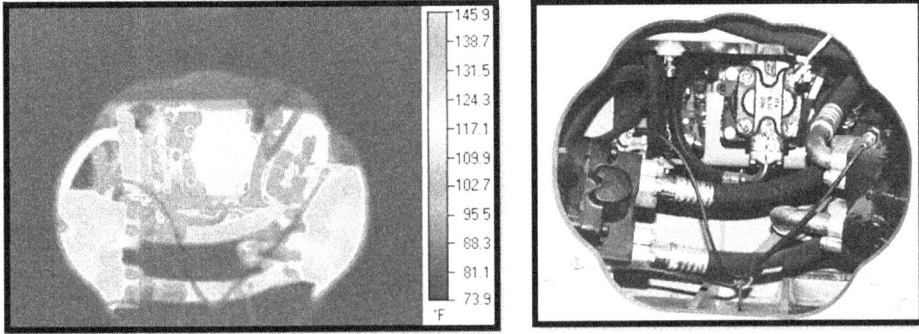

FIGURE 1-14   Hydrostatic crawler heat detection with an infrared camera.

Infrared imaging is widely used in construction, the military, and marine industries. It is just starting to have some use in the mining industry, and there are compelling reasons why it should be useful in mobile equipment maintenance programs. However, the adoption of infrared imaging is lagging in the mobile equipment industry because it is expensive and more complex than fluid analysis.

The ability to see what the naked eye is incapable of seeing is a very suggestive proposition. As shown in Figure 1-15, a large marine engine is having an issue with one injector not firing, which the infrared camera easily detects. Similarly, if a construction machine has an internal leak in a hydraulic cylinder, an infrared camera can detect the resulting increase in temperature.

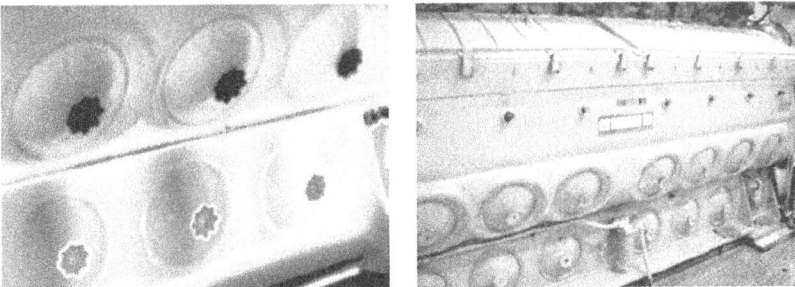

FIGURE 1-15   Injector not firing in a marine diesel engine.

One of the issues with infrared imaging is the cost of the equipment today, but as its use expands, the cost of these cameras is becoming more affordable. Wind is a problem when using an infrared camera in the open, but housing construction already uses infrared cameras with success.

Maintenance Is a Matter of Visibility

Awareness of impending issues in a fleet of equipment is an asset of incalculable value, yet similar machines are not identical. Figure 1-16 helps us to understand what the visibility to conditions mean by showing one of the twins at risk regardless of the similar application and the mechanical similarities of the machines.

Two similar machines work side by side in the same job site, but one is contaminated, so it is only a matter of time until an early failure. Without visibility, the user will always question what caused one to fail while the other continued operating. CBM is about the visibility of conditions and the user's ability to react in time to save the machine from failure.

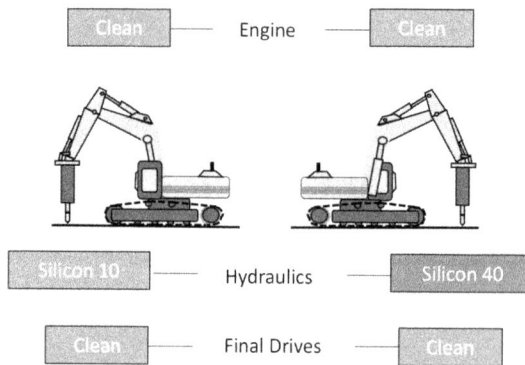

FIGURE 1-16   Twin machines side by side.

When Do the Diagnostics Take Place?

Follow Figure 1-17 as this discussion progresses. The industry diagnoses machines based on pre-established and traditional training approaches. Technicians learn about the machine during their training, and manufacturers do a good job of showing technicians how they build the machine, how to diagnose it, and even how to solve "artificial" problems that perhaps will never happen in real life.

Then the technician may need to fend for himself or herself when a problem arises. This usually happens later in the life of the machine. At that point, if the technician is unable to solve the problem, he or she calls the manufacturer's support center for help. If the problem develops into a failure, the manufacturer or dealer may have to perform the repair.

During what appears to be normal operation of the machine, symptoms may not be visible to the eye, and thus the window to prevent failure is lost. The failure simply happens—because this is how the industry typically operates.

In real life, diagnostics occur when a symptom is already visibly evident. At this point, the operator, technician, or customer has a story to tell regarding machine behavior or poor operation. The machine probably has some kind of fault code present, and the technician is in a hurry trying to figure out what could have gone wrong.

The fact is that technicians are trained to repair but not to prevent failure. They react to symptoms while letting opportunities to prevent failures simply pass by.

A technician may be lucky to pinpoint a problem but also could be detoured to a more complicated diagnosis. If the problem ends up being a faulty connection or a bad sensor, thanks to built-in machine diagnostic capabilities, the technician may be able to resolve the problem. However, if the problem is of a different nature and does not trigger a fault code, the technician faces a much more challenging task.

When wear progresses to failure, there is plenty of secondary damage that can make establishing the real cause much more difficult to diagnose. Catastrophic failures typically branch out, causing so much additional secondary damage that technicians cannot necessarily know where the problem started.

The issue goes back to design. Machines use plenty of sensors and electronic networks that are very good at communicating between each other, but the information they provide to the outside world is not necessarily useful to technicians in a raw view. In summary, diagnostics end up in the failure box when everything becomes more complicated.

When symptoms are so visible that operators complain about performance, this indicates that nobody acted on the messages provided, printed and visual. The only solution left is the practice of forensics (repair), and the cycle starts again.

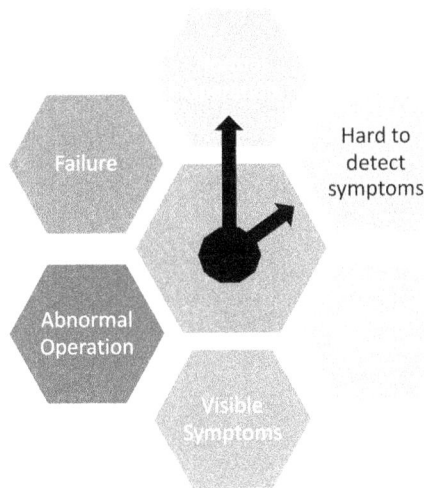

FIGURE 1-17   Timing of diagnostics: machine life watch.

Looking at Wear with CBM Eyes

Wear starts the same day the machine goes to work for the first time. Some wear is detectable via CBM practices, and some wear is simply undetectable. While machines may behave normally as they get older, some types of wear become visible through standard maintenance practices, provided that they are properly interpreted through oil analysis, visual inspections, and other methods.

During the normal wear phase, if problem detection and correction happen in time, the components will continue to operate normally. In this phase, wear can be detected easily, and actions can be taken on the results. However, when a component reaches the point of no return, there is so much wear in it that an overhaul becomes the only solution (see Figure 1-18).

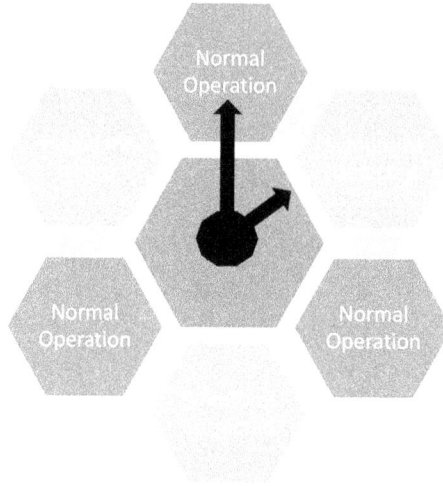

FIGURE 1-18 Extending normal operation and replacing failed components: machine life watch.

When a symptom is visible to an operator, the problem is usually well beyond the initial detection point. The machine may have been sending messages through oil analysis, but nobody paid attention to them, or no one was able to interpret them. In other words, these messages were not visible to the operator or the maintenance crew.

In Figure 1-18, "Normal operation" has replaced "Abnormal operation," and "Fix before failure" has replaced "Failure." This is feasible and allows choosing better times for repairs. Now machines perform longer under normal operation, and symptoms or operator complaints are eliminated.

The task of the maintenance technician is to monitor the condition of the system and modify any conditions in which contamination or wear is present.

The Performance–Failure Curve

The *performance–failure (P-F) curve* bears its name because it represents the life of a component from new or full performance to failure or end of useful life. Traditionally, the use of human senses to determine whether a machine is showing symptoms is a reactive behavior, as Figure 1-19 suggests.

Moving from a reactive to a predictive or proactive mode of maintenance is feasible if we understand the basics. First, this approach is not free, and it may cost extra initially, but once it is established, the returns will be on the order of $5 to $10 for every dollar spent on maintenance.

Changing oil and filters in a machine is not necessarily proactive maintenance unless oil analysis indicates a reason for the change. An oil and filter change does little to improve reliability or uptime. Unless we measure and analyze machine operation, it is simply a change of oil and filters dictated by a routine schedule.

Moving maintenance from a reactive to a predictive level could deliver very different results. However, to get to the predictive maintenance level, certain disciplines and technologies need to be in place, as well as having the proper logistics, processes, and disciplines.

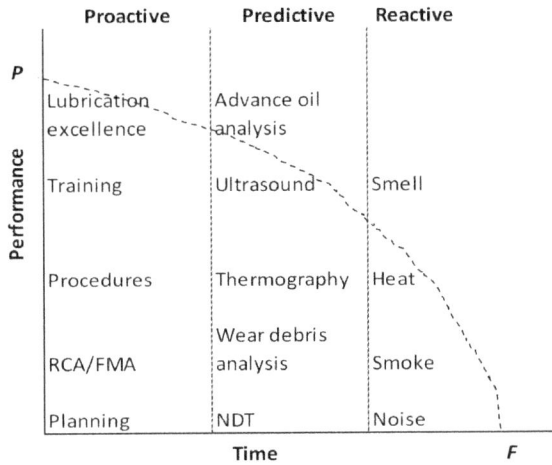

FIGURE 1-19 P-F curve.

CBM versus Scheduled Maintenance: Cultural Differences

CBM is not new, but it has not made progress in the mobile equipment industry at the pace the nuclear or aviation industries have pushed its practice forward. Among the hindrances for CBM application in the mobile equipment industry are the perceived higher cost and especially the human resources to apply the concept. It is also an issue of semantics. What the industry sometimes calls "preventative maintenance" is no more than the regular scheduled services.

The following comparison suggests the benefits of practicing CBM. It also points out that CBM requires more discipline and definitely more knowledge to be effective. Its approach does not become apparent by itself but needs to be learned, especially how to interpret the data that contain the first symptoms of failure in progress and how to react to change the path of failure or the problem detected. In contrast, the advantage of scheduled services, if any, is that anyone can do them. It is just matter of applying what is in the manuals.

Condition-Based Maintenance	Scheduled Services
Reads conditions	Follows the manual
Anticipates problems	Repairs after failure
Prevents potential failure	Cannot plan the results
Requires a higher level of knowledge	Requires no special skills

Impact of CBM on Operating Costs

Operating costs, productivity, and equipment availability will not improve using traditional scheduled service. Greater opportunities for improvement lie in the application of new technologies and an understanding of their results.

We often see users trying to improve a fleet's uptime with traditional scheduled services. However, this task with traditional services is an uphill battle. You cannot expect different results when you apply the same traditional maintenance techniques repeatedly. To ensure a change in results, there needs to be a change in the maintenance approach. CBM provides this opportunity.

Inspections

The aviation industry would not be able to operate without the rigorous and routine inspections that flight crews do before every flight. Inspections are a very important part of CBM (see Figures 1-20 and 1-21).

FIGURE 1-20   Inspection.

FIGURE 1-21   Reportable finding.

The requirements for mobile equipment are not too different from those for aviation, if done properly and routinely. However, the focus needs to be on areas and signals that bring value to the operation of the machine in question. For example, it is important to know if fluid levels are consistently too high or too low, if a leak is present, or if a harness or hose is rubbing against a frame (see Figures 1-22 and 1-23).

FIGURE 1-22   Leaking roller.

FIGURE 1-23   Propel motor cavity.

Inspections require attention to detail and zeroing in on specific trouble areas, but they also require looking at areas that are not easy to reach for lack of direct access. However, executing inspections is the weakest area in mobile equipment. It is never 100% confirmed that the inspection actually occurred and the extent to which the technician inspected the machine.

Another area of concern is whether critical areas were part of the inspection and, if the information was uploaded to a fleet-management application, whether the information will trigger a service order for the machine.

Ultimately, it is important that the information coming from the machine is crossed over to fluid analyses or other forms of prognosis? If this is not happening, many opportunities are being lost in trying to use the information.

The following questions help to uncover the weak areas of inspections:

- Is the operator involved in the inspection?
- Are machine inspections happening routinely?
- Are inspections reaching the hard-to-see areas?
- Are inspections being uploaded to a maintenance application?
- Are inspections crossed over to fluid analysis and telematics data?

Operator

We seldom involve the operator in inspections. The operator is a source of good information about his or her machine. After all, he or she spends the most time with the machine and can catapult the power of inspections to a new level. However, we do not usually talk to operators. Sometimes we arrive for inspections after hours or simply forget to ask the operator simple questions. The questions we need to ask operators are very simple. The list of questions needs to be short and concise. If the machine has operational issues, there is no better person to know this than the operator. Table 1-1 provides a list of questions that can be used as a successful operator questionnaire.

TABLE 1-1 Operator's questionnaire

Question	Y/N	Details
1. Is the machine operating normally?		
2. Have you noticed any leaks or missing hardware?		
3. Is oil consumption abnormal?		
4. Are all grease zerks taking grease?		
5. Do you have any fault codes on your dashboard?		
6. Have you postponed exhaust regeneration lately?		

Operator input, inspections, and data integration from telematics and fluid analysis make the ultimate maintenance application for successful fleet health management.

Telematics

Today's maintenance methods cannot be accomplished without the help of telematics. Every major progressive user is very much tuned to the use of telematics. There are differences among providers, with each somewhat resembling the others but with a slightly different approach to data display and how those data are obtained. All seek to provide the user with what they consider meaningful data to operate a fleet. Telematics become effective when they can transfer operating hours to the maintenance application, provide machine operating data, and interpret fluid analysis data for the user, although few are capable of doing all this. Fluid data analysis is important because interpretation is hard for most people.

Monitoring the health of a fleet and scheduling services or repairs are becoming high-tech activities these days (see Figure 1-24). In terms of speed of service, this suite of applications makes the execution of CBM much more achievable. The snag comes from the ability of a field crew to execute the volume of services generated, for which the industry is ill prepared.

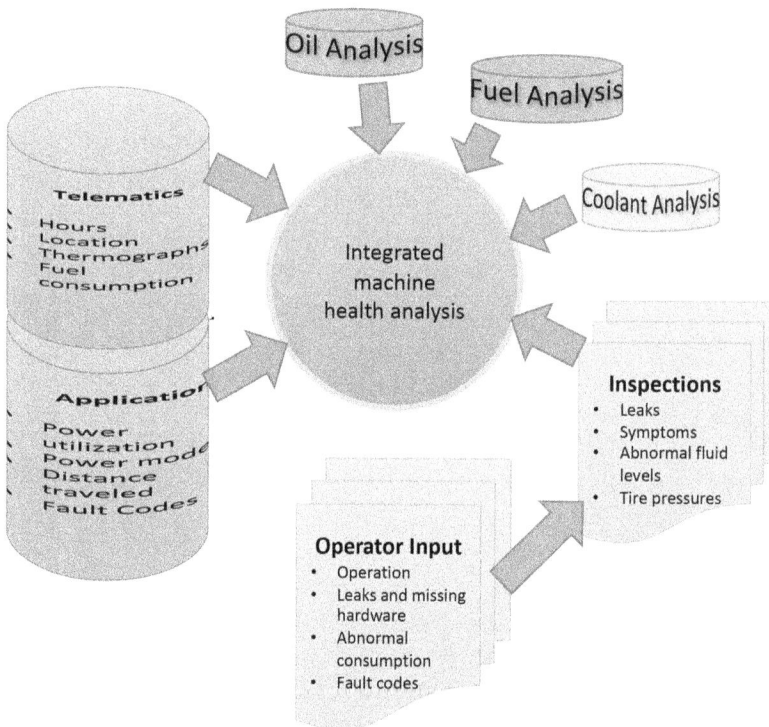

FIGURE 1-24 Integrated machine health analysis.

In contrast, telematics generate an overload of data that makes the information more difficult to grasp. There are pieces of information that are good in the management of a fleet's health, but most data do not contribute to the task. The trick is in the selection of data and their correlations with other parameters, namely fluid analysis.

Displaying, say, pump pressures, although interesting, has little application in the hands of users. A technician, for example, could determine copper-generation correlations in a hydraulic system, but it would require a corresponding graph of temperatures, dirt, water, and particle counts.

Some information coming from telematics is highly valuable. This is the case with idle time, which users can convert into practical operation training on ways to save fuel and injector life.

From a logistics point of view, some telematics information offers great opportunities to locate machines in a vast geographic area and schedule service at the appropriate time. On the utilization side, long idling periods mean fuel dilution or soot generation, whereas high power utilization means higher fuel and oil consumption. For CBM technicians, prolonged idling each month means many technical issues, such as fuel dilution, soot generation, and injector fouling. In contrast, the way inspections are currently used is to diagnose and fix problems, whereas the power in performing inspections is to anticipate impending issues.

Fluid Sensors

Machine sensors are already in use with some degree of success. It is a matter of time before fluid sensors installed on machines will be able to predict changes in the condition of a fluid and produce a recommendation via algorithms. Sensors for viscosity, contamination, dielectric, water, and density exist. Figure 1-25 provides some examples of these sensors.

FIGURE 1-25   Fluid sensors.

The technology is still struggling with the range of sensitivity these sensors need to have, but at present, these sensors can accomplish amazing predictions of fluid quality (see Figure 1-26).

Depending on the speed of changes in the various fluids, whether an uptrend or downtrend, a sensor could indicate several important things about fluid condition. For engines, there is real value if we measure glycol contamination, for example, which would be indicated by a quick uptrend in density and dielectric. By the same token, dilution from a fuel leak could be evident if the viscosity of the fluid goes down together with density and dielectric.

			Soot	Glycol	Water	Fuel	Sulfation	Top-off	Shear
⬆ Up Fast		Density	Slow	Fast	Slow		Slow	Slow	
	Dieliectric		Fast	Fast	Fast		Slow	Very Fast	
Up Slowly	Viscosity		Fast	Slow	Slow		Slow	Very Fast	
Down Slowly	Viscosity					Fast		Very Fast	Fast
Down Fast	Dielectric					Fast		Very Fast	Fast
⬇		Density				Fast		Slow	Slow

FIGURE 1-26 Sensor signals.

Engines are ideal candidates for these types of sensors because engines use several different fluids that need monitoring. The mining industry already uses sensors for contamination, and they work just fine. Many viscosity sensors work well, and it is only a matter of time before these sensors can transmit data through a controller area network (CAN) of a vehicle using telematics. Machines equipped with these sensors will be able to interpret fluid viscosity and density trends and make recommendations for maintenance.

Managing the Data

Having a compressive machine/fleet health analysis would require management of the several results and sources of information in a consolidated way for easier processing, interpretation, and correlation. Given the size of the fleets and the numerous sources of information, it is hard to imagine that a single person could accomplish the processing of these data. It would require a dedicated team that is fully trained in these disciplines.

A comprehensive fluid interpretation plan is a very intriguing concept but extremely hard to accomplish. It requires in-house talent to interpret the different signals, trends, and execution of work orders, but it also requires some sort of programming to merge the data coming from the different inputs.

Telematics is a broadly used technology that is wasted mostly on elementary tasks when in fact it could be used to support the complete health of an entire fleet. A fully trained CBM technician understands that some signals from a machine are important and that these signals can be interconnected. By correlating data from various sources, the technician can anticipate undesired mechanical conditions, improve the timing of service, and ultimately improve uptime.

Machine Health Correlations

With the use of emission engines, the control of regeneration cycles is necessary not only for proper operation of the machine but also for compliance with emissions laws. An operator who continuously cancels the regeneration cycle could take the machine to a full stop and ruin the exhaust filters or catalysts. Telematics allows the technician to see these areas of operation to monitor performance and compliance.

With the power of telematics combined with other tools, such as oil analysis and inspections, data can be pulled from a machine at a distance and used to establish a comprehensive conversation regarding the mechanical issues affecting the machine. However, the huge amount of data collected is beyond the capability of humans to absorb and thus efficiently anticipate the problems affecting a given machine. If the number of messages from a machine is multiplied by the number of machines in a fleet, it is easy to see that the task of anticipating issues is monumental.

Monitoring proper tire pressure is another example of the power of telematics. Proper tire pressure is fundamental in achieving the expected life of tires. Most wheeled machines have a tire pressure differential, which means that the front tires require a different pressure than the rear tires or vice versa. This pressure differential is not always correctly applied in the field, but telematics can provide continuous feedback for easy correction.

In front-end loaders, for example, the front tires could require up to 40 lb/in^2 more than the rear tires. On hard surfaces and fully loaded, this means more tire scuffing, more fuel consumption, and more powertrain strain if the pressure differential is not set correctly. Similarly, lower tire pressures diminish the life of the tires. Having the means to constantly monitor tire pressure is a great addition to CBM activities.

Fluid analysis, telematics, and inspection signals coming from an engine are closely related. For example, a high oil level indicated by telematics could be related to fuel dilution or water contamination. An inspection revealing milky oil would point toward water contamination.

There are many correlations between the signals collected via oil analysis, telematics, inspections and even operator input. Some of the correlations are obvious, such as oil level versus fuel dilution. Others are subtle, like application versus fuel dilution. Still others are difficult to grasp and require expertise to interpret, such as whether iron generation is from use or liner cavitation.

The ideal processing of this amount of data occurs via automated tools that can summarize the results from signals and provide users with a succinct analysis of the situation (see Figure 1-27). Then users can address the issues in order of their importance to make the machine run well most of the time, assuming, of course, that somebody acts on the recommendations produced by the tool.

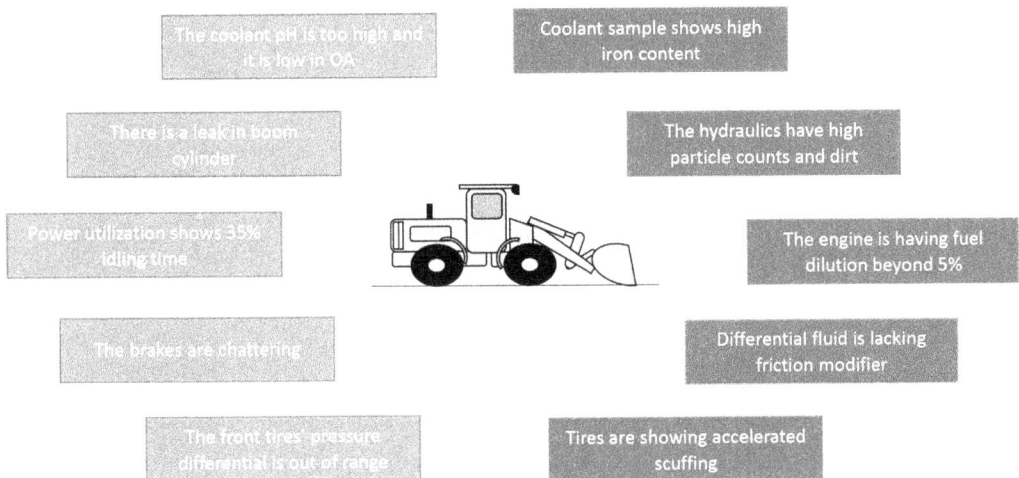

FIGURE 1-27 Machine signal correlations.

Communications with machines are not a project for the future—the technology is here, and the industry is using it. However, the industry is still struggling to make sense of it and put it to use effectively. In addition, technology continues to evolve and will support the industry. However, there needs to be a means to learn how to use the technology properly, but perhaps industry will always be a step behind in trying to make good use of it.

As tablets and smart devices replace computers, medical consultation via the internet is on the rise, and the idea of an automated doctor is just around the corner. Similarly, the secrets behind aging become apparent, and telepathic communication is no longer the monopoly of witchery. So where is the industry that communicates with machines from a distance and at a higher level and the ability to detect impending issues immediately?

Root-Cause Analysis and Failure Scene Investigation

Root-cause analysis (RCA) is an integral part of CBM. It provides the closing loop for failures by diminishing failure repetition. RCA of failures is comparable to the work of a detective. How we handle the case will determine the difference between finding the culprit and never establishing what caused the failure.

RCA does not come by itself. It requires intensive study and practice, especially the methodology. In the practice of RCA, there is a tendency to jump from event to solution, avoiding several steps that are fundamental to solution of the problem. In a crime scene investigation, the detectives install a "Do not cross" ribbon to preserve the evidence. There should be no difference when conducting RCA on mechanical failures. There is a need to preserve the evidence, untouched, to read the message from the failed component.

The most important part of RCA is the disciplined execution of the sequence required by the methodology. The temptation to jump to conclusions is typical and comes from the desire to find a solution fast. This is a dangerous approach because most of the time the culprit is elusive. The discipline to follow the path and investigation sequence is critical (see Figure 1-28).

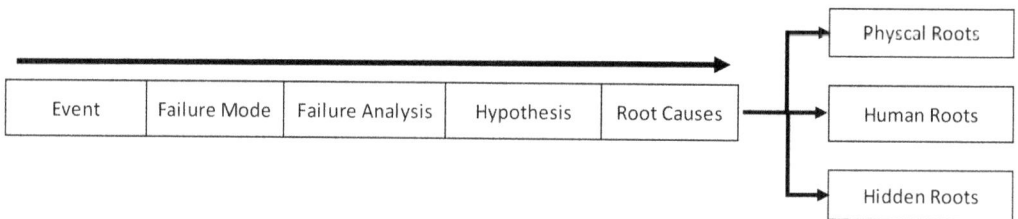

FIGURE 1-28 RCA methodology.

An investigation needs to start with a detailed description of the event, followed by a good understanding of the failure mode. The failure mode is *not* the root cause but the result of the underlying cause. Many investigations go wrong because the failure modes are confused with the root causes (see Figure 1-29).

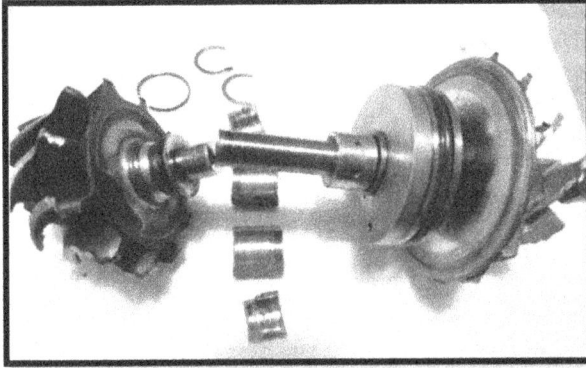

FIGURE 1-29 Turbocharger failure.

There will always be at least three roots. If the investigation fails to uncover credible roots, the case is still unsolved.

RCA Supporting Facts

RCA is feasible thanks to the principles of *order*, *determinism*, and *discovery*:

- Order:
  - We believe in the existence of cause and effect.
  - We believe in the order of things.

    For example, the sun rises every day from the east, and the ocean tides have a defined schedule. If we believe that the error-change exists and follow the logic of cause and effect in reverse, we hope to get to the root causes.

- Determinism:
  - Everything is determinable within a range.

    For example, people react to stimuli within a given range, and parts fail in a defined way with finite modes.

- Discovery:
  - A question generates another question.

    For example, children use this method often. We always look for patterns in chaos.

Most Popular RCA Methodologies

Industry keeps inventing ways to tackle RCA in the best possible way. There are computer programs designed to help with this task, but most people tackle the task by manual means. Ways to do this include a logic tree and the Ishikawa method.

Logic Tree

This is the most used approach in practicing RCA. Although RCA can be applied to anything, from accidents to medicine, the focus in this book is on mechanical and lubrication events that end in failure. Logic trees follow a methodology described in Figure 1-30.

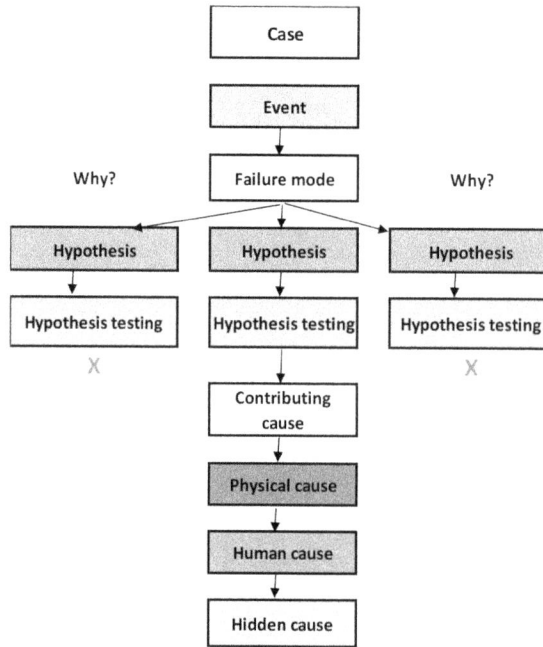

FIGURE 1-30   Logic tree.

An RCA investigation will produce several hypotheses around a single event that need to be tested, as suggested by the logic tree in Figure 1-30, before attempting to determine the final root causes. The logic tree supports the fact that there will always be at least three root causes. There is never a single root cause. If all the root causes are not determined and addressed, the failure mode will repeat.

Ishikawa Methodology

The Ishikawa cause-and-effect diagram, often called a *fishbone diagram*, can help in brainstorming to identify possible causes of a problem and in sorting ideas into useful categories. A fishbone diagram is a visual way to look at cause and effect. The problem, or effect, is displayed at the head or mouth of the fish. Possible contributing causes are listed on the smaller "bones" under various cause categories (see Figure 1-31).

A fishbone diagram can be helpful in identifying possible causes for a problem that might not otherwise be considered by directing the team to look at the categories and think of alternative causes. Always include team members who have personal knowledge of the processes and systems involved in the problem or event being investigated.

The main difference between an Ishikawa diagram and a logic tree is that the Ishikawa diagram lists only the potential causes, whereas the logic tree summarizes the potential causes in logical groups, spells out the hypothesis associated with each group, explains the analysis necessary for testing each hypothesis, and lists the data sources where the information that fuels the analyses can be found.

Advocates of the Ishikawa approach think that their method is superior in the sense that the logic tree could produce different results if a case is given to different individuals to solve. However, both methods require the use of teams that are composed of people with different disciplines, which can provide a much richer analysis.

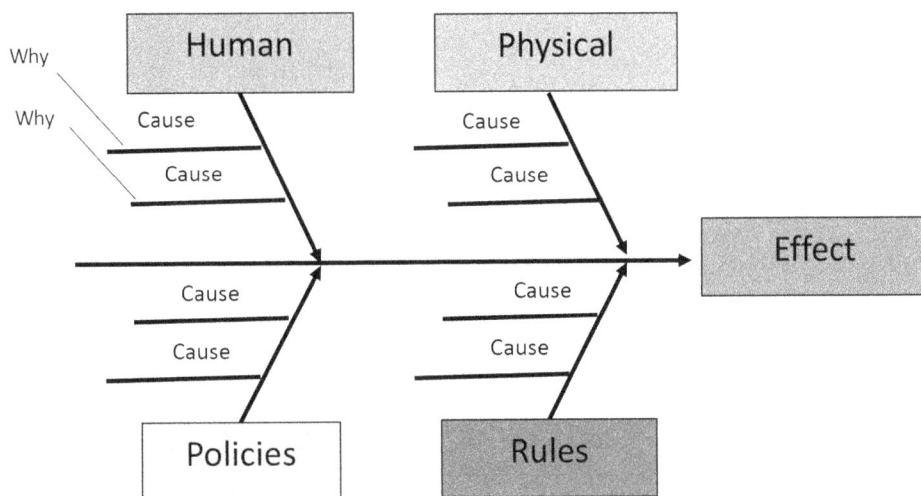

FIGURE 1-31   Ishikawa (fishbone) RCA methodology.

CONCLUSION

This chapter has provided a brief summary of the foundation of this entire book. Fluid analysis relies on all the CBM principles discussed in this chapter, including RCA. As you begin Chapter 2 on lubrication, the RCA methodology will come to bear when we discuss many of the mechanical symptoms of machine failures.

CHAPTER 2

LUBRICATION

LUBRICATION FUNDAMENTALS

Some characteristics of oil are true whether the oil is in an engine, hydraulic system, gearbox, or other application. In this chapter, we will cover some of the background and fundamentals of oil before breaking them down further based on the application in subsequent chapters.

Tribology

Tribology is a division of mechanical engineering involving study of the science and engineering of interacting surfaces in relative motion. In essence, it is the study of friction, lubrication, and wear.

Lubricant Functions

Lubricants do more than prevent direct contact between moving parts. They also cool, clean, transmit power, inhibit corrosion, and seal. When the oil film ruptures, the antiwear and extreme pressure additives lay down a chemical film to prevent wear.

- Think of oil's cooling function as heat transfer. Oil carries the heat away from the piston, wet brake, or gear to the block or housing, where the cooling system can further carry the heat away.
- Lubricants are also great at cleaning components, and they carry contaminants to the filters.
- When lubricants transfer power, as in a hydraulic cylinder or torque converter, we refer to them as *fluids*.
- Lubricants also protect against corrosion by neutralizing acids and employing corrosion inhibitors to set up a chemical barrier.

- Lubricants help to seal components and reduce compression losses, such as around pistons rings.
- Lubricants are great messengers; they carry the operating records from the components. We will concentrate our efforts on interpreting these messages to prolong the useful life of machines.

Petroleum

The theory goes that fossil fuels formed when vegetation and dinosaurs became buried due to catastrophe or a shift in the Earth's tectonic plates. Then, after years of heat and pressure, they became the fossil fuels we have today. Land-based vegetation became coal, and ocean-based vegetation became petroleum (crude oil).

The vegetation in shallow water was different from the vegetation in deep water, which resulted in different types of crude oil. Paraffin crude oil, such as found in the Pennsylvania oil fields, is from shallow-water vegetation. Naphthenic (asphaltic) crude oil, such as that found in Venezuela, is from deep-water vegetation. Most crude oils are a mixture of asphaltic and paraffinic types.

Petroleum has many complex molecules involving carbon and hydrogen, hence the term *hydrocarbon*. Petroleum also contains nitrogen, sulfur, oxygen, and metals such as vanadium. Distillation separates petroleum into its various components, which are primarily different types of fuel—gasoline, diesel fuel, kerosene, and so on. Typically, crude oil yields less than 1% lubricating oil (i.e., base oil, base stock), but this varies with the crude source (see Figure 2-1).

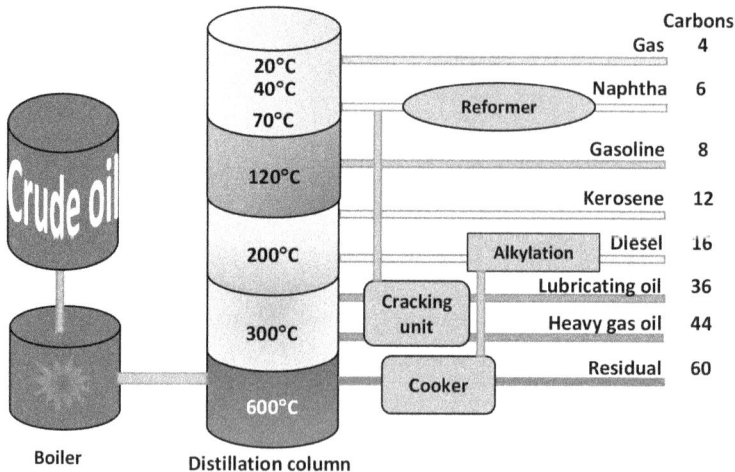

FIGURE 2-1 Crude oil refining.

Hydrocarbons are combinations of carbon and hydrogen (see Figure 2-2). If there are not enough hydrogen atoms to satisfy the carbon atoms, the molecule will be less stable. Therefore, base oil refining includes some type of hydroprocessing to saturate the molecules.

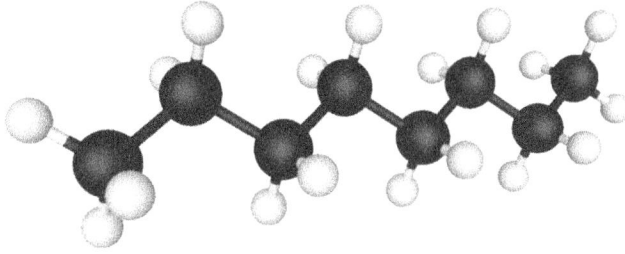

FIGURE 2-2 Hydrocarbon molecule.

You probably can relate to hydrogenation through processed foods. Trans fats are hydrogenated vegetable oils. They are more stable and do not burn as easily as raw vegetable oils when used for frying, but your body has trouble breaking them down for the same reason.

API Base Oil Groups

As refining changed, the properties of base oils changed. Base oil selection affected performance in the engine tests required to meet various American Petroleum Institute (API) specifications. Engine tests are expensive, and lubricant manufacturers needed flexibility in sourcing base oils without running the entire battery of engine tests for every base oil. So the API developed base oil groups based on sulfur content, saturates, and viscosity index, along with a set of rules for base oil interchange and viscosity grade read-across (see Table 2-1).

TABLE 2-1 API Base Oil Categories

Group	Sulfur, mass %	Saturates, mass %	Viscosity index
I	>0.03	<90	≥80 to <120
II	≤0.03	≥90	≥80 to <120
III	≤0.03	≥90	≥120
IV	Polyalphaolefins (PAOs)		
V	All base oils not included in Groups I–IV		

- **Group I** base oils are traditionally refined—solvent extracted, dewaxed, and hydro finished.
- **Group II** base oils are hydrocracked to remove sulfur and increase saturates for greater oxidative stability. Wax molecules (*n*-paraffins) are then broken into isoparaffins and branched-chain paraffins to provide cold-flow performance without physically removing the wax.
- **Group III** base oils are made with a more severe application of the process used to make Group II base stocks. Arguably, the performance rivals that of synthetic oils (Group IV), and most companies call oils made with Group III base stocks *synthetic*. Because some of the molecules are rearranged in the process, the courts decided that this was a step in synthesis, and now most synthetic oils are made with Group III base stock.

- **Group IV** base oils are synthetic hydrocarbon base stocks known as *polyalphaolefins* (PAOs), which are made in a reactor and are synthetic by any measure. Ironically, the feedstocks to manufacture them come from natural gas or petroleum.
- **Group V** is a catch-all for base oils that do not fit into the other groups. They include everything from naphthenic oils that fall below the requirements of Group I to the synthetic esters used in jet engines.

Base oils within each API Group vary significantly in quality, and the lines between them can get somewhat blurry. Some companies use such "terms labels" as "Group II Plus" to differentiate their products, but these are just marketing terms.

A few large gas-to-liquids refineries turn natural gas into fuels and lubricants. The lubricants fall into Group III, where the name *synthetic* clearly applies. Synthetic-blend oils are created by adding some Group III or Group IV oil. Be aware that some manufacturers and marketers recognize the similarity in the process between Groups II and III and call their Group II products *synthetic blend*. Some companies are completely comfortable calling their products synthetic blend because they contain additives that are man-made.

Where the first four API Groups are straightforward, Group V is made up of many extremely different base oils, including:

- Synthetic esters
  - Diesters
  - Phosphate esters
  - Polyol esters
  - Silicate esters
- Polyalkylene glycols (PAGs)
- Alkyl naphthalene (AN)
- Ether polyoils (fire-resistant fluid)
- Naphthenic oils

Esters occur naturally and are the lubricating component in vegetable oils. Synthetic esters are created by the reaction of a fatty acid with an alcohol. Because there are many different fats and alcohols, esters are a large Group with varying properties and include phosphate esters, diesters, polyol esters, and silicate esters.

Overall, esters tend to be high in solvency, which means that they can be aggressive toward paint and elastomers. Water tends to break down molecules of the simpler esters, releasing the acids from which they were made. Synthetic esters are often blended with PAOs, Group IV base stocks, to improve solvency.

Several synthetic esters perform at extremely low temperatures, making them well suited for aviation engines. Some are also fire resistant, providing additional safety where needed, such as hydraulic fluids used in environments with nearby ignition sources.

PAG synthetic oils are mostly water soluble, but the use of oil-soluble types is on the rise. Probably their most common use is in compressors, where their heat resistance, water solubility, and cleanliness make them an excellent choice. Be aware that elastomers must be resistant to this aggressive fluid, and the water-soluble types will not mix with hydrocarbon oils.

Naphthenic base oils have viscosity indexes that are too low to qualify for Group I. Their oxidation stability is too low for them to be used as the primary base oil to meet any modern passenger car or motor oil specification. However, naphthenic oils are still popular in making greases. They are also used in low-quality hydraulic fluids, gear lubes, and tractor fluids, even though their oxidation stability makes it doubtful that they meet current tractor fluid specifications.

As we move from Group I to Group IV, oxidation stability improves, cold flow improves, volatility falls, and solvency declines. The ramifications of these changes are apparent, except for solvency.

Solvency

Solvency of a base oil is important because it relates to the oil's ability to suspend additives and certain contaminants and produce seal swell. Suspending additives is the manufacturer's responsibility, but you may want to be aware of the other performance matters related to solvency.

It is ironic that the oils most resistant to oxidation are the least forgiving once they start to degrade. With a typical Group I hydraulic fluid, for example, an increase in acid number (AN) and viscosity signal oxidation and time for an oil change. These oils typically have adequate solvency to keep the soft by-products of degradation in solution, so they are removed when the oil is drained, assuming that the oil is drained when the signals so indicate. This solvency is also why it is important to drain the oil hot, which keeps more soft-body contaminants in suspension.

Group II and higher hydraulic fluids will run longer before oxidizing. However, if you wait for the same level of acid number and viscosity increase to signal an oil change as with a Group I hydraulic oil, the oil may have already laid varnish down on the surfaces in your hydraulic system.

Solvency is also related to seal swell. Oils with higher solvency penetrate seals more, causing more seal swell. Virtually all oils will penetrate and swell a new seal, so even synthetic oils work well. A problem arises if a seal has been used for a while with a Group I oil and then is switched to an oil with lower solvency. The seal can shrink back somewhat, causing a leak that lets oil out and contaminants in.

Lubricant manufacturers can suspend soft-body degradation products with detergents, and they can increase solvency by adding a synthetic ester. Some hydraulic fluids contain seal-swell agents. Therefore, higher Group numbers do not necessarily create a varnish or seal problem. The point of this discussion is to illustrate why you want to use a consistent product and avoid buying the flavor of the day.

One final example: Remember in the 1990 when a new oil category was created for high-mileage cars? The new oil usually contained something called a "seal conditioner" or some other marketing term to indicate that it dries up seal leaks. What the marketers did not tell you was that most of the leaks were created as the industry migrated from Group I to Group II oils in making passenger car oils.

Viscosity

Viscosity is resistance flow, and it is the most important oil property. There are different methods of measuring and classifying viscosity. *Kinematic* viscosity is measured by the amount of time it takes for the fluid to fall through a thin tube. Test temperatures are typically 100°C or 40°C, and the units of measure are centistokes (cSt). The viscosity of water is approximately 1 cSt. *Absolute* viscosity is measured by the amount of drag on a rotor. The unit of absolute viscosity is the centipoise (cP). Water's viscosity is approximately 1 cP, and water does not shear. Many fluids yield to shear and temporarily lose viscosity. Mayonnaise, toothpaste, latex paint, and multigrade oils are all *thixotropes*, meaning that they thin out from motion; they all temporarily lose viscosity when sheared. Absolute viscosity measures of viscosity with shearing force from the rotor whereas kinematic viscosity does not. Thus:

$$\text{Kinematic viscosity} = \frac{(\text{absolute viscosity})}{\text{density}}$$

For example, a straight-viscosity SAE 40 oil and a multi-viscosity 10W-40 oil can have the same kinematic viscosity, that is, the same measure in centistokes at 100°C. However, the 10W-40 oil will have a lower absolute viscosity, that is, a lower number of centipoises because multigrade oils exhibit temporary shear and straight-viscosity grades do not.

Viscosity Index

Viscosity varies with temperature. Oils become less viscous as temperature is increased. The rate of viscosity change with temperature is indicated by the *viscosity index* (VI), which is a unitless measure of the change in viscosity with temperature. A higher VI indicates an oil that is losing less viscosity as it is heated than an oil with a lower VI.

The scale was developed 100 years ago by taking the best paraffinic oil distilled from Pennsylvania crude, measuring the viscosity loss between 100 and 210°F, and assigning it a VI of 100. Then the same measurement was taken on the worst asphaltic base oil distilled from Texas crude and assigned a VI of zero. Back in the day, a VI of 60 would mean that the oil was roughly 60% paraffinic and 40% asphaltic.

VI is still an important measurement, but refining technique and chemical additives make VIs above 100 prevalent. Multigrade or multi-viscosity oils always have a VI above 100, as do synthetic hydrocarbon oils.

SAE J300 Specification and the Introduction of Additives

The Society of Automotive Engineers (SAE) first met in 1911 and by 1923 had created the SAE J300 specifications, which are still the standard for engine oil viscosity grades, although they have been updated many times. The original version defined the standards for viscosity, flash point, pour point, color, carbon residue, and corrosion resistance.

In those days, oils did not contain additives, even though it was known that some additives reduced friction or improved certain oil properties. By 1939, and possibly forced by the requirements of the military at the beginning of World War II, additives entered the world of lubrication. In 1952, the SAE added the "W" notation to symbolize winter grades, and the low-temperature requirement was actual performance in a low-temperature viscosity test.

Motor Oil Viscosity

The SAE defines viscosity grades for automotive motor oils and gear lubes. Viscosity is always tied to a temperature, which is 100°C for single-grade oils and the second number in a multigrade oil. Winter grades and multigrades also have requirements for cold flow.

Many people have a misconception that multigrade oils become thicker as they are heated. You may have heard that SAE 10W-30 is a 10W oil that "thickens up" to SAE 30 when heated. The fact is that all oils become thinner as temperature increases. For example, SAE 10W-30 oil behaves like SAE 10W in the cold and SAE 30 at 100°C. It still thins out with heat, but it does so at a much slower rate than SAE 10W, as shown in Figure 2-3.

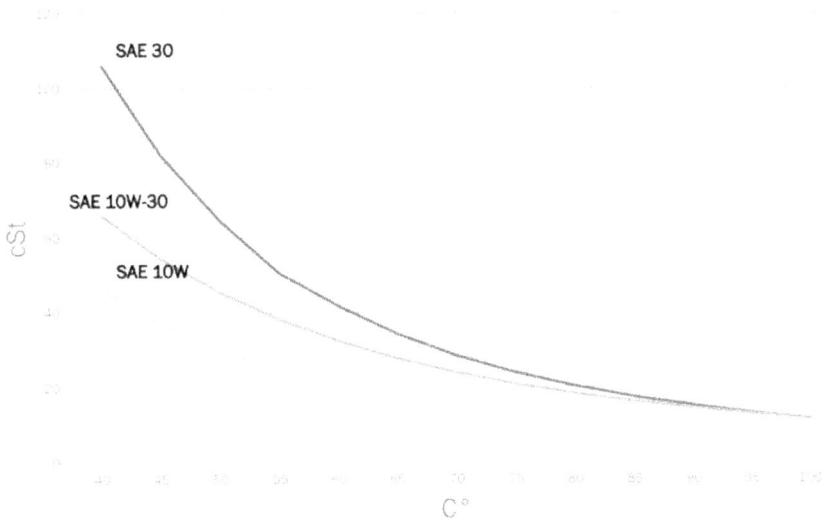

FIGURE 2-3   Effect of temperature on multigrade oils.

Straight SAE viscosity grades and the second number in multigrades are defined by kinematic viscosity at 100°C, as shown in Table 2-2. SAE 5W-40, SAE 10W-40, and straight SAE 40 oils all fall within the same range at 100°C. In 2015, the SAE defined SAE 8 and SAE 12 and apparently made SAE 10W obsolete as a stand-alone grade. No manufacturer recommends them at this time, although some manufacturers are testing them for future uses. It is unlikely that they will ever be recommended for heavy equipment.

Winter grades and multigrades are more complicated. The "W" in an SAE grade stands for "winter," not "weight," and requires cold temperature performance testing. There is a *cold crank simulator test* to ensure that the oil permits winter startup. There is also a *borderline pumpability test* because a worst-case scenario is an engine that starts, but cannot feed the oil pump because the oil in the oil pan has gelled.

TABLE 2-2   SAE J300 Standard

SAE Viscosity Grade	Low-Temp. Cranking (cP)	Low-Temp. Pumping (cP)	Minimum Kinematic Viscosity (cSt) at 100°C	Maximum Kinematic Viscosity (cSt) at 100°C	High-Temp. Shear (cP) at 150°C
0W	6,200 at –35°C	60,000 at –40°C	3.8	—	—
5W	6,600 at –30°C	60,000 at –35°C	3.8	—	—
10W	7,000 at –25°C	60,000 at –30°C	4.1	—	—
15W	7,000 at –20°C	60,000 at –25°C	5.6	—	—
20W	9,500 at –15°C	60,000 at –20°C	5.6	—	—
25W	13,000 at –10°C	60,000 at –15°C	9.3	—	—
8	—	—	4	<6.1	1.7
12	—	—	5	<7.1	2
16	—	—	6.1	<8.2	2.3
20	—	—	6.9	<9.3	2.6
30	—	—	9.3	<12.5	2.9
40	—	—	12.5	<16.3	3.5 (0W-40, 5W-40, 10W-40)
40	—	—	12.5	<16.3	3.7 (15W-40, 20W-40, 25W-40, 40 monograde)
50	—	—	16.3	<21.9	3.7
60	—	—	21.9	<26.1	3.7

The two winter tests are shown in Table 2-2. It should be noted that SAE 5W, SAE 10W, and SAE 20W can be stand-alone grades. SAE 0W, 15W, and 25W are only used as part of a multigrade oil such as 15W-40. The SAE does not define a W grade above 25W, even though some marketing companies incorrectly add the "W" to higher grades.

Multigrade oils must undergo one final test, the *high-temperature, high-shear (HTHS) test.* This is another absolute viscosity test in which the drag on a rotor is measured at 150°C. The HTHS is our best approximation of viscosity in an engine bearing and ensures adequate bearing protection from temporary shear. For limits, see Table 2-2.

Notice that SAE 40 splits based on the first part of the grade. In addition, API FA4 10W-30 provides better fuel economy in diesel engines and is thinner, as measured by the HTHS test, than API CK4 10W-30. Although they are both 10W-30 oils, the HTHS viscosity of FA4 10W-30 is thin enough that it can cause damage in pre-2017 truck engines, and it is not recommended in off-road engines.

You can see how manufacturers use the SAE definitions in making their recommendations in Figure 2-4. Notice that SAE 10W is required to pass the cold crank simulator test at –25°C (–13°F), and SAE 0W must pass the same test at –35°C (–31°F). This coincides with the cold temperature recommendations in the chart, taken from an operator's manual. Similarly, the manufacturer's recommendations for the upper temperature range are based on the SAE's HTHS viscosity.

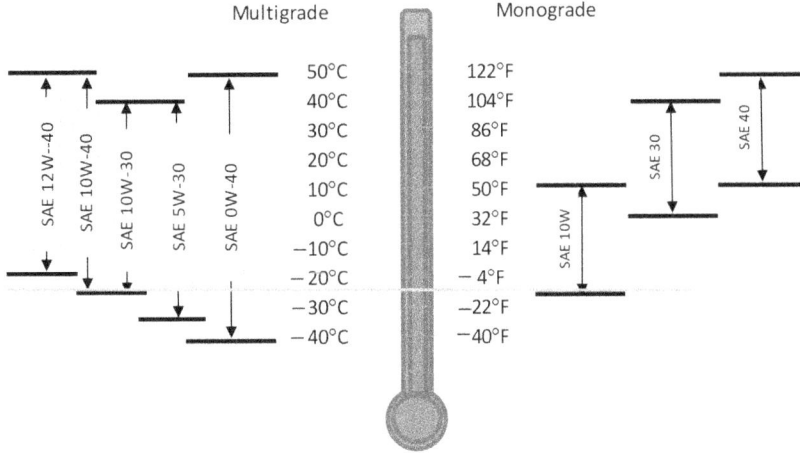

Multigrade | Monograde

50°C   122°F
40°C   104°F
30°C   86°F
20°C   68°F
10°C   50°F
0°C   32°F
−10°C   14°F
−20°C   −4°F
−30°C   −22°F
−40°C   −40°F

SAE 12W–40
SAE 10W–40
SAE 10W–30
SAE 5W–30
SAE 0W–40

SAE 10W
SAE 30
SAE 40

FIGURE 2-4   Lubricant recommendations based on temperature.

SAE J300 Standard for Temperature and Viscosity

Putting it all together, the SAE J300 Standard is the document that rules the viscosity grades for engine oils and helps manufacturers establish their oil recommendations for various operating temperature ranges (as seen in operator manuals). The standard has been updated over the years, at the request of manufacturers, to include more stringent testing or limits to support their recommendations.

Gear Lube Viscosity

The SAE set up the viscosity grades not long after its initial meeting in 1911. Who knows what the members were thinking, but they set up one scale for motor oils and another scale for gear lubes, and the two scales overlap (see Figure 2-5). SAE 90 sounds thicker than SAE 50, but the numbers are misleading.

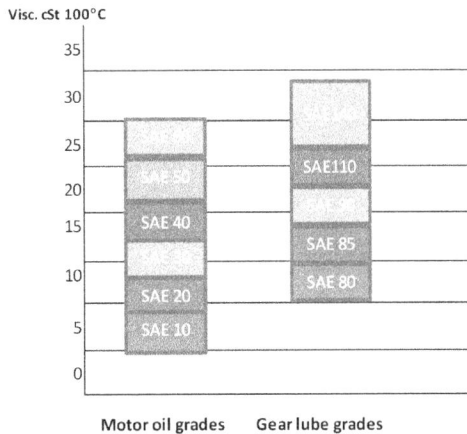

Visc. cSt 100°C

35
30
25
20
15
10
5
0

SAE 50
SAE 40
SAE 20
SAE 10

SAE110
SAE 85
SAE 80

Motor oil grades   Gear lube grades

FIGURE 2-5   Viscosity grade comparison.

Gear lube grades are similar to motor oil grades in the way they are classified (see Table 2-3). Kinematic viscosity at 100°C determines the SAE in straight grade oils and the second number in multigrade oils. As with motor oil grades, the "W" in gear oil grades stands for "winter" and requires a cold flow test. The cold flow test is run at varying temperatures in a Brookfield viscometer with a maximum qualifying viscosity of 150,000 cP.

TABLE 2-3 Viscosities

SAE Viscosity Grade	Max Temperature for Viscosity of 150,000 cP (°C)	Kinematic Viscosity at 100°C (cSt)	
		Min	Max
70W	−55	4.1	—
75W	−40	4.1	—
80W	−26	7	—
85W	−12	11	—
80	—	7	<11.0
85	—	11	<13.5
90	—	13.5	<18.5
110	—	18.5	<24.0
140	—	24	<32.5
190	—	32.5	<41.0
250	—	41	—

ISO Viscosity Grades

Manufacturers of industrial equipment often specify viscosity by International Standards Organization (ISO) grade. Although there are grades lower than those shown in Figure 2-6, they are unlikely to be recommended in construction equipment. Charts that convert ISO grades to SAE grades are easy to come by, but keep in mind that the chart is tied to a VI, usually 95. Here is the weakness in the conversion: a chart for a 95 VI oil will not correctly convert viscosity grades for multigrade or synthetic oils, which have higher VIs.

For a fluid to meet an ISO grade, its viscosity measured in centistokes at 40°C (104°F) must fall within 10% of the ISO number. Therefore, ISO 100 means the fluid is between 90 and 110 cSt at 40°C. Consumers often think that the ISO number is a complete specification, even though it is only viscosity. The ISO grade tells us the viscosity, but the oil could be a non-detergent type, a rust and oxidation inhibited (R&O) hydraulic fluid, an antiwear (AW) hydraulic fluid, gear lube, or something else. Therefore, simply matching ISO grades is an invitation to cross-contamination at best and failure at worst.

SAE 20W and SAE 10W-30 can be formulated to ISO 68, so they are about the same viscosity at 40°C. However, SAE 20W and SAE 10W-30 behave very differently in the cold and at high temperatures. You can match the ISO grade and still have the wrong viscosity if the startup or operating temperature is far from 40°C (104°F).

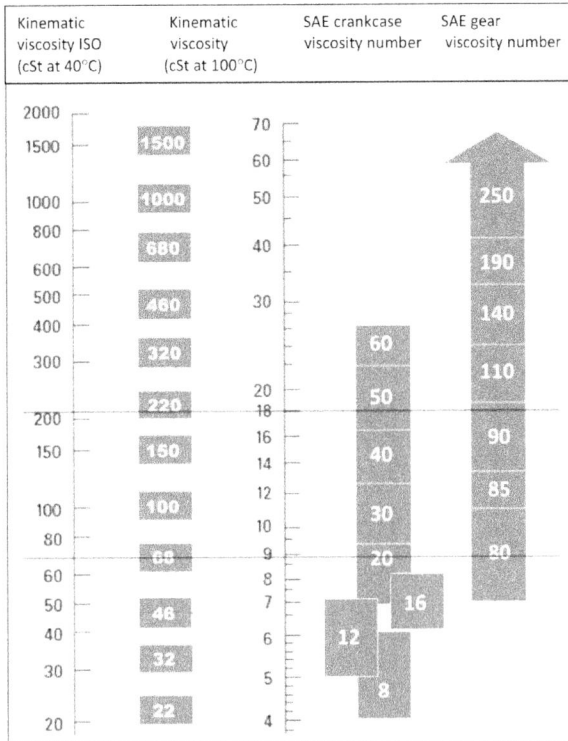

FIGURE 2-6   SAE versus ISO viscosities at 40°C.

API Diesel Engine Categories

The API certifies engine oil specifications beginning with a "C" for compression-ignited fuels or an "S" for spark-ignited fuels. The second letter is just the next letter of the alphabet, so CE is more recent than CD. Starting with CF, the diesel specs started adding a "4" for four-stroke multigrade diesel engine oils to differentiate them from the short-lived two-cycle diesel engine oil specs.

At the time of this writing, the current diesel engine categories are CK4 and FA4. Until 2017, diesel engine oil specs were always backward compatible. In other words, the current specification could be used without harm in older engines, and this is still the case with API CK4 oils. However, FA4 oil is thinner in the bearing, as indicated by the lower HTHS viscosity in Table 2-2, and has extremely limited backward compatibility. The API broke from its traditional C series notation because something different was needed to avoid damage in engines that were not designed for this oil. At the time of this writing, no off-road engine manufacturers are recommending FA4 oil.

CK4 oil and the preceding CJ4 oils are limited in ash, which comes mostly from detergent, to protect emissions equipment. This has raised concerns that they may be inadequate in countries where high-sulfur fuels are still used.

API CF and CD oils are anomalies. The API considers them to be obsolete, and rightfully so, because the hardware to run the tests is no longer available. In contrast, CF is the most recent specification that can be met with straight viscosity oil. Some hydraulic and gear applications may call for API CF or CD in a specified viscosity grade.

Keep in mind that some manufacturers call for additional specifications. Currently, Cummins and Volvo/Mack require the same tests as the API but with limits that are more stringent. Detroit Diesel added an additional engine test requirement to protect against cylinder scuffing. Ford saw additional valve train wear in some of its 6.7 diesels with low-phosphorus oils and developed its own specification to ensure protection.

API diesel specs have become almost a baseline (see Table 2-4). Check the owner's manual to make sure that your oil meets the manufacturer's specifications. Fortunately, the better diesel engine oils meet all these specifications, so it is not necessary to stock multiple products.

TABLE 2-4 Current API Diesel Engine Oil Classifications

Category	Status	Service
FA-4	Current	API Service Category FA-4 describes certain XW-30 oils specifically formulated for use in select high-speed four-stroke diesel engines designed to meet 2017 model year on-highway greenhouse gas (GHG) emission standards. These oils are formulated for use in on-highway applications with diesel fuel sulfur content up to 15 ppm (0.0015% by weight). Refer to individual engine manufacturer recommendations regarding compatibility with API FA-4 oils. These oils are blended to a HTHS viscosity range of 2.9–3.2 cP to assist in reducing GHG emissions. API FA-4 oils are not interchangeable or backward compatible with API CK-4, CJ-4, CI-4 , CI-4 PLUS, and CH-4 oils. Refer to engine manufacturer recommendations to determine if API FA-4 oils are suitable for use. API FA-4 oils are not recommended for use with fuels having greater than 15 ppm sulfur. For fuels with sulfur contents greater than 15 ppm, refer to engine manufacturer recommendations.
CK-4	Current	API Service Category CK-4 describes oils for use in high-speed four-stroke diesel engines designed to meet 2017 model year on-highway and tier 4 nonroad exhaust emission standards as well as for previous-model year diesel engines. These oils are formulated for use in all applications with diesel fuels ranging in sulfur content up to 500 ppm (0.05% by weight). However, the use of these oils with greater than 15 ppm (0.0015% by weight) sulfur fuel may impact exhaust after-treatment system durability and/or oil drain interval.
CJ-4	Current	For high-speed four-stroke diesel engines designed to meet 2010 model year on-highway and tier 4 nonroad exhaust emission standards as well as for previous model year diesel engines. These oils are formulated for use in all applications with diesel fuels ranging in sulfur content up to 500 ppm (0.05% by weight). However, the use of these oils with greater than 15 ppm (0.0015% by weight) sulfur fuel may impact exhaust after-treatment system durability and/or drain interval.
CI-4	Current	Introduced in 2002 for high-speed four-stroke engines designed to meet 2004 exhaust emission standards implemented in 2002, CI-4 oils are formulated to sustain engine durability where exhaust gas recirculation (EGR) is used and are intended for use with diesel fuels ranging in sulfur content up to 0.5% by weight. They can be used in place of CD, CE, CF-4, CG-4, and CH-4 oils. Some CI-4 oils also may qualify for the CI-4 PLUS designation.
CH-4	Current	Introduced in 1998 for high-speed four-stroke engines designed to meet 1998 exhaust emission standards, CH-4 oils are specifically compounded for use with diesel fuels ranging in sulfur content up to 0.5% by weight. They can be used in place of CD, CE, CF-4, and CG-4 oils.

Gasoline Engine Oil Classifications

For gasoline engine oils, the categories start with an "S" for "spark-ignited fuels." The tendency with gasoline engine specifications has been to improve fuel economy. In trying to do this, viscosities have been reduced while still protecting engine parts. However, if the oil is too thin, damage will occur. For this reason, a new API "shield" logo was developed for engines calling for SAE 0W-16 (GF-6B), which is not backward compatible and not compatible with all new engines. The shield differentiates such oils from thicker oils bearing the API "starburst" logo (see Figure 2-7). It appears that manufacturers are looking for the absolute thinnest oils that will work because some are testing SAE 0W-12 and SAE 0W-8 oils for future models. Some manufacturers of passenger vehicles have their own lubricating oil specifications that are beyond the API specifications, particularly General Motors and some of the European manufacturers.

FIGURE 2-7 SAE 0W-16 shield and API starburst.

As oils have become thinner, volatility has been a challenge because thinner oils evaporate at lower temperatures. This makes perfect sense when you think about how base oils are produced. Different viscosities of base oil are separated by distillation, and the most volatile oils, the thinnest, rise to the top. Higher volatility can mean more oil consumption, which is why manufacturers have embraced synthetic oils and synthetic blends.

A limit on phosphorus was put in place starting with the API SH specification, tightened with API SJ and tightened again with API SM, and this level carries through the current (at the time of this writing) SP category. Many in the industry consider these antiwear-restricted specifications to be inadequate for older engines with flat tappet cams and prefer to use diesel engine oils with higher phosphorus and zinc levels (see Table 2-5). Some small engines, such as those used in lawn equipment and concrete saws, specify API SG and older classifications or recommend their own brand for additional cam/lifter protection.

TABLE 2-5   Current API Passenger Car Oil Classifications

Category	Status	Service
SP	Current	Introduced in May 2020 and designed to provide protection against low-speed preignition, timing-chain wear protection, improved high-temperature deposit protection for pistons and turbochargers, and more stringent sludge and varnish control. API SP with Resource Conserving matches International Lubrication Specification Advisory Committee (ILSAC) GF-6A by combining API SP performance with improved fuel economy, emission control system protection, and protection of engines operating on ethanol-containing fuels up to E85.
SN	Current	For 2020 and older automotive engines.
SM	Current	For 2010 and older automotive engines.
SL	Current	For 2004 and older automotive engines.
SJ	Current	For 2001 and older automotive engines.

Oil Breakdown

You may have heard the myth that oil never breaks down, it just gets dirty. Hydrocarbon molecules actually do break down with heat and time. Although oil can thermally degrade in the absence of oxygen, oxidation is normally the mode of oil breakdown in construction equipment.

According to the Arrhenius rule, oil life is cut in half for every 18°F (10°C) of temperature increase after the base activation temperature is reached. A turbine oil that runs at about 100°F (37.7°C) will last for years. However, once an oil is above 140°F (60°C), every increase of 18°F (10°C) cuts its life in half. Synthetic (hydrocarbon) oils provide a longer oxidative life than petroleum oils, but the Arrhenius rule still applies.

As straight viscosity grade oils break down, they darken, get thicker, and create acids. This is why we are interested in viscosity increase and acid number in oil analysis.

Soft degradation by-products accumulate in the oil, causing the darkening, and are dissolved in the oil at first. As these contaminants build up, the oil eventually reaches its saturation point, and the soft dissolved solids begin to fall out. Just as more sugar can be dissolved in hot coffee than cold coffee, temperature is a factor in the ability of oils to solubilize the degradation solids and why it is best to drain oil while it is still hot. The soft solids from oil degradation will first fall out of solution in low-flow areas and where there is a temperature drop, such as in oil coolers and under valve covers. At first, the deposits are soft and gooey, like sludge, and can be removed with a rag. Given time, they harden into solids that resemble coffee grounds and varnish and become very difficult to remove.

Once varnish formation takes hold, it can cause hydraulic valves to stick, lubrication systems to fail, and equipment overall failure. The machine becomes unreliable and potentially unsafe.

The degradation impurities also tend to catalyze further degradation. Using viscosity as a measure of breakdown, viscosity will rise gradually at first. As oxidation progresses, the rate of oxidation accelerates exponentially. Once you begin to lose oxidation control, it is as if you have knocked over the first domino.

Most impurities tend to catalyze oil breakdown. Water, air bubbles, and metal particles all speed up the process of oxidation.

The discussion of breakdown until now has been about straight viscosity oil. Multigrade oils experience viscosity breakdown to varying degrees. The chemical additives that are VI improvers are massive molecules that can break down because of mechanical shearing action. These big molecules are broken into smaller pieces, and viscosity drops.

During the life of a multigrade oil, viscosity will drop at first as the VI improver shears. With continued use, the base oil begins to break down and starts pushing viscosity back up.

Contaminants and Filtration

Most contaminants accelerate wear and oxidation. It is easy to visualize how dirt and wear particles increase wear but do not overlook the damaging effects of water, air bubbles, coolant, and degradation solids.

It is generally much easier and more cost effective to exclude dirt and other environmental contaminants than to remove them after they have entered the component. Store lubricants in a clean, dry place. In high-pressure hydraulic systems, it may be necessary to filter new oil when dispensing it into a hydraulic system to achieve the desired level of cleanliness.

Make sure that breather filters are functioning and properly sized. Desiccant filters may be useful to control moisture in humid environments. Evaluate any leaky seals to determine whether they are a potential entry point for dirt. A leaky seal on a hydraulic cylinder is more than a nuisance. As the cylinder extends, dirt sticks to the excess oil that has leaked onto the rod. As the cylinder retracts, dirt is dragged back under the faulty seal to contaminate the hydraulic fluid and increase wear throughout the system.

Bypass filters and filter carts can be used to remove contaminants within the component or system if the fluid is still serviceable otherwise. Properly sized, these filters can remove much smaller contaminants than the full-flow filter.

Then there are detergent and dispersant additives that surround contaminants. Using water as an example, the detergent and dispersant additives surround and suspend the water droplets. Then, if a filter removes the water, the detergent attached to the droplet is also removed, lowering the detergent level.

The important thing is to remember we are looking for trends. What is normal for a given component? If we see readings outside the normal trend lines, it is a reason for further investigation.

TYPES OF LUBRICATION

Boundary lubrication occurs at contact points. This is where additives play a role in preventing scuffing between two moving parts. *Hydrodynamic lubrication*, in contrast, implies lubricant under flow and pressure separating the two parts with a moving cushion of laminar fluid (see Figure 2-8).

Figures 2-9 and 2-10 depict the wonders of boundary lubrication in protecting parts in motion. Figure 2-9 shows two contacting points with no antiwear (AW) or extreme pressure (EP) protection where the parts scuff each other under load and movement produces material transfer. There is no protection to avoid damage between them.

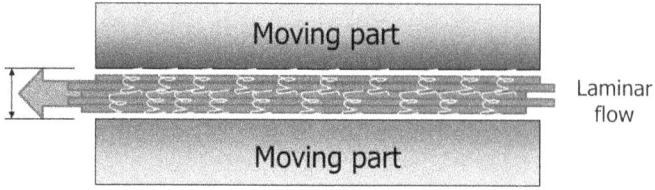

FIGURE 2-8 Laminar flow.

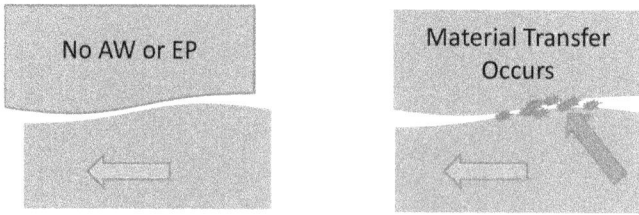

FIGURE 2-9 Without EP or AW.

Figure 2-10, in contrast, how the AW or EP additive separating the moving parts inhibits them from damaging each other.

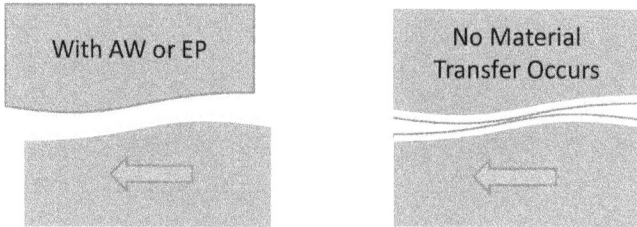

FIGURE 2-10 With EP or AW.

Hydrodynamic Lubrication Example

High-pressure flow inside an axial pump keeps moving parts from physical contact, and the higher the pressure and flow, the higher is the laminar flow keeping the moving elements apart. This is the case with valve plates and slippers. They actually roll on a cushion of oil (see Figures 2-11 and 2-12).

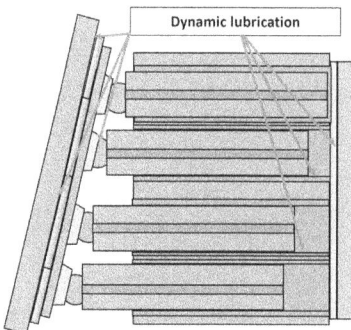

FIGURE 2-11 Axial pump.

FIGURE 2-12 Pump pistons and slippers.

Boundary and Dynamic Lubrication

With boundary lubrication, only the antiwear additives protect the parts. As the moving parts start turning, lubrication moves into the hydrodynamic realm, and friction reaches its minimum point. At the transition point, the load can be increased and still have the same level of protection. With a sufficient increase in speed and load, lubrication enters the elastohydrodynamic realm, where the metal surfaces flex under the pressure without rupturing the oil film. The lubricant temporarily becomes a solid under the pressure in ball bearings and heavily loaded gears (see Figure 2-13).

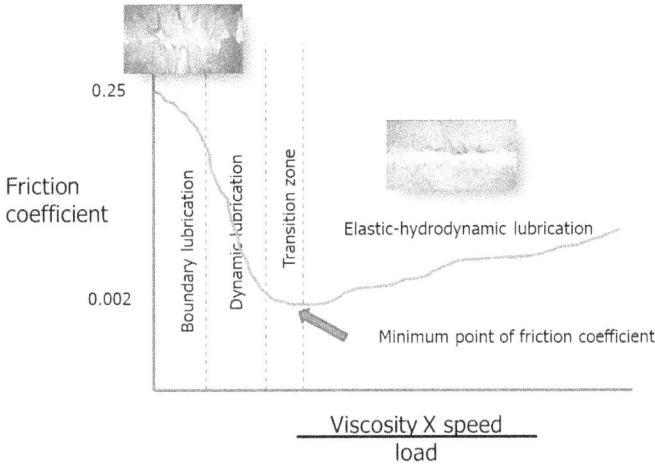

FIGURE 2-13 Boundary and hydrodynamic lubrication.

Boundary and hydrodynamic lubrication are easier to understand in an engine, where both types of lubrication are present. Let us take the case of engine piston rings. A piston reaches its top and bottom dead centers, and the piston comes to a full stop for a short moment in both positions. This is when hydrodynamic lubrication ends and boundary lubrication begins. It is also where the most engine wear takes place.

The top ring dead-stop zone shows more wear than the lower dead-stop zone because in an engine the maximum temperature is at the top of the cylinder, affecting the quality of the oil film (see Figure 2-14).

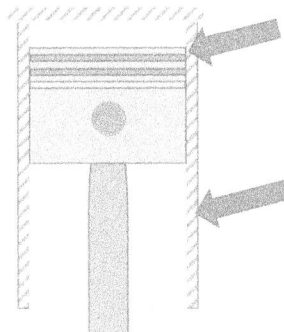

FIGURE 2-14 Area of maximum wear.

Herztian Forces

The *Herztian force* describes the loads that ball and, to a lesser degree, roller bearings stand and still survive. They manage to survive, thanks to the increase in pressure of the trapped lubricant in the race producing an increase in viscosity and providing protection to rolling elements from coming into contact with the race.

The pressure on the point of contact can be as high as 300,000 lb/in^2 and still avoid metal-to-metal contact. Together with the increase in viscosity, the race also deforms to form what we know as *elastohydrodynamic lubrication*. The viscosity of the oil can increase as much as 300 times under load, which explains the wonders of ball bearings.

Elastohydrodynamic Lubrication

Elastohydrodynamic lubrication is part of the wonders of lubrication by complementing hydrodynamic lubrication with actual bearing race deformation (see Figure 2-15).

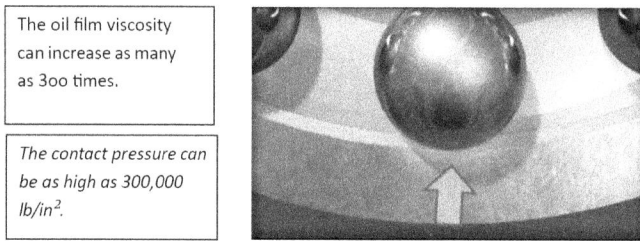

The oil film viscosity can increase as many as 3oo times.

The contact pressure can be as high as 300,000 lb/in^2.

FIGURE 2-15   Elastohydrodynamic lubrication and Herztian forces.

ADDITIVES AND THEIR FUNCTIONS

Table 2-6 lists the types of additives used in different kinds of lubricants and their functions.

TABLE 2-6   Additive Types and Their Functions

	Additive Type	Action	Typical Compounds Used	Functions
1	Antifoaming	Prevent foam formation	Silicone polymers and organic copolymers	Reduce surface tension to speed collapse of foam
2	Antioxidant	Slow lubricant oxidation	Zinc dithiophosphates, hindered phenols, aromatic amines, sulphurized phenols	Decompose peroxides and terminate free-radical reactions
3	Antiwear Agent	Reduce boundary wear, reduce thin-film	Zinc dithiophosphates, organic phosphates and acid phosphates, organic sulfur and chlorine compounds, sulfurized fats, sulphides and disulfides	Chemical reaction with the metal surface to form a film with lower shear strength than the metal, thereby preventing metal-to-metal contact

	Additive Type	Action	Typical Compounds Used	Functions
4	Corrosion and Rust Inhibitor	Protect surfaces against chemical attack	Zinc dithiophosphates, metal phenolates, basic metal sulfonates, fatty acids and amines	Preferential adsorption of polar constituent on metal surface to provide a protective film and/or neutralization of corrosive acids
5	Demulsifier	Facilitate water separation from oil	Polyalkoxylated phenols, polyols, polyamines	Concentrate at water–oil interface and promote coalescence of water droplets
6	Detergent	Keeps surface clean	Metallo-organic compounds of barium, calcium and magnesium phenolates, phosphates and sulfonates	Chemical reaction with sludge and varnish precursors to neutralize them and keep them soluble
7	Dispersant	Suspend and disperse undesirable wear and combustion by-products	Polymeric alkylthiophosphonates and alkylsuocinimides, organic complexes containing nitrogen compounds	Contaminants are bonded by polar attraction to dispersant molecules, prevent from agglomerating and keep them in suspension due to solubility of dispersant
8	Extreme Pressure	Increases the load at which scuffing occurs	Sulfurized esters, olefins, diarl disulfides, organic sulfur, phosphorus compounds, lead naphthenate, bismuth naphthenate	Reacts chemically with the metal surface to form a layer, which reduces friction at high loads and temperatures
9	Friction Modifier	Change coefficient of friction	Organic fatty acids and amines, lard oil, high molecular weight organic phosphorus and phosphoric acid esters	Preferential adsorption of surface-active materials
10	Metal Deactivator	Reduce catalytic effect of metals by passivating surfaces	Organic complexes containing nitrogen or sulfur, amines, sulphides and phosphites	Form inactive film on metal surfaces by complexing with metallic ions
11	Pour Point Depressant	Enable lubricant to flow at low temperatures	Alkylated naphthalene and phenolic polymers, polymethacrylates	Modify wax crystal formation to reduce interlocking
12	Seal Swell Agent	Swell elastomeric seals	Organic phosphates, aromatics, halogenated hydrocarbons	Chemical reaction with elastomer to cause slight swell
13	Viscosity Improver	Reduce the rate of viscosity change with temperature	Polymers and copolymers of methacrylates, butadiene olefins and alkylated styrenes	Polymers expand with increasing temperature to counteract oil thinning

Foam Inhibitors

Foam is an undesirable phenomenon that is present in most systems to some degree. Foam could affect proper lubrication. It is also detrimental to pump life and the temperature of the whole system. Foam is sometimes the triggering agent for varnish formation.

There are additives whose function is to control foam. They are huge molecules that create a weak spot in the surface of the tiny bubbles that form or inhibit foam from forming in the first place. There are also precursors to foam, which could be, but are not limited to, contamination, mixing, insufficient time in the reservoir, too much foam inhibitor, or a high-pressure internal leak (see Figure 2-16).

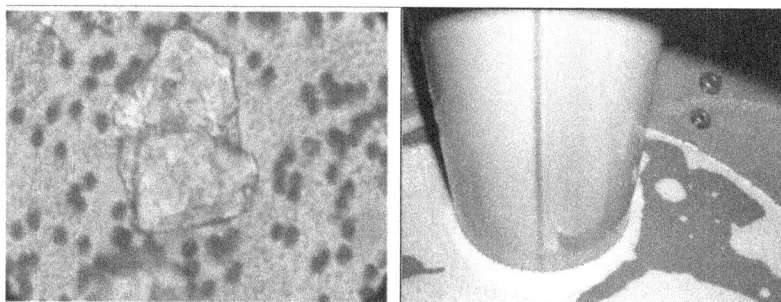

FIGURE 2-16 Foam inhibitor additive (*left*); foam in a reservoir (*right*).

Additive Antioxidant Synergies

Another aspect of additives is their synergy with the base stock. Figure 2-17 shows that inhibited phenols and aromatic amines (antioxidants) produce much longer oxidation resistance when combined with Group III and IV base stocks. This suggests that lubricants made of higher than Group I base stocks are more responsive to typical oxidation inhibitors.

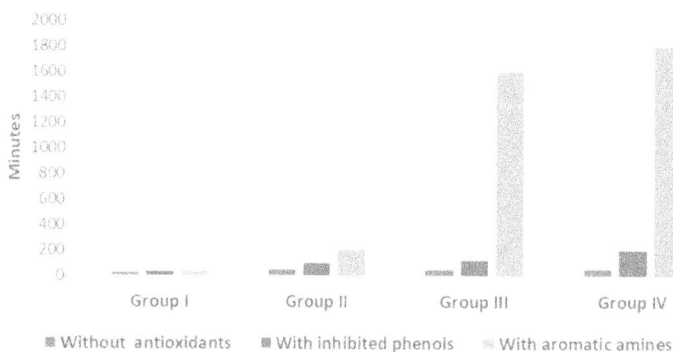

FIGURE 2-17 Antioxidant synergy with base oils.

Zinc Dialkyl Dithiophosphate and Tricresyl Phosphate

Zinc dialkyl dithiophosphate (ZDDP) is a wonder additive that has saved the world since 1944, when Hebert Frauler invented it while working for the Union Oil Company in Los Angeles.

ZDDP is a sulfur compound made of zinc and phosphorus in a strong bond that controls not only wear, but also oxidation and corrosion.

The other famous antiwear additive is tricresyl phosphate (TCP), which is popular in hydraulic fluids. Although older than ZDDP, it has not been used as widely. There are many other antiwear additives on the market, but the ones discussed here are the most common and most likely to be used in mobile equipment lubrication (see Figure 2-18).

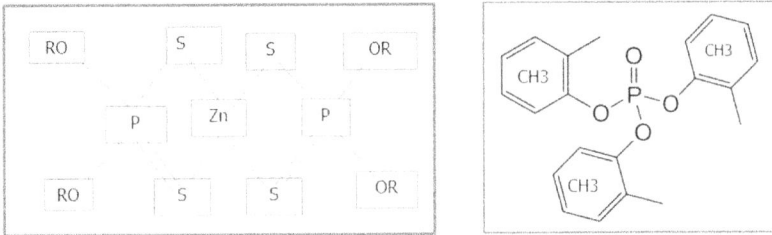

FIGURE 2-18   Zinc dialkyl dithiophosphate (ZDDP) (*left*); tricresyl phosphate (TCP) (*right*).

Dispersants and Detergents

Dispersants and detergents work similarly, and the only difference between them is the strength and size of the micelles. When the core is polar, they surround the particle in a three-dimensional way, keeping the particle from joining other particles. When the isolated particles are solid, the result is called a *colloid*. When the isolated particle droplets are liquid, they are called an *emulsion*. Micelles behave differently in water because the nonpolar tails engulf the particles (see Figure 2-19).

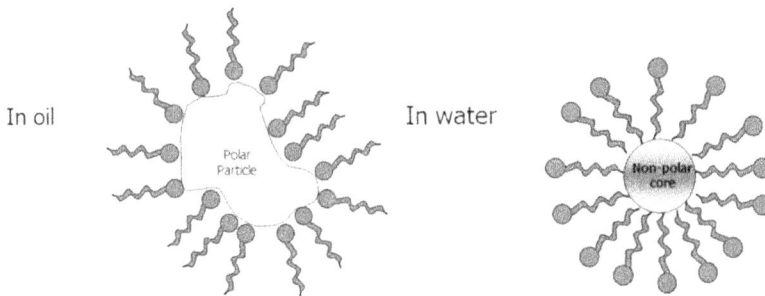

FIGURE 2-19   Dispersants and detergents.

Additives versus Load

Figure 2-20 shows the increase in lubricity (reduction in friction) on contact surfaces when different antiwear additives or friction modifiers are used and the ways the load-carrying ability is improved. Additives turn base oils into final products for their specific application. When oxidation occurs, these additives turn into different compounds, losing their ability to reduce friction.

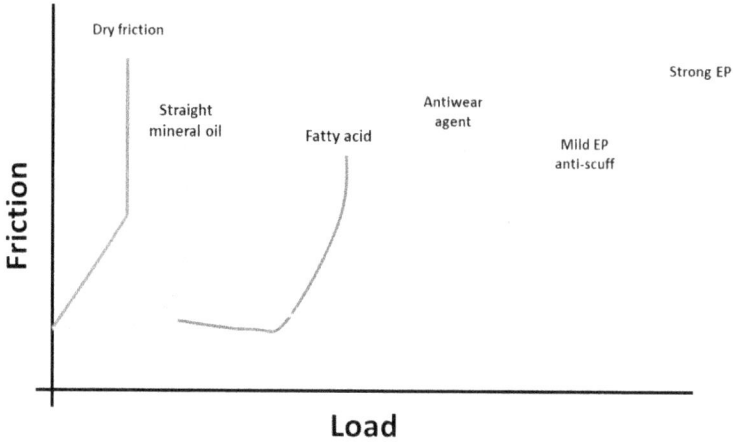

FIGURE 2-20 Friction versus load (Courtesy Noria Corp.).

Additive Synergies versus Temperature

Different additive families have different characteristics and behave differently, depending on the temperature at which they operate. Mineral oil loses its antifriction protection as the temperature goes up, as shown in Figure 2-21. Sulfur-based compounds, however, have an interesting range of load protection that decreases friction when the oil reaches a given temperature and keeps its antifriction protection up to very high temperatures. Phosphorus, in contrast, is activated and provides a constant reduction in friction at lower temperatures than sulfur, but its protection ceases and sulfur takes over when temperatures continue to rise. Sulfur and phosphorus combined provide a synergistic protection over a wider temperature range. In all cases, temperature activation is a fact. This shows that all components need to be at the right operating temperature to get the best additive protection offered.

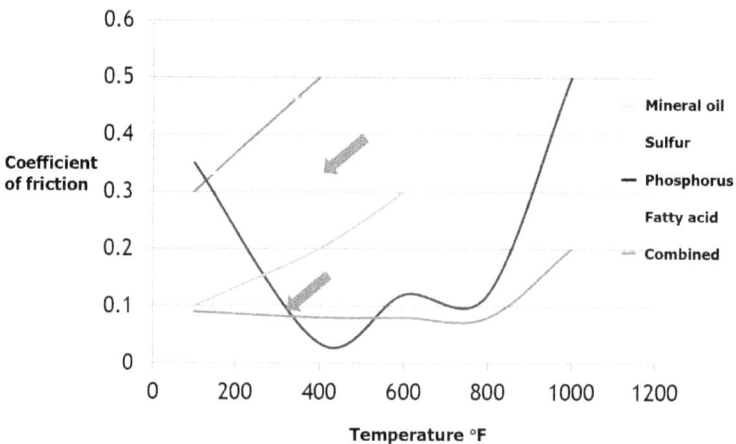

FIGURE 2-21 Additive synergies.

Additives Fight for Surface

Extreme pressure (EP) additives, namely sulfur and phosphorus, react to metal contact differently. They provide EP and AW characteristics to the base oil, respectively. Sulfur absorbs into the surface and is more stable because it is imbedded into the metal. Phosphorus, in contrast, only clings to the surface and needs to fight for surface with corrosion inhibitors (see Figure 2-22).

When the wrong formulation exists, AW additives may lose their space and allow metal-to-metal contact. This could be the case when different fluids are mixed. When mixing occurs, the AW and corrosion-inhibition capabilities are affected.

A similar situation arises with glycol contamination because the corrosion inhibitor from the coolant breaks the lubrication film to take possession of the surfaces. Glycol is not a good lubricant; rather it is a surfactant, and metal-to-metal friction will occur, and bearing material will show up in the oil analysis.

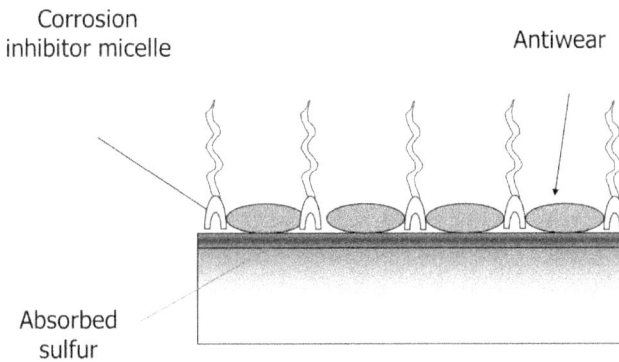

FIGURE 2-22 Additives fight for space.

Water and Rust Inhibitors

Water is a contaminant that attacks iron and steel. It consumes the additives and rust inhibitors. Rust inhibitors are little molecules that cling to the metal surfaces and provide protection, thanks to their polarity. When these additives are exhausted, they can no longer offer protection, and corrosion takes over. Iron oxides accelerate fluid aging because they catalyze chemical reactions and consume additives. Water level control is essential (see Figure 2-23).

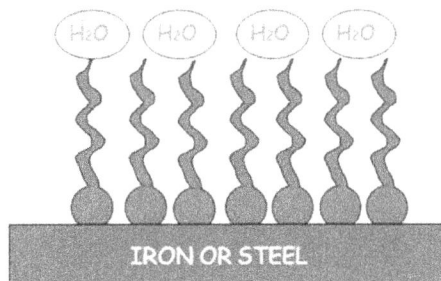

FIGURE 2-23 Rust inhibitors (Courtesy Noria Corp.).

Dispersants for Soot

Dispersants, as found in engine oils, are hydrocarbon micelles with a polar head made of oxygen or nitrogen. Their function is to separate soot (a by-product of combustion) from resins (by-products of oil degradation). In doing so, they keep soot from forming big coagulations that obliterate filter function and clog lubrication galleries (see Figure 2-24).

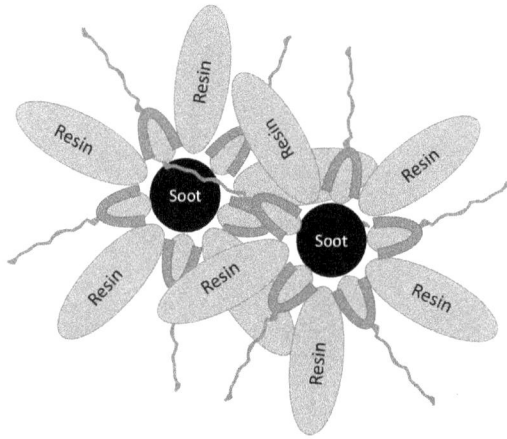

FIGURE 2-24 Dispersants.

Copper Passivators

Copper passivators appear in both engine oils and hydraulic systems. In engine oils, they protect the oil cooler. They came late to the market as a requirement for emission engines that run hotter and produce more corrosion metals. They protect yellow metals by corroding the surface and sealing it from further attack. However, given the numerous types of additive packages and concentrations available in the industry, using different oils may cause the new product to create its own corrosion crust, increasing copper readings until the oil cooler is passivated again. Copper readings from cooler passivation in engines mask the copper coming from bearings or bushings, so you cannot use them to determine bearing wear (see Figure 2-25).

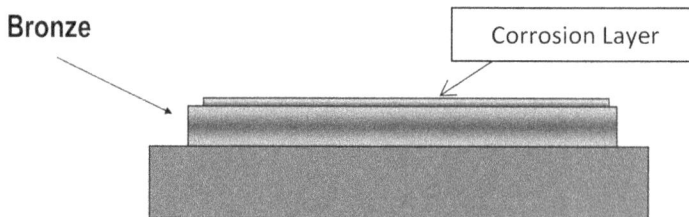

FIGURE 2-25 Copper passivators.

Etching versus Physical Erosion

Copper etching is one of the least understood phenomena. It happens by chemical attack on bronze alloys. Erosion and pitting, in contrast, could be the direct result of water in the fluid, or cavitation. Water under high pressure is in a liquid state. When it escapes from the high-pressure port, it changes to vapor rapidly, taking away little chunks of metal. When the vapor is under pressure again inside the pump, it will cause vaporous cavitation, damaging the bronze alloy, especially in rotary group plates and sleeves (see Figure 2-26).

FIGURE 2-26   Bronze etching (*left*); bronze erosion (*right*).

Friction Modifiers

Friction modifiers are typically phosphorus compounds that sacrifice themselves to protect surfaces during engagement. By "burning" during engagement, they protect the surfaces from scuffing and promote a modulated and progressive engagement. These applications are typical in wet brakes and power shift transmissions as well as in limited-slip differentials. Some manufactures may opt for a more abrupt and direct engagement and may recommend a lubricant without these additives (see Figure 2-27).

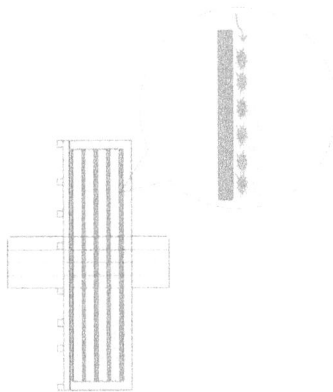

FIGURE 2-27   Friction modifiers.

How Additives Show Up in Oil Analysis

An additive package allows users to make educated guesses as to the type of oil/fluid in use. The three groups of additives show up here. AW additives can be present as boron or molybdenum, but also as sulfur, zinc, or phosphorus. ZDDP shows up in many oils and fluids and is recognized by the presence of zinc and phosphorus. ZDDP is an antiwear, antioxidant, and corrosion-inhibitor additive. Sulfur is usually not in the oil analysis report because it is not common in oil additives analysis. TCP is also an antiwear additive, but it does not have zinc. Fluids containing TCP should only have it in small amounts. Silicon alone is typically a foam inhibitor, but it also be could the result of dirt, silicate from coolant, or silicone from sealant (which will be discussed in Chapter 3). Detergents are recognizable for the readings in calcium, magnesium, barium, and very seldom sodium (see Figure 2-28).

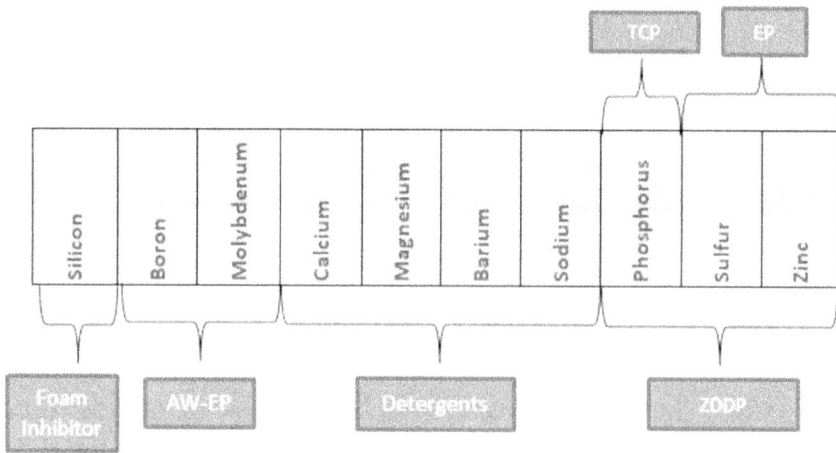

FIGURE 2-28 How additives show up in oil analysis reports.

Lubricants and Additives Used in Mobile Equipment

See Table 2-7.

TABLE 2-7 Additives Seen in Oil Analysis Reports

Component	Type of Lubricant	Class	Typical Additives	Percent of Volume
Engines	Engine oil	CK4, CJ4, CI4	Detergents/dispersants, antiwear agents, antifoam agents, alkalinity improvers, antioxidants	10%–30%
		SP, SN, SM, SL	Detergents/dispersants, antiwear agents, antifoam agents, alkalinity improvers, antioxidants	10%–30%

Component	Type of Lubricant	Class	Typical Additives	Percent of Volume
Power shift transmissions	Tractor fluid	Zinc based	ZDDP, friction modifiers, antiwear agents, antioxidants, fatty acids, detergents/dispersants	10%–30%
		Zinc free	Phosphates, friction modifiers, antiwear agents, antioxidants, fatty acids, detergents/dispersants	10%–30%
	Automatic transmission fluid	Dexron or Mercon	Friction modifiers, antiwear agents, antioxidants, foam inhibitors	2%–10%
		Specialty	Dedicated friction modifiers, antiwear agents, antioxidants, foam inhibitors	2%–10%
Axles, final drives, gearboxes	Tractor fluid	Zinc based	ZDDP, friction modifiers, antiwear agents, antioxidants, fatty acids, detergents/dispersants, foam inhibitors	10%–30%
		Zinc free	Phosphates, friction modifiers, antiwear agents, antioxidants, fatty acids, detergents/dispersants, foam inhibitors	10%–30%
	Gear oil	Standard	Extreme pressure agents, antiwear agents, antioxidants, antifoam agents	3%–10%
		Limited slip	Extreme pressure agents, friction modifiers, antiwear agents, antioxidants, antifoam agents	3%–10%
Hydraulic and hydrostatic	Antiwear (AW) oil	AW zinc based	ZDDP, antiwear agents, antifoam agents, corrosion inhibitors, antioxidants, pour-point depressants	2%–10 %
		AW zinc free	TCP, antiwear agents, antifoam agents, corrosion inhibitors, antioxidants, pour-point depressants	2%–5%

TYPICAL SIGNATURES OF VARIOUS LUBRICANTS

Lubricant signatures are important in the world of oil analysis. They help in establishing whether the correct product matches the application. They also help in recognizing elements that do not belong to the original formulation, thus confirming contamination or mixing.

Hydraulic Fluids

Hydraulic fluids are great for transmitting power, and their name derives from this fact. However, they also need to lubricate parts in relative movement, and they need to keep a very stable viscosity. In most cases, hydraulic fluids shed water rather than keeping it in suspension.

Two major tendencies are seen today in hydraulic fluid additive packages. One is zinc based, and the other is zinc free. Zinc-based fluids depend heavily on ZDDP, and zinc-free fluids depend on TCP. TCP fluids are more environmentally friendly, but they are not biodegradable (as advertisements sometimes try to suggest).

Hydraulic fluids for mobile equipment are usually available in ISO 22, 32, 46, and 68 viscosities to adapt to different seasons and requirements. Some hydraulic fluids come in Group II base oils and some in Group III oils. The latter are more oxidatively stable, providing long service hours. Buyer beware: many AW hydraulic fluids are formulated with Group I base oils and naphthenic base oils of insufficient quality to qualify for Group I. There are other types of fluids, such as synthetic, fire resistant, and biodegradable fluids, for specific applications. In some instances, an original equipment manufacturer (OEM) may choose an engine oil or a tractor fluid as the hydraulic fluid, but most OEMs prefer a straight zinc-based AW fluid.

Polyalkylene glycol (PAG) and naphthenic hydraulic fluids are not popular in mobile equipment because the pressure and temperature requirements render them less desirable. Some additional hydraulic fluids worth mentioning are the John Deere Hydrau and Hydrau XR and Cat HYDO fluids, which have higher levels of ZDDP and are designed for heavy equipment. John Deere's Hydrau is a multigrade fluid with 850 ppm zinc and is designed to demulsify water. Cat HYDO, in contrast, is a single-grade fluid with 900 ppm zinc (see Table 2-8).

TABLE 2-8 Hydraulic Fluid Signatures

Hydraulic Fluids	Additives								
	Sodium	Silicon	Magnesium	Calcium	Barium	Phosphorus	Zinc	Molybdenum	Boron
AW 46 Plus Brand A	0	0	3	74	0	412	293	0	1
AW 46 Brand B	0	0	0	67	0	396	369	0	0
AW 32 Plus Brand B	0	0	3	74	0	412	293	0	1
AW 32 Brand A	1	0	1	29	0	268	474	0	0
Zinc-Free AW 46 Brand A	0	0	0	0	3	106	0	0	0
Zinc-Free AW 46 Brand B	3	0	0	0	0	560	<1	<1	1
Zinc-Free AW 46 Brand C	1	0	17	5	0	245	4	0	0
Zinc-Free AW 46 Brand D	1	0	6	23	0	336	17	0	1
Zinc-Free AW 46 Brand E	0	1	1	0	0	319	0	0	0
Zinc-Free HVI 46 Brand A	0	2	1	18	0	91	27	2	1
Zinc-Free AW 32 Brand A	0	0	2	11	—	653	7	0	1
Zinc-Free AW32 Brand B	0	0	0	4	0	456	5	0	1

Multi-Viscosity Fluids

The need for multiseason hydraulic fluids that can perform year round is a wish never fulfilled for the spectrum of temperatures. However, some multi-viscosity (MV) fluids are available that can perform reasonably well during the summer and still have excellent pour points in winter. Manufacturers can reach these requirements using two different approaches or a combination of them. One approach is by means of synthetic base oils, and the other is by using viscosity improvers in a relatively low-viscosity base oil. Figure 2-29 provides two examples of these kinds of products.

Kinematic viscosity	Synt. 46	Synt. 68
cSt at 40°C	46	68
cSt at 100°C	9	11.7
Viscosity index	180	170
Flash point °C(°F)	221(430)	215(419)
Pour point °C(°F)	−51(−60)	−48(−54)
Brookfield viscosity ASTM D2983 cP at −30°C	-	5100

Kinematic viscosity	Multiseason
cSt at 40°C	32.5
cSt at 100°C	6.9
Viscosity index	180
Flash point °C(°F)	190(374)
Pour point °C(°F)	−50(−58)
Brookfield viscosity ASTM D2983 cP at −20°C	1040
Brookfield viscosity ASTM D2983 cP at −30°C	3310
Brookfield viscosity ASTM D2983 cP at −40°C	14800

FIGURE 2-29 Synthetic fluid ISO 46 and ISO 68 (*left*); MV fluid ISO 32 (*right*).

As you can see from the charts, the synthetic in ISO 46 has a pour point of −51°C (−60°F), whereas the MV fluid has a pour point of −50°C (−58°F), which is very close. However, at 100°C, the synthetic IOS 46 has a 9.0-cSt viscosity compared with only 6.9 cSt for the MV ISO 32 fluid. Another point is the viscosity index (VI). A true synthetic base oil is going to have a high VI and requires less viscosity improver to achieve the same formulated VI, making it less susceptible to viscosity loss.

Tractor Fluid and Automatic Transmission Fluid Signatures

Fluids used in transmissions, and sometimes in hydraulics, have their own signatures and purposes. A universal tractor fluid is available for hydraulics, transmissions, wet brakes, and differential locks. Therefore, it is imperative that the fluids for these systems do not deviate from the manufacturers' specifications. There are also zinc-free tractor fluids. such as Case IH HI-TRAN.

Automatic transmission fluids (ATFs), in contrast, are mainly for automatic transmissions with specific friction modifier needs and oxidation stability. They tend to have more friction modifiers than tractor fluids and are thinner. Tractor fluids are typically SAE 10W-30, whereas ATFs generally range from SAE 0W-8 to SAE 0W-20.

Cat TO-4 is in its own category. As with most tractor fluids, it meets the requirements of API GL4 for gear protection, and it is a capable hydraulic fluid. TO-4 fluids are usually recommended in straight grade SAE 10W, SAE 30, and SAE 50. They are also set apart in that they contain no friction modifiers. Substituting TO-4 fluid in a tractor hydraulic system will cause additional brake chatter and wear, and substituting tractor fluid in a Cat drivetrain may cause slippage and poor brake performance (see Table 2-9).

TABLE 2-9   Universal Tractor Fluid, ATF, and TO-4 Signatures

Tractor Fluids, ATF and TO4	Additives							
	Silicon	Magnesium	Calcium	Barium	Phosphorus	Zinc	Molybdenum	Boron
Tractor Fluid ISO 46 Brand A	0	145	3570	<1	1290	1640	<1	6
Tractor Fluid ISO 46 Marca B	0	22	2975	0	1135	1380	0	89
Tractor Fluid ISO 46 Marca C	19	17	3493	0	1320	1666	0	118
Tractor Fluid ISO 46 Marca D	1	60	3200	0	1193	1330	1	2
Tractor Fluid ISO Bio ISO 46	0	<1	656	<1	899	486	<1	97
Tractor Fluid ISO Low Visc. ISO 32	0	145	3570	<1	1290	1640	<1	2
Tractor Fluid ISO Synthetic ISO 46	9	13	3192	0	1173	1287	0	76
ATF Zinc-Free Ecofluid Plus A	0	0	24	0	196	7	0	120
ATF Transynd	4	0	21	0	331	0	0	153
TO4-30 SAE 30	7	14	2011	0	948	1085	0	0
TO4-10W SAE10-W	0	544	334	0	1072	1225	0	26

Gear Oils Signatures

Gear oils are simpler in term of additive content, as their signatures suggest. They do not use sulfonates as such, and their signatures show heavy doses of phosphorous. Some may show boron, whereas others with limited-slip (LS) additives show much higher phosphorous numbers (see Table 2-10).

TABLE 2-10   Gear Oil Signatures

Gear Oils	Additives							
	Silicon	Magnesium	Calcium	Barium	Phosphorus	Zinc	Molybdenum	Boron
SAE 75W-80 LS	0	<1	<50	<1	2445	<50	<1	300
SAE 80W-90	0	9	50	0	950	40	9	131
SAE 80W-90 LS	0	<1	<1	0	1842	<1	0	<1
SAE 85W-140	0	0	3	6	750	3	<1	120

Oils with high phosphorous content are designed for special applications, such as an acute-angle hypoid gears or brake or clutch materials requiring that protection. In some cases, a concentrated supplemental additive is recommended by the equipment manufacturer to provide LS performance.

Engine Oil Signatures

Signatures in engine oil are also important, especially in the era of low-emission engines. The lubricant signature allows quality checks, such as making that sure top-offs are of the same type of lubricant. Ensuring the use of the same type of engine oil diminishes the risk of spiking copper readings from cooler passivation and guaranties that there is no additive conflict. Table 2-11 shows typical signatures of different products on the market. The list goes from CI4 through CK4 products. It is interesting to notice the drastic reduction in calcium, phosphorous, and zinc content from the old CI4 oils to current-specification CK4 brought about by limitations to protect emissions equipment. New products are retro-compatible with engines of previous generations, but older specification products are not necessarily compatible with modern engines given their emission control accessories.

TABLE 2-11   Engine Oil Signatures

Engine Oils	Additives							
	Silicon	Magnesium	Calcium	Barium	Phosphorus	Zinc	Molybdenum	Boron
0W-40 Synthetic CI4 Brand A	0	0	3950	0	1420	1560	110	160
0W-40 Synthetic CI4 Brand B	6	14	1710	0	1385	1275	0	0
10W-30 CJ4 Brand A	4	267	2514	0	1246	1213	0	0
10W-30 CJ4 Brand B	8	16	1770	0	1328	1199	2	1
10W-30 CI4 Brand C	0	21	1480	0	1110	1280	90	51
10W-30 CI4 Brand D	0	21	2360	0	1110	1220	90	51
10W-30 CI4 Brand E	5	961	415	0	1110	1409	59	3
15W-40 CI4 Brand A	6	16	3073	0	1234	1300	107	126
15W-40 CI4 Brand B	7	9	1953	0	1071	1276	1	6
15W-40 CI4 Brand C	0	21	1480	0	1110	1280	90	51
15W-40 CI4 Brand D	0	21	2360	0	1110	1220	90	51
15W-40 CI4	9	925	391	0	1442	1218	47	1
15W-40 CK4 Group III	0	0	2360	0	1110	1212	0	9
15W-40 CK4 Group II	0	0	2425	0	1141	1298	0	106

Viscosity Decline at 40°C

Hydraulics, Hydrostatics, and Power Shift Transmissions

Viscosity is a very important property of fluids and oils. Knowing their shear-down numbers is important when deciding on a product for a machine and the season in which it will be working. Viscosity in AW hydraulic fluids is very stable if they do not depend on viscosity improvers as engine oils and tractor fluids do. Viscosity loss also varies depending on the shear stress of the component the fluid/oil is serving. For example, in power shift transmissions equipped with torque converters, the viscosity shear-down on a tractor fluid is going to be more noticeable because of the sharp edges of the torque converter fins. Table 2-12 shows typical changes in the viscosity of different fluids commonly used in mobile equipment. As the chart indicates, SAE 0W-40 engine oils used as hydraulic fluids shear down substantially. Engine and tractor fluid figures for >2,000 hours are truncated because these products usually are not used for >3,000 hours.

TABLE 2-12 Hydraulic Fluid Viscosity Shear-Down

Hours ⟶	0	300	500	1000	2000	3000	4000	Decline
AW 46 Sample A	44.1	41.8	40.57	39.7	38.8	37	36	−18%
AW 46 Sample B	46	43	41	38	37	36	35	−24%
AW46 Average	45	42	41	39	38	37	36	−21%
Zinc-Free Fluid AW 46 Sample A	47	46	45	44	43	41.7	40	−15%
Zinc-Free Fluid AW 46 Sample B	45	44	44	43	42	41	40	−11%
Zinc-Free Average	46	45	45	44	43	41	40	−13%
SAE 15W-40 Hydraulics/ Hidrost	117	95	74	73	72	71	70	−40%
SAE 10W-30 Hydrostatics	67.4	60	54	52	50	49	49	−27%
SAE 10W-30 Crawler Hydraulics	67.4	61	58	55	53	52	51	−24%
SAE 10W-30 Power Shift with Torque Converter	67.4	59	48	47	46	NA	NA	−32%
SAE 10W-30 Power Shift W/O Torque Converter	67.4	61	58	55	53	NA	NA	−21%
SAE 0W-40 Hydraulics Sample A	85.7	60	56	53	50	49	48	−44%
SAE 0W-40 Hydraulics Sample B	95	54	53	52	51	50	49	−48%
0W-40 Hydraulics Average	90	57	55	53	51	50	49	−48%
THF ISO 68 Hydraulics	57	52	48	40	39.9	NA	NA	−30%
THF ISO 68 Axles	57	55	54	53	45	NA	NA	−21%
THF ISO 68 Power Shift with Torque Converter	57	50	45	39	38	NA	NA	−33%
Low Visc THF ISO 32 Power Shift W/O TC	33.7	28	27	26	24	NA	NA	−29%
Bio-THF ISO 46 Hydraulics	46	40	36	31	30	NA	NA	−34%

Gear Oil Viscosity Decline

Gear oils also lose some viscosity with use, but it is not significant. However, knowing their viscosity at a given point when the hours of use are not recorded could be of great value in estimating the age of the oil. Table 2-13 gives approximate viscosity values calculated from averages of many product brands. Users need to develop their own viscosity timelines.

TABLE 2-13 Gear Oil Viscosity Shear-Down

	OH	300H	500H	1000H	2000H	3000H	Decline
75W-90 Synthetic	118	114	112	110	109	108	–8%
80W-90	136	132	129	128	126	126	–8%
80W-90LS	165	163	162	160	160	159	–5%
75W-90HD	148	145	144	143	142	141	–5%
75W-140 Synthetic	184	177	172	168	167	166	–10%
80W-140	271	260	252	247	245	243	–10%
85W-140	335	339	329	322	319	317	–10%

Additive Decline

All additives are sacrificial, which means that they lose their original chemical structure over time. Although the oil analysis results still show their individual constituents, it does not mean that those constituents are still providing protection. For this reason, we measure oxidation in long-term-use fluids to understand their degree of change. However, some additives decrease in numbers over time, and we need to be aware of this fact. The additives that typically decline in concentration are the coarse additives such as calcium and magnesium, which, when attached to a contaminant, become trapped in the filters. We can blame this additive loss on the high filter beta rating and the hours of operation. Other additives, such as limit-slip (LS) additives, that contain phosphorus decrease more rapidly when the application involves wet clutches and brakes. Phosphorus is especially sacrificial, and in some LS gear oils, it may be necessary to bring its level back by adding a concentrated supplement in a specific percentage (see Table 2-14).

TABLE 2-14 Additive Decline

	Calcium	Phosphorous	Zinc
Decline ppm/h	0.25–0.5	0.025–0.1	0.025–0.1

Physical Properties

A good maintenance program keeps a strict eye on lubricant physical properties through oil analysis. We have already covered the viscosity aspects of oils, and we still need to cover physical properties such as total acid number (TAN), total base number (TBN), and oxidation.

We can say that there are three main groups of TAN limits: those from oils with low calcium content, those for oils with high calcium content, and those for gear oils. Among the gear oils, there are two subgroups: standard gear oils and heavily loaded LS gear oils.

The oils with high calcium content are going to have high TANs and high TBNs as a rule. This is the case for engine oils and tractor fluids. In contrast, fluids with low calcium contents have lower TANs and TBNs and are designed for applications where acid formation is less of an issue. However, it is always advisable to adjust these numbers based on results because some fluids fall outside the norm.

With regard to gear oil TANs, certain products specifically designed for high-friction applications have heavy doses of LS additives that elevate the TAN above that of standard gear oils. It is important to know this to avoid false alarms when the lab results highlight high TANs on a given product. In general, we are more interested in TAN increases than in the actual number. Therefore, it is important to know the initial TAN.

TBNs in engines using ultra-low-sulfur diesel fuel (<15 ppm of sulfur) can start out lower and safely be allowed to go lower than with higher-sulfur fuels because the fuel does not pose a threat to the engine and the alkaline reserve.

Oxidation is a measure of how much the additive package has degraded in use. Once again, we are looking for the change rather than the number. For oxidation measurement to be effective, the lab requires a sample of the new product for comparison with the results of the used sample. Otherwise, getting oxidation results without this reference is a waste of time (see Table 2-15).

TABLE 2-15   Lubricant Physical Properties

Used Oil Limits			
Physical Properties	**Normal**	**Abnormal**	**Critical**
TAN Tractor Fluids	<5	5.0–6.0	>6
TAN AW Fluids	<1	1.0–2.0	>2
TAN ATF's	<1.5	1.5–2.0	>2.0
TAN Gear Oil	<3.0	3.0–4.5	>4.5
TAN Gear Oils LS	3.5–5.9	6–7.0	>7
TAN Engine Oils CJ4	<5	2.9–2.0	>2
TBN CJ4 Engine Oil with "15 PPM Sulfur Fuel"	≥3	2.0–2.9	<2.0
TBN CK4 Engine Oil with "15 PPM Sulfur Fuel"	≥3.5	2.5–3.4	<2.5
Oxidation	<25	25–30	>30
Other Contaminants	**Normal**	**Abnormal**	**Critical**
Sulphation (Diesel Engines)	<25	25–40	>40
Nitration (Engines/Hydraulics)	<25	25–30	>30

Lubricant Compatibility

Lubricant compatibility for hydraulic and powertrain applications is an important topic that is poorly discussed in many manuals. It is important from the standpoint of both lubricant performance and component longevity. It takes years of development to produce dedicated fluids and oils to suit specific applications.

It is then reasonable to let those fluids/oils perform in their original formulations to achieve their highest potential. After a new machine goes to work for the first time, it does not take long before its main systems contain a mixture of lubricants. It is the norm of the industry, after all, that users are rarely going to perfectly abide by the manufacturer's recommendations for a given brand and type of lubricant. It is reasonable to think that mixed fluids are part of the daily life in this industry, and the limitations and consequences of mixing fluids should be understood. Figure 2-30 illustrates the compatibility of a limited number of lubricants used in hydraulic systems. The most likely negative results of mixing fluids are high particle counts, foaming, and reduced anti-wear capabilities.

HYDRAULIC FLUIDS COMPATIBILITY MATRIX

Compatible | Marginal/acceptable in certain applications | Non-compatible/not allowed

E Engine oil | H Hydraulic fluid | T Tractor fluid | ZF Zinc-free | MS Multiseason | Bio Biodegradable | ATF Automatic | AW Antiwear

	E_5W30_CI4	E_10W30_CJ4	E_10W30_CI4	E_10W40_CI4	E_15W40_CJ4	E_15W40_CI4	E_0W40_CJ4	E_0W40_CI4	H_ZF46_AW	H_ZF32_AW	H_68_AW	H_46_AW	H_46_AW_EUR	H_32_AW	H_22_AW	H_BIO46_AW	H_BIO32_AW	H_MS43_AW	T_68_J20C	T_32_J20D	T_MS_J20C/D	T_ZF68_J20C	T_BIO_J20C	T_ATF	T_10W_TO4	T_30W_TO4	T_50W_TO4
E_5W30_CI4		1	1	1	1	1	1	1	2	2	3	3	3	3	3	4	4	3	5	5	5	7	7	8	10	10	10
E_10W30_CJ4	1		1	1		1	1	1	2	2	3	3	3	3	3	4	4	3	5	5	5	7	7	8	10	10	10
E_10W30_CI4	1	1		1	1		1	1	2	2	3	3	3	3	3	4	4	3	5	5	5	7	7	8	10	10	10
E_10W40_CI4	1	1	1		1	1	1	1	2	2	3	3	3	3	3	4	4	3	5	5	5	7	7	8	10	10	10
E_15W40_CJ4	1	1		1		1	1	1	2	2	3	3	3	3	3	4	4	3	5	5	5	7	7	8	10	10	10
E_15W40_CI4	1		1	1	1		1	1	2	2	3	3	3	3	3	4	4	3	5	5	5	7	7	8	10	10	10
E_0W40_CJ4	1	1	1	1	1	1		1	2	2	3	3	3	3	3	4	4	3	5	5	5	7	7	8	10	10	10
E_0W40_CI4	1	1	1	1	1	1	1		2	2	3	3	3	3	3	4	4	3	5	5	5	7	7	8	10	10	10
H_ZF46_AW	2	2	2	2	2	2	2	2			2a	2a	2a	2a	2a	4	4	2a	2	2	2	2	2	8	12	12	12
H_ZF32_AW	2	2	2	2	2	2	2	2			2a	2a	2a	2a	2a	4	4	2a	2	2	2	2	2	8	12	12	12
H_68_AW	3	3	3	3	3	3	3	3	2a	2a						4	4		11	11	11	11	11	8	10	10	10
H_46_AW	3	3	3	3	3	3	3	3	2a	2a						4	4		11	11	11	11	11	8	10	10	10
H_46_AW_EUR	3	3	3	3	3	3	3	3	2a	2a						4	4		11	11	11	11	11	8	10	10	10
H_32_AW	3	3	3	3	3	3	3	3	2a	2a						4	4		11	11	11	11	11	8	10	10	10
H_22_AW	3	3	3	3	3	3	3	3	2a	2a						4	4		11	11	11	11	11	8	10	10	10
H_BIO46_AW	4	4	4	4	4	4	4	4	4	4	4	4	4	4	4			4	4	4	4	4	4	8	4	4	4
H_BIO32_AW	4	4	4	4	4	4	4	4	4	4	4	4	4	4	4			4	4	4	4	4	4	8	4	4	4
H_MS 43_AW	3	3	3	3	3	3	3	3	3	3						4	4		7	7	7	7	7	8	7	7	7
T_68_J20C	5	5	5	5	5	5	5	5	12	12	7	7	7	7	7	4	4					12	7	8	10	10	10
T_32_J20D	5	5	5	5	5	5	5	5	12	12	7	7	7	7	7	4	4					12	7	8	10	10	10
T_MS_J20C/D	5	5	5	5	5	5	5	5	12	12	7	7	7	7	7	4	4					12	7	8	10	10	10
T_ZF68_J20C	6	6	6	6	6	6	6	6	6	6	6	6	6	6	6	4	4	6	6	6	6		6	8	6	6	6
T_BIO_J20C	7	7	7	7	7	7	7	7	7	7	7	7	7	7	7	4	4	7	7	7	7	7		8	7	7	7
T_ATF	8	8	8	8	8	8	8	8	8	8	8	8	8	8	8	4	4	8	8	8	8	8	8		8	8	8
T_10W_TO4	10	10	10	10	10	10	10	10	12	12	10	10	10	10	10	4	4	10	10	10	10	10	10	8			
T_30W_TO4	10	10	10	10	10	10	10	10	12	12	10	10	10	10	10	4	4	10	10	10	10	10	10	8			
T_50W_TO4	10	10	10	10	10	10	10	10	12	12	10	10	10	10	10	4	4	10	10	10	10	10	10	8			

As a general rule, mixing should never happen because fluids are designed for specific application purposes.

1	Mixing is permissible but the system will run with high particle counts. Viscosity will be difficult to trend
2	Zinc-free hydraulic fluids are not allowed to be mixed with engine oils which are zinc-based. Consequences of this mixing have not been measured. When mixed with engine oil it
2a	Zinc-free fluids are allowed, but not encouraged, to be mixed with zinc-base AW hydraulic fluids. Intervals needs to be reduced by 50%. Expect high particle counts.
3	Zinc-based hydraulic fluids blend with engine oils and zinc-free hydraulic fluids. Expect high particle counts.
4	Biodegradable hydraulic fluids should never be mixed with anything else. Mixing can cause gels and loss of biodegradability properties.
5	Tractor fluids should never be mixed with engine oils because serious foaming could develop.
6	Zinc-Free tractor fluids should never be mixed with any tractor fluid J20C or TO4
7	Biodegradable tractor fluids should never be mixed with non-bio product or bio-product of different type, as their biodegradability and performance will be hindered.
8	ATF Automatic transmission fluids are designed for a very specific application and should never be blended with anything else.
9	NOT USED
10	CAT's TO4 fluids are not part of the John Deere tractor fluids but they are popular fluids in mixed fleet. There is little application interchangeability between these lubricants but if
11	Zinc-based hydraulic fluids are simply not compatible with tractor fluids. Serious foaming and damage to hydraulic pump could occur.
12	Mixing of zinc-free hydraulic fluids with tractor fluids or TO4 should never happen at any time.

FIGURE 2-30 Hydraulic fluids compatibility matrix.

Where Lubrication Happens

Most of the lubrication requirements for mobile equipment are met at the job site in a less-than-clean environment. However, many companies address this challenge by performing these activities with well-equipped lubrication trucks that deliver lubricants from a tank directly to the machine. Others companies that are less well equipped use manual deliveries from a pail or drum with funnels or hand pumps, resulting in lesser quality of the lubrication service.

In terms of space, lubrication trucks need to consider, not only new and diverse oils, but also returning used oils. Therefore, space is an issue in lubrication trucks. For this reason, lubricant types need to be minimized so that the trucks can carry enough volume to service large fleets. Users need to optimize the proliferation of lubricants for their mixed fleets and find ways to service their fleets without having to worry about "black sheep" machines that require specialized lubricants (see Figure 2-31).

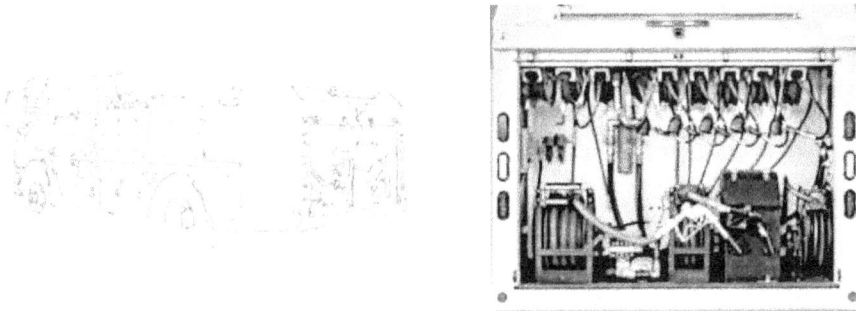

FIGURE 2-31 Limited number of tanks in a lubrication truck.

Lubricant Optimization

Most mobile equipment fleets are made up of different brands and types of equipment. Each brand has its own lubricant requirements, which sometimes makes compliance difficult. For example, a given manufacturer could be adamant that only zinc-free fluids can be used in its machines' hydraulics, whereas other manufacturers may require a certain amount of zinc in hydraulic fluid formulations. Users end up with hard decision to make:

- How to reduce the number of engine oils, hydraulic fluids, and transmission fluids to suit most of the fleet?
- How to decide which machine brand will lead fluid selection?
- How to determine the warranties, if any, that manufacturers will put on the user for not sticking to their lubrication recommendations?
- How to determine the fluid variations the user can accommodate based on lubrication truck cargo space?

Let us address the two first considerations using an imaginary construction company. Let us assume that the company has 45 machines distributed as shown in Table 2-16.

TABLE 2-16 Fleet Composition

No. of Units	Type	Brand
4	4WD Loaders	A
3		B
2		C
4	Excavators	A
4		B
3	Bulldozers	A
4		B
3	Backhoe	A
2	Loaders	B
5	Compactors	A
3	ADTs	A
3	Dump Trucks	A
3		B
2	Scrapers	A
45		

There are eight types of equipment in this imaginary company, which involve 18 different types of fluids, among oils, coolant, diesel exhaust fluid (DEF), grease, and fuel—all this without considering fluids and oils for seasonal changes. The task for the fleet manager is to decrease the proliferation of different fluids to ensure that the lubrication truck will be able to carry all the needed fluids, but also ensuring that the fleet gets the proper lubricants. In a fleet the size of this exercise, the demand for lubricants is going to be on the order of 500 gallons per month without counting fuels, DEF, and coolants.

The task starts with an analysis of which systems could use a single type of lubricant. Possibly the easiest choice is engines because most brands of engines can run very well on one type of oil. The second decision involves hydraulics. In many cases, the hydraulic pumps of different brand machines are similar and come from the same manufacturer. With rare exceptions, a hydraulic fluid won't work for all hydraulic systems of a given brand. There will be instructions from manufacturers to use their lubricants with their machines. Still, consulting with machine dealers is always a good practice, and very possibly the operator's manuals of the individual machines will display the options available.

There are 16 types of lubricants required for this fleet without counting fuel, coolant, DEF, and grease—three types of engine oils, six types of hydraulic fluids, four transmission fluids, and three gear oils (see Table 2-17).

TABLE 2-17 Fleet Lubrication Diversity

Type	Brand	Engine	Gal/h	Hydraulics	Gal/h	Transmission	Gal/h	Axles/Final Drives	Gal/h
4WD Loaders	A	Tier 4-CK4 15W40/10W30	0.133	AW46	0.080	J20C	0.008	J20C	0.036
	B	Tier 3-CJ4 15W40/10W30	0.120	CAT 10	0.040	TO4	0.006	GL5	0.069
	C	Tier 2-CI4 15W40/10W30	0.080	10W30	0.067	ATF C	0.004	GL5 LS	0.02
Excavators	A	Tier 4-CK4 15W40/10W30	0.160	Zinc-Free AW46	0.113	NA	0.008	GL5	0.02
	B	Tier 3 CJ4 15W40/10W30	0.147	AW46	0.088	NA	0.008	GL5	0.036
Bulldozers	A	Tier 4-CK4 15W40/10W30	0.100	TO4	0.060	TO4	0.006	GL5	0.0105
	B	Tier 3 CJ4 15W40/10W30	0.160	10W30	0.147	10W30	0.008	GL5	0.02
Backhoe Loaders	A	Tier 4 CK4 15W40/10W30	0.080	10W30	0.015	J20C	0.006	J20C	0.0105
	B	Tier 2-CI4 15W40/10W30	0.047	CAT 10	0.009	TO4	0.004	J20C	0.007
Compactors	A	Tier 3-CJ4 15W40/10W30	0.117	Zinc-Free AW46	0.020	Zinc-Free AW46	0.01	GL5	0.0125
ADTs	A	Tier 3 CJ-4 15W40/10W30	0.120	CAT 10	0.031	CAT 10	0.006	GL5	0.027
Dump Trucks	A	Tier 4 CK-4 15W40/10W30	0.100	10W	0.010	ATF ZF	0.006	GL5	0.0075
	B	Tier 4-CK4 15W40/10W30	0.090	10W	0.010	ATF	0.006	GL5	0.0075
Scrapers	A	Tier 2 CI4 15W40/10W30	0.080	10W	0.057	TO4	0.004	GL5	0.012
Hour	3		1.533	6	0.747	4	0.09	3	0.2955
Day			12.267		5.973		0.72		2.364
Month			294.4		143.36		17.28		56.736

Engines

Because the oldest machines in the fleet are equipped with tier 2 engines, we can assume that these machines will run well with CK4 oils. However, the fleet also has tier 4 engines, which also require CK4 oil. Because CK4 oils can be run in older engines, it makes sense to choose CK4 oil to run the entire fleet, which helps consolidate the number of oils (see Figure 2-32).

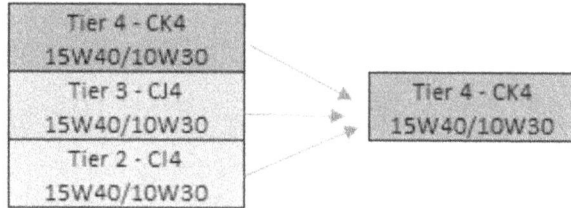

FIGURE 2-32 Engine oil optimization.

Hydraulics

The optimization of hydraulic fluids can take different approaches depending on the economy of scale and operating costs. The approach shown in Figure 2-33 is one option, but it is not the only one.

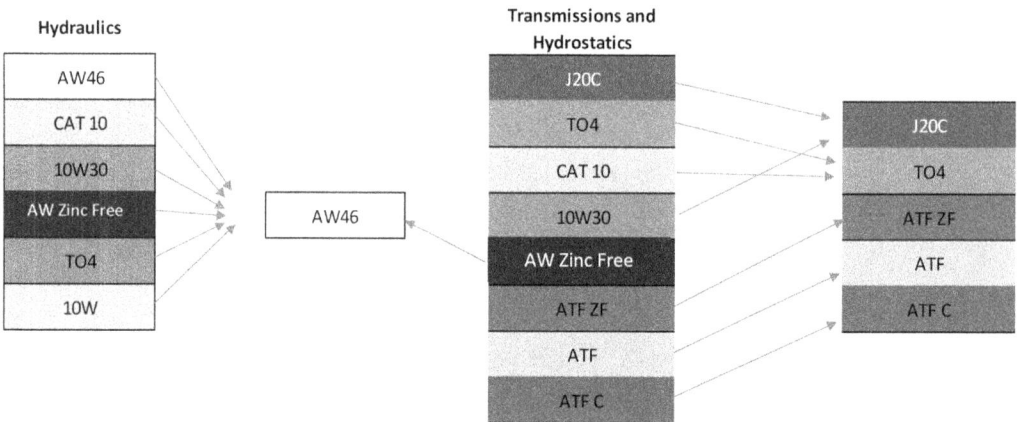

FIGURE 2-33 Hydraulic and transmission fluid optimization.

When deciding on which product to use, try to determine the best fluid or the one best suited to your operation. For example, John Deere and Caterpillar each recommend hydraulic fluids with much higher levels of zinc than typical AW hydraulic fluids, so one of these fluids likely would benefit other brands of equipment as well. John Deere's Hydrau, Hydrau XR, and Cat's HYDO disperse water. If you do not periodically purge water from the hydraulic system or clean your hydraulic fluid with a filter cart, as you should, dispersing water is safer for a piston pump with water-contaminated fluid.

Transmissions

The optimization of transmission fluids is more complex and limited. The materials used in clutches are very specific and need the dedicated friction modifiers required by the manufacturers. The options are very limited. J20C, for example, has little interchangeability with TO-4 and vice versa. However, they both meet the obsolete API GL4 standard, and they can be used in some applications without clutches or wet brakes, assuming that their viscosity matches the application.

Axles and Final Drives

Similarly, there is very limited overlap with gear oils, and some require LS additives.

In summary, we started with 16 different oils, including engine oils, and ended up with six. This is a reduction of 37.5% in lubricant proliferation. Every fleet is going to be different, and every fleet will offer optimization possibilities.

On switching from one hydraulic fluid to a different one, plenty of the original fluid remains in the machine's system. Getting rid of this remaining fluid takes great effort. Figure 2-33 offers some guidance in establishing the outcomes of various fluid mixtures. In the preceding case example, the most typical adverse reaction would be high particle counts in mixtures with zinc-free fluid, which are no more than gels that form when the two different chemistries comingle.

Flushing Hydraulic Systems

When it comes to changing 100% of the fluid in a machine, it could take up to six reservoir volumes to achieve 90% new fluid. Please refer to Chapter 4 for more details on how to successfully flush a system.

LUBRICANT STORAGE

Storing lubricants requires consideration of shelf life to additive separation. It is important to consider where to keep lubricants. The following list deserves discussion, so let's discuss each point:

- Lubricant shelf life (temperature stability)
- Additive separation
- Labeling (proper identification)
- First in, first out (FIFO) arrangement (easier access with high-activity lubricants)
- Lubricant cleanliness (filtration warnings)
- Desiccant filters for totes
- Assigned pumps by type of lubricant
- Intelligent filter caddies

Lubricant Shelf Life and Storage

Any active company that uses mobile equipment stores lubricants with enough inventory to ensure continuous supply for maintenance operations, but avoids overstocking unless those purchases enjoy a good discount when done in large quantities. Stable temperature has a direct influence on lubricant shelf life. To achieve a stable temperature, pails, drums, and totes need protection from the elements. If the temperature of oil totes remains stable, there is less opportunity to draw in humid air, thus keeping moisture contamination to a minimum. Because stored lubricants have no activity and do not generate heat, they cannot eliminate humidity by themselves. To achieve dry air inside totes, desiccant filters offer a solution (see Figure 2-34).

FIGURE 2-34 Poor handling of lubricants.

Additive Separation

Be aware that certain lubricants that stay in inventory for long periods could suffer from additive separation, which is expected and is not an abnormal condition. Heavily loaded tractor fluids or engine oils could present this phenomenon, and the solution is to agitate the lubricant. With pails, shaking and turning them can help, whereas totes and barrels benefit from recirculation via a dedicated pump.

Labeling (Proper Identification)

Labeling and proper identification help to avoid serious mixing issues. With inadequate labeling, mistakes can happen even with fresh deliveries from the supplier. For lubrication technicians, it is important that they have no doubts about the product they are about to pour into a component. The same care applies when filling the tanks in the lubrication truck. Mistakes are very expensive and sometimes disastrous.

First In, First Out

Some oils are used with great frequency, which is the case for engine oils and hydraulic fluids. It makes business sense to get rid of the oldest inventory first to always keep a good balance of lubricants with extended shelf lives. Inventory should be rotated when taking delivery of pails and totes so that the oldest inventory is the easiest to reach.

Lubricant Cleanliness

New lubricants are not necessarily clean from a particle count point of view. There are soft particles and unblended additives that could give high particle counts. The natural tendency is to connect a filter caddy and try to reduce particle counts by circulating the oil though a filter. However, the chances of affecting the additive balance in the oil are real because the filter can trap coarse additives easily when the oil is cold. This is even more risky when bulk tanks are located underground and temperatures are low. Attempting filtration without a heat source will produce a reduction in coarse additives, especially calcium and magnesium. As an option, it is better to send a sample to the lab to confirm the nature of the particle counts. If the results show no significant metallic particles, dirt, or water, it is better to leave the oil cleaning to the machine's filtration system.

Tote Air Desiccant

For operations that need to stock oil in bulk tanks or totes, the use of desiccant filters is a sound idea. Some oils are more hygroscopic than others, but regardless, keeping the airspace free of humidity contributes to the well-being of the fluid. This is especially true for hydraulic fluids, which by nature need to be dry.

Pumps and Mixing

For smaller operations where lubricant transfers are made with hand or electric pumps, it is imperative that these pumps are always dedicated to one fluid. An ordinary barrel pump can hold more than a quart of oil, so using the pump from the oil X barrel or tank to pump oil Y, or vice versa, is a clear path to cross-contamination.

Intelligent Filter Caddies

Intelligent filter caddies have become popular and have several uses within a mobile equipment fleet. An intelligent filter caddy can:

- Pump new fluid through filtration to a machine
- Polish the fluid while the machine is in operation
- Remove free water from the fluid
- Check for particle counts
- Measure water saturation
- Transfer fluids without filtration

Intelligent filter caddies should have storage for several filters so that they can filter more than one type of fluid. However, ideally, under the proposed fluid consolidation scenario, the caddy will be dedicated to filtering only one type of fluid (see Figure 2-35).

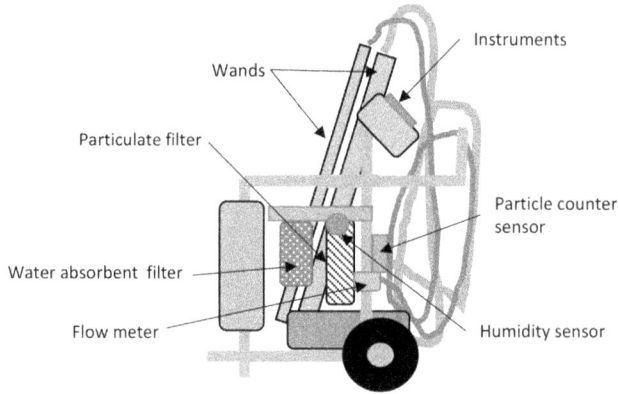

FIGURE 2-35 Intelligent filter caddy.

GREASES

In the world of mobile equipment, grease is the most common lubricant, and in many cases it is a daily lubricant. With advances in manufacturing and lubricant quality, greasing intervals have been extending. This is not to say that in special applications where water or extreme dust is present, long intervals can still stand.

The word *grease* comes from the Greek word *crassus*, which means "animal fat."

Grease has been used since 1400 BC, and records show that it was already being used by the Greeks and Egyptians, who used grease to lubricate chariot wheels and to facilitate the transportation of wooden sleds carrying large stones cut from the quarry.

In 1845, a grease made from animal fat, mineral oil, and limestone was produced in United States. Shortly afterwards, the British came up with a beef tallow and sodium grease. Sodium and calcium soap-based greases were in use by the turn of the twentieth century, but both were needed because sodium greases suffered from poor water tolerance, and calcium greases cannot withstand high temperatures. Multipurpose lithium grease was developed in 1938 and eliminated the need for most users to carry more than one type of grease. In 1954, aluminum complex grease came to the market, followed by lithium complex grease in 1959 that could withstand much higher temperatures than conventional greases.

We need thick lubricants in places where other lubricants leak. We also need the sealing power of grease to keep water or other contaminants out of critical external moving parts. Grease is used where additional wear protection is needed from friction and corrosion.

To produce grease, three basic components are needed: a base oil, a thickening agent, and an additive package. Given the large number of base oil and thickener options, many types of grease are feasible, which allows us to cover many different needs and applications.

To achieve the desired grease properties, many types of base oil come into consideration. Therefore, paraffinic, naphthenic, polyalphaolefins (PAOs), esters, and glycol base oils could end up in grease. The same principles we used in selecting oils apply to selecting greases for different applications in mobile equipment. If we need grease for arctic use, then lower-viscosity base oil goes into the formulation. Alternately, if we need a high-temperature grease, then a heavier-viscosity base oil goes in.

Two categories of thickeners are commonly used by grease manufacturers: organic and inorganic. Organic thickeners can be soap based or non–soap based, whereas inorganic thickeners are always non–soap based. Soaps come from the saponification of fatty acids or esters, which can be vegetable or animal based, combined with an alkali or an alkaline earth metal. They are reacted under heat, pressure, and agitation. The earth metals (i.e., lithium, sodium, and calcium) provide the stability and physical properties of the product.

To improve the dropping point and load-carrying capability of a grease, an agent goes into the mix for secondary reactions to convert the thickener structure to a soap–salt complex thickener. These final products are the greases we call *complex* and can include lithium, aluminum, sodium, and calcium. Extreme pressure (EP), antiwear (AW), and friction-reduction additives also go into greases to give them the enhanced performance needed by different applications, in similar fashion to engines oils and hydraulic fluids. Requirements for compatibility, environmental regulations, color, and cost also figure into the grease manufacturing equation (see Figure 2-36).

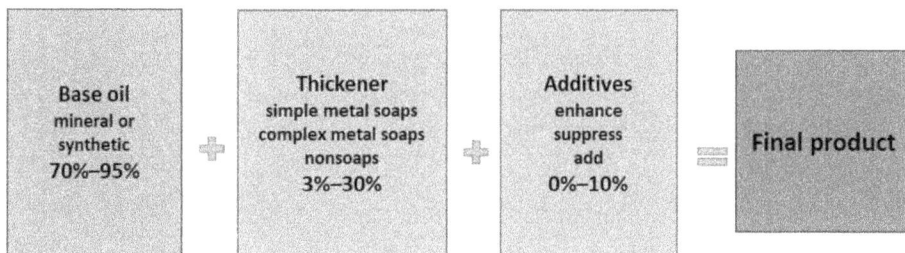

FIGURE 2-36 Grease manufacturing equation.

Grease Classification

The National Lubricating Grease Institute (NLGI) established nine grades of thickness or consistency, from grade 000 to 6. Grade 000 grease is very fluid and pourable inside a component compartment, whereas grade 6 is more like a brick of cheese. The most common grease used by the mobile industry is NLGI 2, in many cases of the EP type. However, if the application requires performance in arctic weather, NLGI 1 would be a better option. Harder grades are good to avoid leakage or water washout.

Grease Temperature Performance

One aspect that should help users with selecting the right grease for their application has to do with temperature performance. Table 2-18 helps in understanding the range and limitations of commercial greases in deciding what is best for a fleet. In an equipment fleet, there are going to be different grease requirements, the same as with fluids and oils, and optimizing the selection of greases is the task of the maintenance manager in making logistics and maintenance decisions.

TABLE 2-18 Grease Operating Temperatures

NLGI Consistency Grade	Worked Penetration Range ASTM 217 at 77°F	Consistency
000	445–475	Fluid, like ketchup
00	400–430	Fluid, like yogurt
0	355–385	Very soft, like mustard
1	310–340	Slightly soft, like tomato sauce
2	265–295	Most common, like peanut butter
3	220–250	Stiff, like margarine
4	175–205	Hard, like frozen yogurt
5	130–160	Hard, like fudge
6	85–115	Very hard, like brick cheese

Use Table 2-19 to define the temperature ranges of various grades of grease.

TABLE 2-19 Temperature Ranges of Various Grease Grades

Thickener	Dropping Point at Temp. °F	Max Temp. in Use °F
Aluminum Soap	230	175
Calcium Soap	270–290	250
Sodium Soap	340–350	250
Lithium Soap	390	275
Calcium Complex	>500	350
Lithium Complex	>500	350
Aluminum Complex	>500	325
Polyuria (Nonsoap)	>450	350
Organic Clay Nonsoap)	>500	350

Grease Colors

Although some initiatives were posited for grease color coding, there are no fixed rules for grease colors. It was suggested that the industry use the color coding shown in Table 2-20.

TABLE 2-20 Grease Color Coding

Color	Intended Purpose
White	Food machinery
Black	Heavy-duty applications
Red	High temperature
Blue	Cold temperature
Yellowish	Chassis lubrication
Green	Environmentally friendly

Keep in mind that color does not add value to any lubricating property of a grease but that excess amounts of dye may deteriorate some properties, such as copper corrosion resistance. The addition of titanium or zinc oxides (TiO_2 and ZnO), which impart a white color to greases, also does not add value to any grease properties, but may increase wear and oil bleeding when used in higher concentrations. Additionally, some grease manufacturers may not follow the loosely established color rule, and this could cause misinterpretation of the applicability of the product.

In summary, grease color coding was intended to help technicians use the right product while doing their routine fleet lubrication, but now it is used in marketing instead of identifying types of grease. After use and time, the grease will change color anyway, but the coding provides an initial visual that helps technicians recognize which grease they are using, assuming that they do not have two grease products that are the same color.

Performance Specifications for Automotive Use (Mobile Equipment Included)

The Classification and Specification of Automotive Service Greases Standard (ASTM D4950) sets performance specifications for automotive chassis (LA, LB) and wheel-bearing greases (GA, GB, and GC), with NLGI GC-LB representing the highest performance levels (see Tables 2-21 and 2-22).

TABLE 2-21   ASTM D4950 Chassis Grease Categories

Test	Property	LA Limit	LB Limit
D217	Consistency, worked penetration, mm/10	220–340	220–340
D566 or D2265	Dropping point, °C, min	80	150
D2266	Wear protection, scar diameter, mm, max	0.9	0.6
D4289	Elastomer SAE AMS 3217/3B compatibility		
	Volume change %	0 to 40	0 to 40
	Hardness change, durometer A-points	–15 to 0	–15 to 0
D1742	Oil separation, mass %, max		10
D1743	Rust protection, rating, max		Pass
D2596	EP performance		
	Load wear index, kgf, min		30
	Weld point, kgf, min		200
D4170	Fretting protection, mass loss, mg, max		10
D4693	Low-temperature performance, torque at –40°C, N·m, max		15.5

TABLE 2-22 ASTM D4950 Wheel Bearing Grease Categories

Test	Property	GA Limit	GB Limit	GC Limit
D217	Consistency, worked penetration, mm/10	220–340	220–340	220–340
D566 or D2265	Dropping point, °C, min	80	175	220
D4693	Low-temperature performance, torque at −40°C, N·m, max	15.5	15.5	15.5
D1264	Water resistance at 80°C, %, max		15	15
D1742	Oil separation, mass %, max		10	6
D1743	Rust protection, rating, max		Pass	Pass
D2266	Wear protection, scar diameter, mm, max		0.9	0.9
D3527	High temperature life, hours, min		40	80
D4289	Elastomer SAEA AMS 3217/3B compatibility			
	Volume change, %		−5 to +30	−5 to +30
	Hardness change, durometer-A points		−15 to +2	−15 to +2
D4290	Leakage tendency, g, max		24	10
D2596	EP performance			
	Load wear index, kfg, min			30
	Weld point, kfg, min			200

- *LA.* For chassis and universal joints in passenger cars, trucks, and other mild-duty vehicles with frequent relubrications in noncritical applications
- *LB.* For chassis and universal joints in passenger cars, trucks, and other mild- to severe-duty vehicles with prolonged relubrication intervals, high loads, severe vibration, and exposure to water and other contaminants
- *GA.* Wheel bearings in passenger cars, trucks, and other mild-duty vehicles that require frequent relubrications in noncritical applications
- *GB.* Wheel bearings in passenger cars, trucks, and other mild- to moderate-duty vehicles operated under normal urban, highway, and off-road service
- *GC.* Wheel bearings in passenger cars, trucks, and other mild- to severe-duty vehicles operated with high bearing temperatures (includes vehicle with frequent stop-and-go service, such as buses, taxis, and urban police cars, or under severe braking service, such as trailer towing, heavy loading, and mountain driving)

NLGI HIGH-PERFORMANCE GREASE SPECIFICATIONS FOR MOBILE EQUIPMENT (2021)

As of 2021, new high-performance multiuse (HPM) grease specifications for mobile equipment became available from the NLGI. The new specifications are much more stringent and are suited for diverse mobile equipment. The new specifications offer a core specification and four performance tags. There is no limit to the number of performance tags for each certified grease as long as they meet the specifications.

HPM Core

In the HPM core specifications, seven tests are common to the GC-LB specifications but may have more restrictive limits:

- Cone penetration (ASTM D217)
- Elastomer compatibility (ASTM D4289)
- Water washout (ASTM D1264)
- Oil separation (ASTM D1742)
- Four-ball wear (ASTM D2266)
- Four-ball EP (ASTM D2596)
- Corrosion prevention (ASTM D1743)

Other tests, not included in the GC-LB specification, include two mechanical stability tests:

- Extended worker penetration (100,000 strokes by ASTM D217)
- Roll stability (ASTM D1831)

There are two more corrosion tests:

- EMCOR rust test (distilled water by ASTM D6138)
- Copper corrosion (ASTM D4048)

Two more temperature tests have been added:

- Oxidation stability (ASTM D942)
- High-temperature oil bleed (ASTM D6184)

Finally, low-temperature torque of ball bearing grease (ASTM D1478) replaced low-temperature torque of wheel bearing grease (ASTM D4693). All these tests raise the level of performance compared with the GC-LB specifications. The new specifications are more relevant to multiuse in industrial applications (see Table 2-23).

TABLE 2-23 HPM Specifications

Property	Test Conditions	Test Method	Units	Min	Max
Cone penetration of lubricating grease	Worked 60 strokes	ASTM D217	cmm	220	340
Cone penetration of lubricating grease	Worked 60 strokes penetration (Δ100K)	ASTM D217	cmm	−30	30
Elastomer compatibility of lubricating grease and fluids (Using NBR standard reference elastomer ISO 13226)	168 hours at 125°C	ASTMD4289	ΔHardness (Shore A points)	−15	2
			Volume percent	−5	30
Oxidation stability of lubricant greases by the Oxygen Pressure Vessel Method	Pressure drop after 100 hours at 100°C	ASTM D942	kPa (psl)		35 (4.9)
Water washout characteristics of lubricant greases	60 minutes at 79°C	ASTM D1264	wt%		10
Low temperature torque of ball bearing grease	−20°C	ASTM D1478			
Starting torque			nMn (g-cm)		1000 (10,200)
Running torque at 60 minutes			nm (g-cm)		100 (1,020)
Oil separation from lubricating grease during storage	24 hours at 25°C	ASTM D1742	wt%		5.0
Oil separation from lubricating grease (conical sieves method)	30 hours at 100°C	ASTM D6184	wt%		7.0
Roll stability of lubricant grease (using 1/2 scale penetration)	2 hours at room temperature	ASTM D1831	dmm	−10%	10%
Wear preventive characteristics of lubricating grease (Four-Ball Method wear scar diameter)	75°C, 1200 RPM, 60 minutes	ASTM D2200	mm		0.6
Measurement of extreme-pressure properties of lubricant grease (Four-Ball Method), weld point	1770 rpm at 27°C	ASTM D2596	kgf	250	
Corrosion preventive properties of lubricant grease	48 hours at 52°C	ASTM D1743	rating	Pass	
Corrsion-preventive properties of lubricant grease under dynamic wet conditions (Emcor Test)	Distilled water, 2 bearings	ASTM D6138	rating		0.1
Copper corrosion from lubricant grease	24 hours at 100°C	ASTM D4048	rating		1B

HPM + WR (Water Resistance)

The HPM+WR specifications include three tests intended to demonstrate an increased level of performance over the HPM core specifications in wet or water-wash environments. Water washout (ASTM D1264) is the same test as used in the HPM core specifications, but it has a more restrictive limit. Water spray off (ASTM D4049) demonstrates a grease's ability to resist water spray, whereas wet roll stability (ASTM D8022) evaluates the effect of water on grease mechanical stability (see Table 2-24).

TABLE 2-24   HPM+WR Specifications

Property	Test Conditions	Test Method	Units	Min	Max
Determining the water washout characteristics of lubricant grease	60 minutes at 79ºC	ASTM D1264	wt%		5.0
Determining the resistance of lubricating grease to water spray	5 minutes at 38ºC	ASTM D4049	wt%		40
Roll stability of lubricant grease in presence of water (10% by wt distilled water) (using $^1/_2$ scale penetration)	2 hours at room temperature	ASTM D8022	dmm	−15%	15%

HPM+CR (Saltwater Corrosion Resistance)

The HPM+CR specifications include three tests intended to demonstrate improved corrosion resistance over the HPM core specifications in saltwater environments. Saltwater rust (ASTM D5969) is similar to ASTM D1763 in the HPM core specifications but uses 10% synthetic seawater. Two versions of EMCOR rust (ASTM D6138) evaluate corrosion protection in both 100% synthetic seawater and 0.5 N sodium chloride solution (see Table 2-25).

TABLE 2-25   HPN+CR Specifications

Property	Test Conditions	Test Method	Units	Min	Max
Corrosion-preventive properties of lubricant grease in presence of dilute synthetic sea water environments	10% synthetic seawater (as in ASTM D665)	ASTM D5969	Rating	Pass	
Determination of corrosion-preventive properties of lubricating grease under dynamic wet conditions (Emcor Test)	100% synthetic seawater (as in ASTM D665)	ASTM D6138	Rating		1.2
Determination of corrosion-preventive properties of lubricating grease under dynamic wet conditions (Emcor Test)	0.5 N solution (3% NaCl solution)	ASTM D6138	Rating		2.3

HPM+LT (Low Temperature)

The HPM+LT specifications include three tests intended to demonstrate improved low-temperature performance over the HPM core specifications. Low temperature torque of ball bearing grease (ASTM D1478) is the same test as used in the HPM core specifications, but it runs at a lower temperature. Grease mobility (U.S. Steel method) demonstrates grease resistance to flow at low temperatures, whereas flow pressure (Kesternich method, DIN 51805) is another way to look at flow at low temperatures (see Table 2-26).

TABLE 2-26 HPM+LT Specifications

Property	Test Conditions	Test Method	Units	Min	Max
Low temperature of ball bearing grease	−30°C	ASTM D1478			
Starting torque			mMm (g-cm)		1000 (10,200)
Running torque at 60 minutes			mNm (g-cm)		100 (1,020)
Grease mobility	−20°C	U.S. Steel LT-37	g/min	10	
Determining of flow pressure of lubricating greases according to Kesternich Method	−30°C	DIN 51805			1400

HPM+HL (High Load)

The HPM+HL specifications include five tests intended to demonstrate improved load-carrying capability over the HPM core specifications. Both the four-ball wear (ASTM D2266) and four-ball EP (ASTM D2596) tests are the same tests as in the HPM core specifications, but they have limits that are more challenging. These properties over the fretting wear by FAFNIR test (ASTM D4170) are also included with a tighter limit than the same test in the LB specifications (see Table 2-27).

TABLE 2-27 HPM+HL Specifications

Property	Test Conditions	Test Method	Units	Min	Max
Wear preventive characteristics of lubricanting grease (Four-Ball Method) wear scar diameter	75 C, 1200 rpm, 60 minutes	ASTM D2266	mn		0.50
Measurement of extrme-pressure properties of lubricating grease (Four-Ball Method), wels point	1770 rpm at 27°C	ASTM D2596	kgf	400	
Determining extreme pressure properties of lubricant greases using a high frequency, linera-oslillation (SRV) test machine	Procedure B at 80°C	ASTM D5706	N	800	
Fretting wear protection by lubricating grease	Average of 2 runs, 22 hours at room temperature	ASTM D4170	mg		5.0

Property	Test Conditions	Test Method	Units	Min	Max
Determining fretting wear resistance of lubricanting greases under high hertzian contanct pressures using a high frequency, linear-oscillating (SRV) test machine	50°C, 100N, 0.300mm, 4 hours	ASTM D7594	mm		0.500

Grease Selection

Grease selection for mobile equipment requires some basic criteria to make the best decision—to start with, operating temperature, penetration, dropping point, water spray-off resistance, washout resistance, four-ball testing, Timken load testing, and thickener compatibility. Table 2-28 can help in selecting the type of grease thickener based on application and the advantages and disadvantages each type. For most mobile equipment, *lithium complex* grease in NLG1 and NLG2 thickness is the most popular. In certain areas of machines that receive shock loads (e.g., booms and arms) or have large swing gears and bearings, a grease with EP additives is the right call.

TABLE 2-28 Grease Selection/Applications

Thickener	Advantages/Disadvantages	Applications
Aluminum soap	Low dropping point, good water resistance.	Low speed bearings, wet applications.
Calcium soap	Low dropping point, good water resistance.	Bearings working exposed to water, railroad.
Sodium soap	Poor water resistance. Good adhesion.	Old equipment with frequent lubrication needs.
Lithium soap	High dropping point.	Chassis, wheel bearings, general purpose.
Calcium complex	Good water resistance, Good EP lubricant.	Automotive and heavy equipment, high temp.
Lithium complex	Resistant to leaking, moderate resistance to water.	Automotive and heavy equipment, high temp.
Aluminum complex	Excellent water resistance, resistance to softening.	Industrial bearings at high temp.
Polyurea	Good water resistance and oxidation. Tends to soften and leak.	CV Joints and drive shafts.
Organic clay	Good water resistance. Thickener does not liquefy.	High temp. bearings that require frequent lubrication.

In a fleet, choosing the grease that will work in every moving part is a matter of economics and logistics. Having one grease that satisfies all the needs of the machines is highly attractive if the cost of doing so makes sense.

Grease Compatibility

Perhaps one of the lubrication activities that has lesser scrutiny in a fleet is grease management. After all, the types of greases used are of limited variety, and sometimes a single good-for-all grease is adopted without major consideration of the impact on the specific components that depend on it.

Mistakes in mixing greases happen all the time. The same as with oils and fluids, mixing of greases is very common and sometimes leads to disastrous wear. An infamous example is the case of an Alaskan Airlines accident in 2001, where the elevator worm-type screw mechanism wore out because incompatible greases were used. Let us use the compatibility chart in Figure 2-37.

Perhaps the most important consideration with grease mixing is thickener compatibility. Some mixes will cause the grease to soften to the point of migrating out of the area of mechanical contact, whereas other mixes may harden, limiting the protection the grease is supposed to give. Extreme cases can result in the oil separating from the thickener mixture. Compatibility charts abound, but you should take them with a grain of salt because they often contradict each other, and some authors have not updated the information for a long time.

The oils that go into grease manufacturing could be mineral or synthetic. Some of these products are incompatible and cause the grease to underperform. The same care we take to avoid oil mixing needs to be applied to grease.

We also need to consider the compatibility of the additive package. If the additives start fighting among themselves, this will not cause foaming or settling as in oils, but the antiwear protection they are supposed to give could suffer. *At best, we need to consider the following initial steps if mixing is going to occur (as in changing from one type of grease to another):*

- Make sure that the dropping point of the mixture is not significantly lower than that of the individual greases.
- Check that the mechanical stability of the mixture is within the range of consistency of the individual greases.

	Aluminum complex	Barium	Bentonite clay	Calcium	Calcium 12-hydroxy	Calcium complex	Calcium sulfonate	Lithium	Lithium 12-hydroxy	Lithium complex	Polyurea
Aluminum complex											
Barium											
Bentonite clay											
Calcium											
Calcium 12-hydroxy							NA				
Calcium complex											
Calcium sulfonate					NA						
Lithium											
Lithium 12-hydroxy											
Lithium complex											
Polyurea											

Green = Compatible   Yellow: Marginal   Black= Incompatible

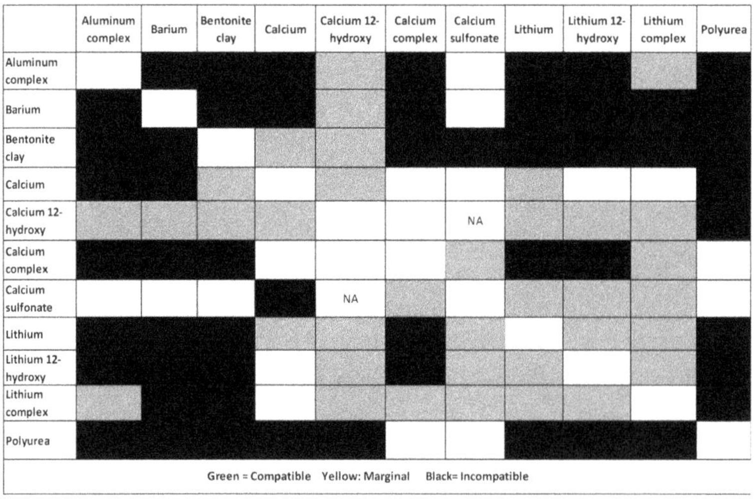

FIGURE 2-37   Testing grease compatibility.

The most incompatible greases are barium- and clay-based greases, followed by aluminum complex and polyurea-thickened greases. The most compatibles greases are calcium sulfonate, calcium 12-hydroxystearate, and all lithium-based greases. In mixtures with barium greases, oil often separates easily.

Not all thickeners of the same group are compatible with each other. Polyurea grease is an example of this because two polyurea grease formulations in specific cases may not be compatible with each other. Sometimes thickeners are generally compatible, but two greases may contain clashing base oil or additive formulations.

Choosing the Best Option for a Fleet

Based on the information in Figure 2-37 it is reasonable to consider the use of the most compatible greases to service a fleet to achieve two objectives: limit the proliferation of types of greases and get the best protection at the same time. This does not negate the need for specialty greases in specific applications.

Calcium sulfonate greases exhibit superior mechanical and shear stability over lithium complex greases, indicating less leakage and run-out during operation. The dropping point and high-temperature life of calcium sulfonate greases are also better, allowing these greases to be used at higher temperatures.

Calcium sulfonate thickeners have inherent EP and AW properties but also provide excellent water-resistance properties, and they do not break down, even in the presence of water. As a comparison, lithium complex greases usually require tackifiers, which are prone to deplete quickly in the presence of water. Calcium sulfonate greases are also compatible with lithium and lithium complex greases.

Example of a Greasing Lubrication Decision

There are four basic considerations when choosing the right grease for a fleet: application, environment, season, and economy. A subset of these considerations involves compatibility. Under application, there are shock loads and work in water, such as excavators with long fronts. Some machines operate under constant shock loads that benefit from greases with at least 3% of molybdenum disulfide (moly). And some machines/applications are going to require a grease that is superior at withstanding operation in water (see Figure 2-38).

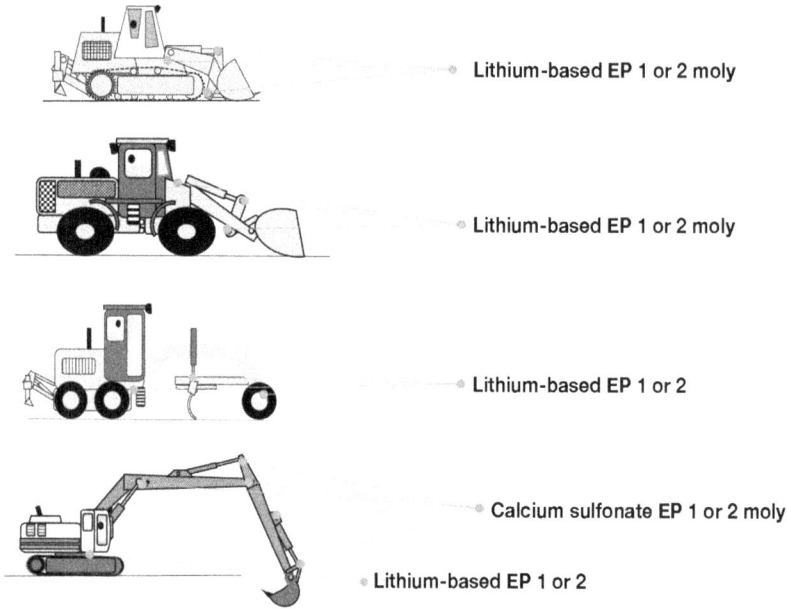

Lithium-based EP 1 or 2 moly

Lithium-based EP 1 or 2 moly

Lithium-based EP 1 or 2

Calcium sulfonate EP 1 or 2 moly

Lithium-based EP 1 or 2

FIGURE 2-38 Grease selection example.

Greases for High-Load and High-Shock Applications

Heavy-duty sliding applications, such as the arm of a construction excavator or the boom pivots of a four-wheel-drive loader, are going to require the help of heavy-duty grease in NLGI grades 1 or 2 that contains moly. Moly is a solid additive that is used most commonly in EP applications.

The moly particles have a laminar structure that slides apart to reduce friction and provide passive EP protection. The particles fill in tiny imperfections in the machined surfaces and protect against "welding" during shock loading and heavy loads. However, the material tends to stick to surfaces and cause a coating to build up. For this reason, there are limitations on its use, such as not to exceed 5% moly content in automatic lubricators because it could cause seizing of the lubricator pump pistons (see Figure 2-39).

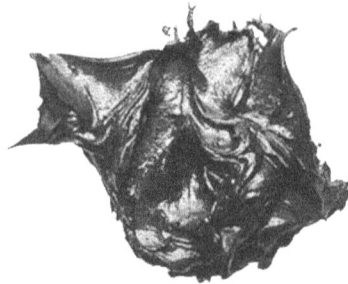

FIGURE 2-39 Molybdenum disulfide (moly) grease.

Automatic Lubricators

Automatic lubrication solves the problem of inconsistent lubrication by reducing the human element. Automation makes the lubrication events more consistent and avoids waste. Automatic systems apply a small replenishing volume of grease, keeping a constant film of lubricant between the contact surfaces of the bearing or bushing. With automatic lubrication, there may not be the same need for moly grease because every bearing will receive the grease it requires to keep it lubricated and purged of contaminants (see Figure 2-40).

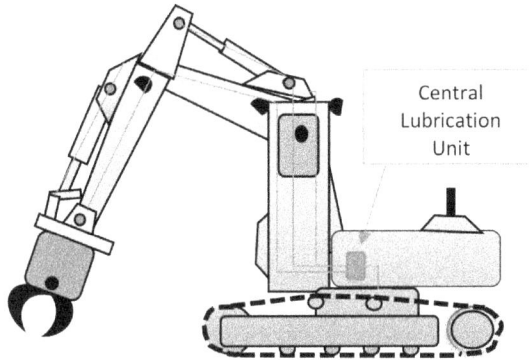

FIGURE 2-40   Centralized lubrication.

The advantages of centralized lubrication are easy to understand by examining Figure 2-41. Automatic and centralized lubrication ensures that the components get grease all the time with no dry periods and no waste. The challenge becomes the automated system itself, which requires maintenance and monitoring to ensure that it never goes dry and that the lines do not plug.

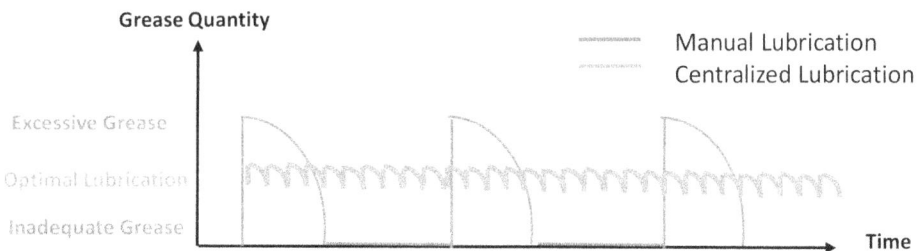

FIGURE 2-41   Centralized greasing.

CONCLUSION

This chapter covered various aspects of lubrication together with additives and types of oils and greases. The chapter also covered physical properties and compatibility. Now it is time to move to Chapter 3, Contamination, and learn how to measure and control various contaminants.

CHAPTER 3

CONTAMINATION

Contamination is the presence of minor and unwanted constituents or impurities that could be solid, gaseous, or fluids in a host fluid. The presence of these contaminates impairs lubrication by third-body contact, increases two-body contact, or causes pitting by vapor cavitation and/or corrosion (see Figure 3-1).

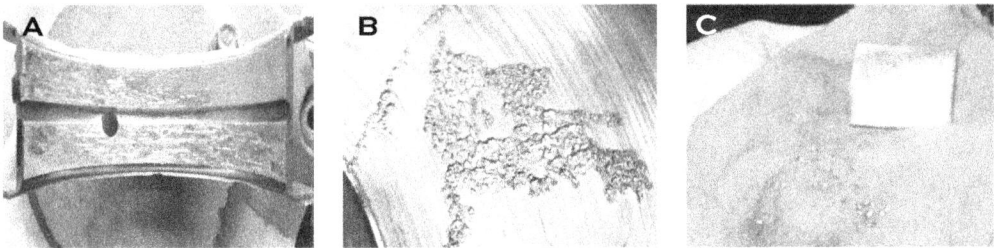

FIGURE 3-1 (A) Bearing failure; (B) pump cavitation; (C) Pump failure.

What Are Those Contaminants?

Contaminants come in three large categories: gases, liquids, and solids. Most gaseous and liquid contaminants have aggressive effects on wear, as do some solid contaminants. However, some solids have less harmful effects, such as soft metals, fibers, carbon, and rubber. During this discussion, we will cover most of the aggressive contaminants in more detail (see Figure3-2).

Gases	Liquids	Solids
Air Nitrogen Hydrogen	Water Glycol Fuel	Dirt Metals Oxides
		Fibers Rubber Carbon

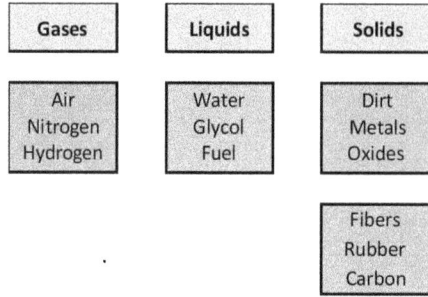

FIGURE 3-2 Categories of contaminants.

Particle Sizes and Visibility

Most machines are fluid-dependent systems, just like the human body. Lubricants, hydraulic fluids, coolants, fuels, and air bring contaminants into the system and transport the contaminants within the system. When too much contamination is in a system, the components wear, and machine performance drops. This could lead to shorter service life or catastrophic failure, just like a massive heart attack.

Humans tend to believe only in what they see. Humans understand that a microscopic world and an outside world exist. They accept big unknowns based on faith, culture, and beliefs. However, because they do not normally carry a microscope or a telescope with them, they accept that these worlds exist but ignore the intimacies that occur within those worlds. In fact, there is very little general knowledge about these worlds and a lot of skepticism.

Particles that cause harm are usually smaller than our eyes can see. So let us start the discussion with this premise in the forefront. The range of visibility for the human eye is around 40 μm, which is about half the diameter of a human hair. Nevertheless, to say that we will see those sizes is very optimistic. The fact is that a particle of 40 μm or even 100 μm will not be visible if the background is similar in color to the particle. Take the case of a large particle in oil; very possibly, we will not be able to see it.

Particle classification comes from the wear pattern generation or morphology. Figure 3-3 can help us identify some of the particles we must deal with. Rubbing particles are very small, whereas fatigue and sliding wear particles are generally large. In our discussion, we will be dealing with particles measured in microns. These particles are the ones that cause most of the wear issues.

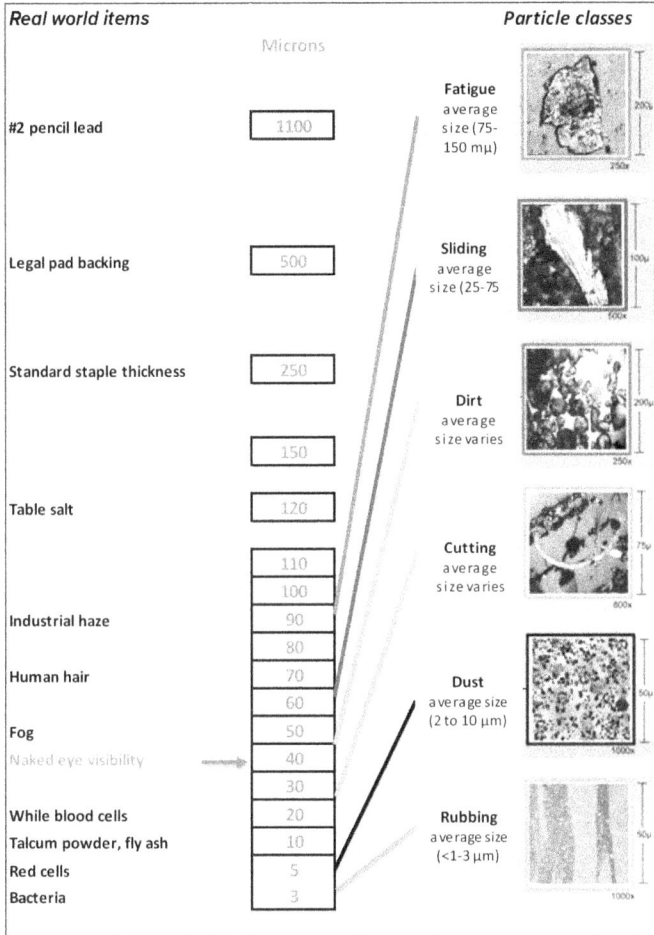

Real world items	Microns		Particle classes
#2 pencil lead	1100		**Fatigue** average size (75-150 mμ)
Legal pad backing	500		**Sliding** average size (25-75
Standard staple thickness	250		
	150		**Dirt** average size varies
Table salt	120		
	110		**Cutting** average size varies
	100		
Industrial haze	90		
	80		
Human hair	70		**Dust** average size (2 to 10 μm)
	60		
Fog	50		
Naked eye visibility	40		
	30		
While blood cells	20		**Rubbing** average size (<1-3 μm)
Talcum powder, fly ash	10		
Red cells	5		
Bacteria	3		

1 millimeter	=	1000 microns (μm)
1 micron	=	1000 nanometers
1 nanometer	=	1000 picometers
1 picometer	=	1000 femtometers
1 femtometer	=	10000 attometers

FIGURE 3-3 Particle contaminant classes.

How Small Is One Part per Million, Billion, and Trillion?

Table 3-1 provides some perspective on the sizes of real-world items and their potential concentrations. In general, this industry does not use concentrations in parts per billion very often, but concentrations in parts per million (ppm) are the most common units to describe contamination content in a sample.

TABLE 3-1 Perspectives on Real-World Items and Concentrations

Item	1 Part per million (ppm)	1 Part per billion (ppb)	1 Part per trillion (ppt)
Money	1 cent in $10,000	1 cent in $10,000,000	1 cent in $10,000,000,000
Time	1 minute in 2 years	1 second in 32 years	1 second in 3,200 years
Length	1 inch in 16 miles	1 inch in 16,000 miles	1 inch in 16,000,000 miles

The sources of contamination in hydraulic systems are common to other systems as well and behave in the same fashion. The primary sources are illustrated in Figure 3-4 and listed below:

1. **Residue from fabrication**, as in the case of reservoirs that have not been cleaned properly, leaving welding slag or dirt behind.

2. **Leaks through the cover or filler of a reservoir**, as in many types of reservoirs where an inspection cover depends on a gasket and a set of bolts. Water is the most common contaminant through the cover of a reservoir.

3. **Leaks through a breather.** Not all breathers can hold all types of contaminants. For instance, breathers allow some fine dust and humidity to go through. In addition, while pressure washing a machine, introducing water with the high-pressure spray is feasible.

4. **When topping off a machine**, if the new fluid or the container is not clean, the consequences are obvious. This is a very common mistake.

5. **Pump wear particles.** Although most pump wear is the results of prevalent contamination, material fatigue also can make a pump produce metals. Still, there are many reasons a pump produces metals, ranging from humidity to high total acid number (TAN). In addition, pump seals allow some dust ingression.

6. **Corrosion in tubing and reservoirs** is sometimes the major source of iron readings. Humidity, an aggressive fluid, or a high TAN could be the culprit.

7. **Cylinders are by far the most important source of dirt and humidity entry** into a hydraulic system. When leaks occur through seals, dirt and humidity will go in, and even with perfectly good seals, you can expect some contaminants to go in as well.

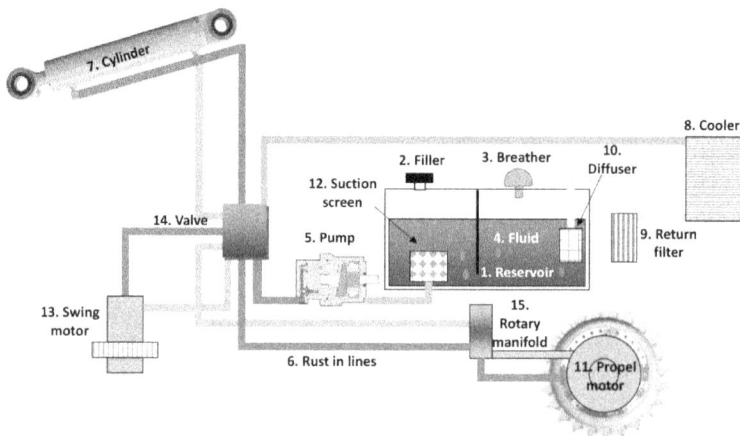

FIGURE 3-4 Sources of contamination.

Various industry statistics show that most failures in hydraulic systems do not happen because of design flaws but rather by contaminants such as water, dirt, or lubricant degradation. Typically, hydraulic systems provide many years of trouble-free service. However, on many occasions, the conditions in which these components work are changed by the actions of some contaminants, which the system's own filtration is unable to contain.

A second series of events occurs when the technicians repairing a unit after a catastrophic failure ignore the micro world on which the hydraulic system is so dependent. Their actions are to repair and replace the most affected components and flush the fluid from the reservoir, leaving behind millions of destructive particles lodged in lines, valves, coolers, and crevices.

The results are always the same: premature failure of the main component and even greater levels of contamination. The story repeats itself, and some may blame the manufacturer for the poor quality of the design. Figure 3-5 shows how contamination might occur through seals.

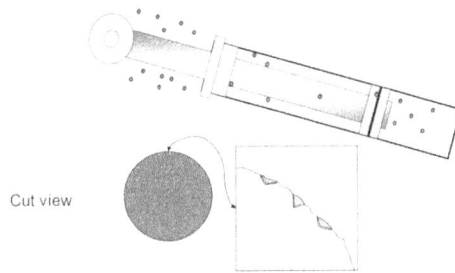

Cut view

FIGURE 3-5 Contamination through seals.

Figure 3-6 shows examples of poor maintenance. The first cylinder belongs to a bulldozer and the second to a marine crane, and both are leaking fluid. If cylinders leak, contamination occurs from dirt and from salty water, respectively.

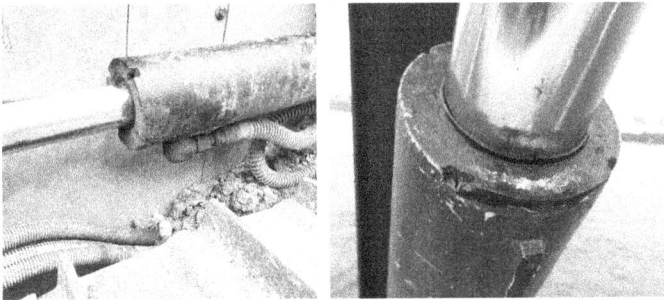

FIGURE 3-6 Leaking hydraulic cylinders.

In most applications, the elements present in the material the machines handles will show up in the oil analysis. For example, a machine working in a fertilizer plant will produce potassium and sodium readings in fluid analysis results.

The aggressiveness of these contaminants varies according to their type; dirt by itself is nasty because it contains very hard elements such as silicon and aluminum oxides. The maximum amount of dirt allowed in a fluid, without causing premature failure, is equivalent to one baby aspirin per 100 gallons of fluid. This amount is extremely small considering that a baby aspirin is perhaps 20 mg in 110 gallons, or the equivalent of 5 ppm (see Figure 3-7). The guidelines for silicon (constituent of dirt) need to be tight and in the range of 10 to 15 ppm.

FIGURE 3-7   One baby aspirin.

The Rossin–Rammler Distribution Graphic

Figure 3-8 shows that the majority of the small particles concentrate in sizes below 15 μm and that the hardest ones are in the range of 5 to 10 μm. The figure also shows that the smaller particles are the hardest ones.

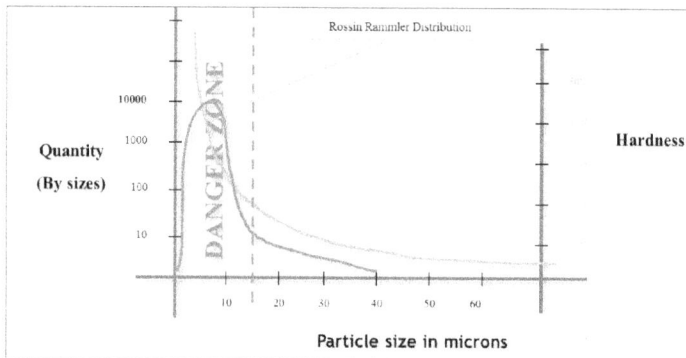

FIGURE 3-8   Rossin–Rammler particle distribution. (Courtesy of Noria Corporation.)

This brings us to some additional facts about contaminants:

1. Systems wear out silently right in front of us unless we do regular oil analyses.
2. Most filters in hydraulic systems handle small particles poorly unless the system uses low-micron bypass filtration.
3. The hardest and most dangerous particles are smaller than 15 μm, and we hardly control them.

Silica (SiO_2) is the oxidative form of silicon, a metalloid. Alumina (Al_2O_3) is the oxidative form of aluminum, a metal. Silica and alumina, as found in dirt, are among the hardest particles on Earth.

Silica and alumina will show up as silicon and aluminum in most oil analyses given their abundance in the Earth's crust. It is important to distinguish them from other forms that not are from dirt, namely oil foam inhibitors, sealant material (silicone), and wear from aluminum parts. When determining whether a sample has dirt based on silicon-aluminum ratio readings, we face the challenge of isolating dirt from other possible combinations that are not dirt and establishing whether the actual readings are in fact dirt. Keep in mind that the readings could be the result of silicates from coolant, foam inhibitor additives, or "gasket maker."

In the case of dirt, the typical silicon-aluminum ratios are between 4:1 and 6:1, depending on geographic location and whether the compartment has aluminum elements, as in the case of pistons in engines. Coolant is also a source of silicon readings because silicates, an additive that protects aluminum from corrosion, could be present. Note that:

- Ninety percent of the Earth's crust is made of silicon-containing materials.
- Silicon is not to be confused with silicone, a synthetic compound.
- Silicon has a hardness of 7 on the Mohs scale.
- Aluminum is the most abundant metal in the world.
- Aluminum is the third most abundant element on Earth.
- Alumina has a hardness of 9 on the Mohs scale (see Figure 3-9).

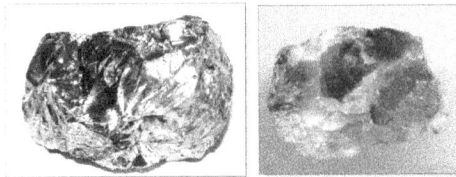

FIGURE 3-9 Silicon (*left*); aluminum/corundum (*right*).

Foam inhibitors made out of silicone compounds also could be present, depending on the type of fluid. Such readings may not show up in the typical silicon-aluminum ratio of 4:1 to 6:1, making the segregation of this reading from dirt easier. Silicone as sealant material tends to be easy to detect because of the lack of corresponding metals such as aluminum and because it typically appears in a new machine or a component that have been recently repaired and sealed. In summary:

- Silicon/aluminum as dirt → 6:1 to 4:1 ratio
- Silicon as silicates → as found in coolants
- Silicon as polydimethyl siloxane foam inhibitor → as found in oils
- Silicon as silicone sealant → as found in gasket maker

The presence of dirt leaves unequivocal marks that make identification of the issue much easier. Axial pumps and swash valve that have hardened surfaces can show scratches caused by hard contaminants, typically dirt (see Figure 3-10). Dirt also leaves unequivocal signs of its high hardness by denting roller races that end up with surface fatigue.

FIGURE 3-10   Dirt impact on a pump swash plate.

The example from a construction machine in Figure 3-11 shows several trending graph lines that react to the presence of silica. When silica (silicon) is present, everything wears out. Silica is the nastiest of all contaminants. It is fine, it is hard, and it is abundant. The different readings in the figure show that every time the silica was high, other elements, such as copper and iron, had a corresponding episode of trending up.

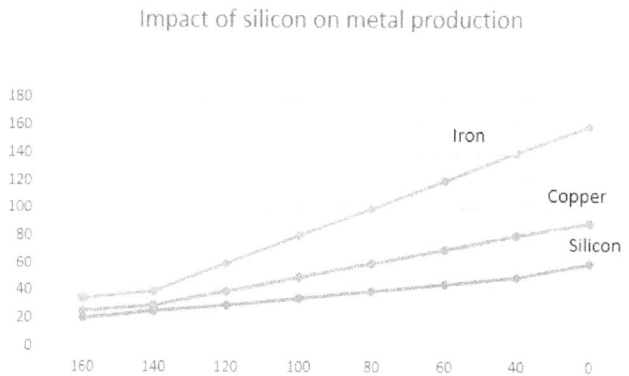

FIGURE 3-11   Trending silicon.

FILTRATION

Filters exist to capture contaminants, and there are many designs and uses. Surface filtration is a mesh that takes out coarse contaminants, and it is typically used in front of a pump. Deep filtration implies several layers of different media, synthetic or otherwise, that capture various types and sizes of particles, typically in the return flow to a tank. Deep filtration media also can be seen in the feeding line before a pump, as in hydrostatic pumps (see Figure 3-12).

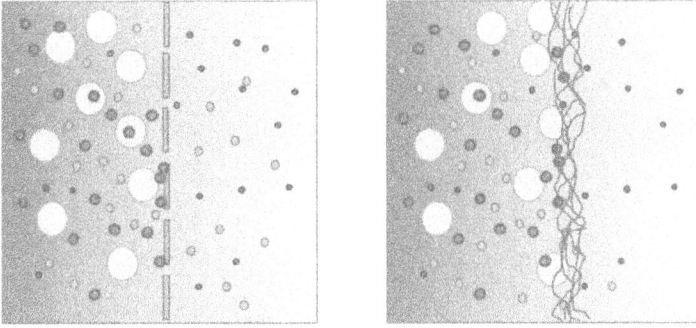

FIGURE 3-12 Surface filtration (*left*); deep filtration (*right*).

Filter Beta Efficiency

The *beta ratio* is a measure of a filter's efficiency in removing particles of a stated size. A single multipass test goes through many small time segments. During each of these counting periods, the number of particles of a specific size—size X—and greater upstream of the filter are totaled, and the number of size X and greater particles downstream of the filter are totaled. The number of particles found upstream of the filter divided by the number of particles found downstream of the filter equals the beta value of the filter element at the given particle size (see Figure 3-13 and Table 3-2).

$$\beta x = \frac{n'_{\text{ upstream} \geq X \, \mu m}}{n'_{\text{ downstream} \geq X \, \mu m}}$$

FIGURE 3-13 Filter beta rating.

Beta ratios for filter elements are determined during a multipass test (ISO 16889). The 1999 standard for multipass testing states that this test is applicable to filter elements that exhibit an average beta ratio ≥ 75. Individual element manufacturers determine the beta ratio specification for their elements. Most manufacturers currently use a minimum beta ratio of 200 for a particular micron rating.

TABLE 3-2 Beta Ratings

Beta	Efficiency	Particles upstream	Particles Downstream
2	50	100,000	50,000
4	75	100,000	25,000
10	90	100,000	10,000
20	95	100,000	5,000
40	97.5	100,000	2,500
60	98.33	100,000	1,667
75	98.66	100,000	1,333
100	99	100,000	1,000
125	99.2	100,000	800
150	99.33	100,000	667
200	99.5	100,000	500
300	99.66	100,000	333
500	99.8	100,000	200
1,000	99.9	100,000	100
2,000	99.95	100,000	50
4,000	99.97	100,000	25
5,000	99.98	100,000	20
10,000	99.99	100,000	10
20,000	99.995	100,000	5
50,000	99.998	100,000	2

Table 3-2 helps in understanding the beta efficiency of each category. A good beta filter is not always the answer on high-flow applications such as an excavator. The high flows require large filters with low beta ratings to avoid crushing the filters. In the case of hydrostatic systems, however, where the flow is between the pump and motor with no filter in between, the beta ratio for the charge flow filter could run as high as 200.

To grasp the idea of beta ratios in filters, some examples are useful. A filter cart, for example, could use a beta 1000 at 7-μm filter that is extremely capable of cleaning systems very fast, whereas a hydraulic bypass filter could have a beta 2000 at 1 μm equivalent, although bypass filters are not subject to this measurement and only get an estimation of their beta. Modern Tier 4 diesel engines use filters with betas around 5 at 7 μm, but previously had a beta of 1000 at 20 μm, which is a very impressive change with regard to the past.

Bypass Filtration

To understand bypass filtration and its benefits, we need to understand contamination and the impact of particles on a system first. Bypass filtration takes a small flow, typically around 1 gallon per minute from the main feeding line, and passes it through a very restrictive medium where very small particles are removed, sometimes down to 1 μm in size. With so little flow, it takes time for bypass filtration to achieve a good cleanliness level, but when it reaches equilibrium, that component will have a much better chance of exceeding its projected life than if it is only equipped with standard filtration.

An engine bypass filter, as manufactured by Puradyn, uses long strand cotton to capture particles and neutralize sulfur, but it also has an evaporator to get rid of water (see Figure 3-14). New models replacing this type of evaporator are already on the market. The new models use water-absorbing material to deal with water.

A hydraulic bypass filter, as manufactured by Stauff, has an even more restrictive medium than an engine bypass filter. The Stauff bypass filter has an equivalent beta of 2000 at 1 μm, which is a very impressive number (although there is no official testing for beta rating for these types of filters). This filter also captures water with water-absorbing material (see Figure 3-15).

FIGURE 3-14 Puradyn engine bypass filtration.

FIGURE 3-15 Stauff hydraulic bypass filter.

There are nonconventional bypass filter designs, such as the Balance Charge Agglomeration (BCA) filter, where the flow passes through electrodes, as shown in Figure 3-16. The system uses static electricity to attract varnish and separate it from larger particles. Larger particles are then removed downstream with another high-efficiency filter.

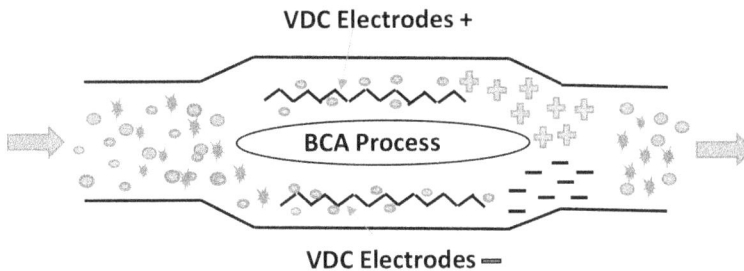

FIGURE 3-16 BCA electrostatic filtration for varnish.

Desiccant Filters

Desiccant filters, which remove humidity from incoming air, can be especially handy in applications where humidity needs to be low, as in the case of biodegradable hydraulic fluids, where the water content needs to be <500 ppm. These filters have silicone desiccant gel inside and some valves that allow the reservoir to inhale through the desiccant medium, but expel air directly. The filter typically has a window to show the amount of filter life remaining (see Figure 3-17).

FIGURE 3-17 Desiccant filter.

Rare Earth Magnets

Rare earth magnets are sometimes useful where filtration is not feasible. Given their strong magnetic field, they attract iron in a very efficient way. Users need to be aware that the ferrous debris, the nest of filings, could cause alarm to a casual observer. With experience, this point comes under control, and the rare earth magnet can be helpful in removing iron particles that otherwise would reduce bearing life (see Figure 3-18). However, rare earth magnets do not capture nonferrous materials, dirt, or fibers.

FIGURE 3-18 Rare earth magnet.

Heat and Air

Some authors consider heat and air to be contaminants because they contribute to fluid oxidation. Air accelerates the oxidation rate of oil, especially if aeration is present. Heat accelerates oxidation, with every 10°C (18°F) increase over 60°C (140°F) cutting lubricant life in half.

Overheating hydraulic systems and overheating axles are examples of systems that can run hotter than expected for numerous reasons. The fluid in an axle working at 20°C (36°F) above the ideal temperature will be shorter by three-fourths. Therefore, instead of changing oil every 2000 hours, a 500-hour oil change would be appropriate to remove the oxidized fluid. Knowing the cause of the overheating might solve the problem.

Some oils resist heat oxidation better than others, and this is the case with synthetic oils. As Figure 3-19 suggests, synthetic oils can run hotter and longer than conventional hydrocarbon oils if oxidation is the limiting factor.

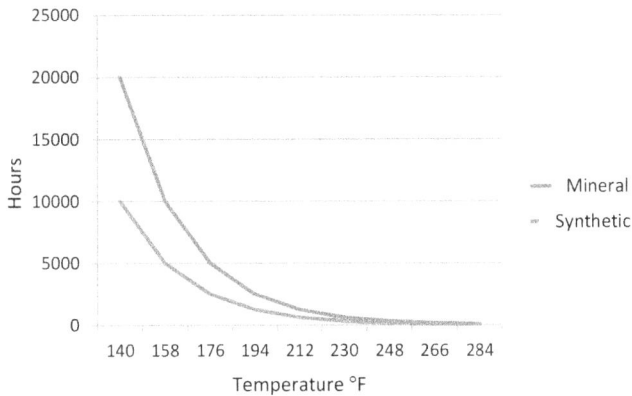

FIGURE 3-19   Heat and oxidation.

High compression of air bubbles generates high temperatures. This physical phenomenon occurs in hydraulic cylinders when they are assembled empty. The high temperature causes *microdieseling*, where the oil vapor inside air bubbles combusts, burning seals and turning fluid blackish. This phenomenon can occur in new machines and recently repaired cylinders. The carbon itself is not an issue for the proper operation of the machine but can be alarming due to the color of the fluid. Therefore, when repairing hydraulic cylinders, it is always a good idea to fill them with clean fluid before operating the machine to avoid microdieseling (see Figure 3-20).

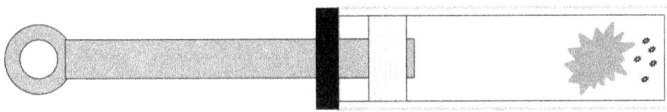

FIGURE 3-20   Cylinder microdieseling.

Air not only causes cavitation and microdieseling in hydraulic systems, but it also accelerates oxidation. Oxidation, in turn, causes an increase in acidity of the fluid, as represented in Figure 3-21. The main source of air is pump cavitation. Air can cause pump cavitation failure when the bubbles expand during the suction phase and then collapse violently in the pump pressure phase.

When cycling hydraulic cylinders that contain air, the fluid becomes foamy. This foamy condition dissipates with use. When using on-site particles counters, such as with a filter caddy, keep in mind that these air bubbles could give misleading particle readings because the particle counter interprets the tiny bubbles as particles.

% AIR	TAN
0	0.1
3	0.15
6	0.25
9	0.5
12	0.6

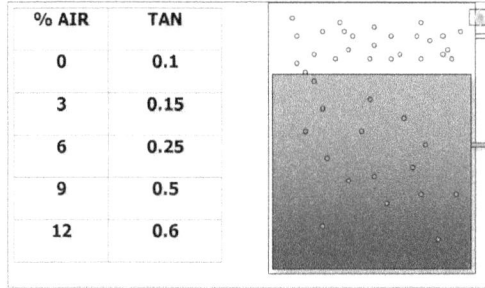

FIGURE 3-21 Air and total acid number (TAN) (Courtesy Noria Corp.)

Things to Look for during a Visual Inspection of a Machine

The educated eye pays more attention to the appearance of a fluid than to the actual level of the fluid. Both are important, but appearance will show whether there is an issue with the system. Items that we can see during an inspection of a machine, even without an inspection form, are listed in Figure 3-22.

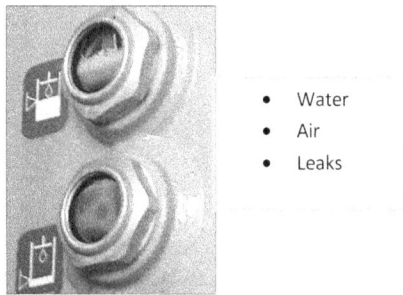

- Water
- Air
- Leaks

FIGURE 3-22 Air in a reservoir.

Metals as Catalysts

Metals can be catalysts that accelerate the oxidation process, which is the precursor of varnish. Copper is by far the most active in this sense. However, copper is typically not the most abundant metal in fluid samples—iron is. As shown in Figure 3-23, iron is still very active as a catalyst and generally is the catalyst involved in varnish generation.

Copper	100
Lead	75
Steel	60
Iron	45
Zinc	25
Tin	8
Aluminum	4

FIGURE 3-23 Oxidation by catalysts (%) (*left*); varnish (*right*).

Static Current

When two dissimilar materials come into contact and are then separated in a sliding motion, electron transfer occurs between the two materials. This is knowns as *triboelectric charging*. Frictional contact between a fluid and a filter medium in a lubrication or hydraulic system results in triboelectric charge generation that, if not dissipated, accumulates and discharges to a lower-potential surface, usually the filter housing. This is like a lightning bolt in a filter housing. Static current generated by high-velocity hydraulic fluid is sometimes the culprit in filter damage. If iron or other fluid oxidation catalysts are present, the static discharge may lead to varnish deposits.

Water Content: Parts per Million and Percentage of Saturation

Water can be present in hydraulic fluids and other lubricants as dissolved water, emulsified water, and/or free water. One drop of free water can damage a hydraulic pump, so free water is of primary concern. The insidious thing is that harmless dissolved water can fall out of solution as temperatures fall, becoming damaging free water.

The point at which a fluid cannot hold any more dissolved water is the *saturation point*. If more dissolved water is present than the fluid can hold, the excess water can be present as either separated free water or an emulsion. Note that the saturation curve is always at 100%. At low temperatures, oil cannot hold a great deal of water, but it can still be 100% saturated at that temperature.

Typically, oversaturated fluids appear cloudy. The amount of water a fluid can hold at saturation strongly depends on the type of base stock, the additive package, temperature, and pressure. For instance, highly refined mineral oils with few additives hold very little water before becoming saturated—about 100 ppm at 70°F. At the other extreme, engine oil and tractor fluids can have saturation levels of >3000 ppm at 70°F and even higher levels at higher temperatures (see Figure 3-24). Water-absorbing filters can collect free water. Ideally, the oil should not be too hot to capture the maximum amount of free water.

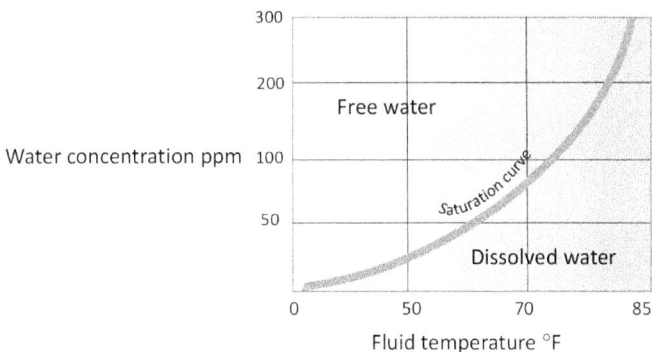

FIGURE 3-24 Water saturation curve.

When water contamination in oils exceed the saturation level of the specific oil, it becomes visible as a milky solution. When a fluid reaches this point of contamination in mobile equipment, there is little value in trying to save it. Using vacuum and heat techniques to remove water works for industries that use huge reservoirs, but these methods nearly always exceed the cost of fluid replacement and disposal in mobile equipment.

Visibility of Water

The visibility of water varies by the method used to measure it and whether we are measuring total water content or only free and emulsified water. Figure 3-25 provides a perspective on these differences.

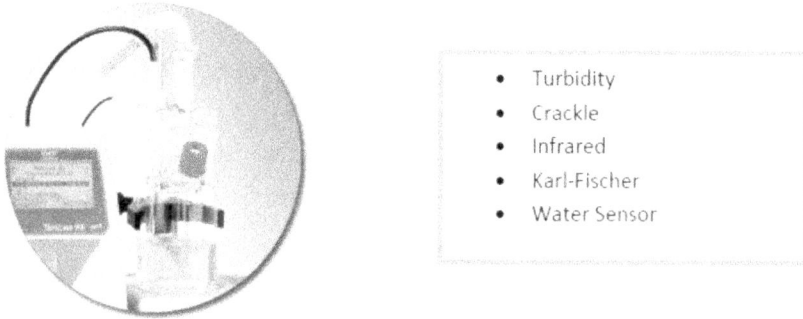

- Turbidity
- Crackle
- Infrared
- Karl-Fischer
- Water Sensor

FIGURE 3-25 Titration (*left*); methods (*right*).

The methods used to measure water in oil are always challenging for labs and users. Most of labs will test water by Fourier transform infrared (FTIR) spectroscopy and also by the crackle test. However, one should never forget the easiest way—visual observation of turbidity. When water contamination in oils exceeds the saturation level for the specific oil, the oil becomes milky and is very identifiable. The absence of a Karl Fischer codistillation test limits our ability to examine water, leaving a dangerous hidden zone, which for hydraulics is critical. The crackle test is effective, but it is very dependent on the human eye, and as we discussed earlier, the human eye is extremely limited (see Figure 3-26).

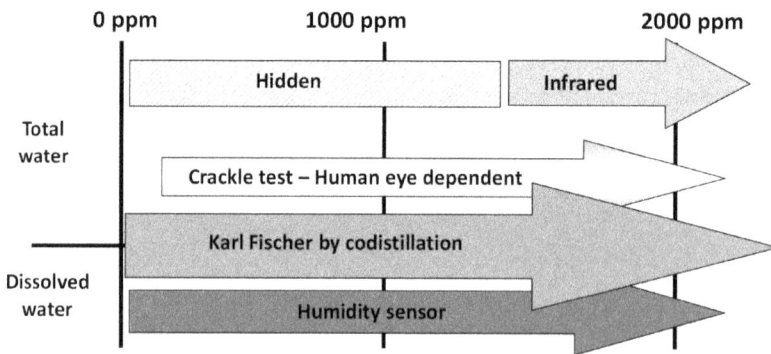

FIGURE 3-26 Visibility of water.

The infrared test is effective only for water concentrations that exceed 1600 ppm in oils that have a high calcium content, such as engine oils and tractor fluids. The infrared test potentially can detect water in hydraulic fluid at around 1200 ppm, but this is beyond the critical point, which typically ranges from 750 to 1000 ppm. The Karl Fischer test, for which there are several standards, offers high sensitivity in detecting water from 50 ppm and up, whereas a less sensitive test detects water at around 1000 ppm. The water sensor, as used in some field tools, provides a measurement of dissolved water in addition to providing a constant and cheap way to detect harmful free water.

When a fluid reaches this point of water saturation, there is little value in trying to save it. Removing the water requires expensive methods of vacuum dehydration and heat that are not justifiable in the relatively small volumes seen in mobile equipment (see Figure 3-27).

FIGURE 3-27   Water saturation.

Examples from Real-Life Situations

Figure 3-28 shows a maintenance flaw in the marine industry where a reservoir is lacking a cap on the filler neck. Allowing this condition to occur ensures contamination not only with salty water but also with dirt and particulate.

FIGURE 3-28   Marine application mishap.

In dredging operations, which are common for excavators, water ingression through cylinders is a real possibility. Figure 3-29 shows the rust generated in this water-contaminated machine.

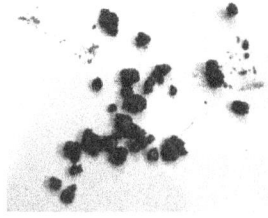

FIGURE 3-29 Rust from the reservoir of an excavator.

Visibility of Particles

Elemental analysis by inductively coupled plasma mass spectrometry (ICPMS) or other commonly reported methods show particle sizes up to 6–7 µm. Therefore, if you can see a particle or a big chunk of metal after a failure, do not expect the oil analysis to show it. This 6- to 7-µm size limit of elemental analysis leaves valuable information behind. Particle counts, as we will see later in this book, are essential to understanding total cleanliness of a fluid. Not having particle counts is equivalent to getting only half the story (see Figure 3-30).

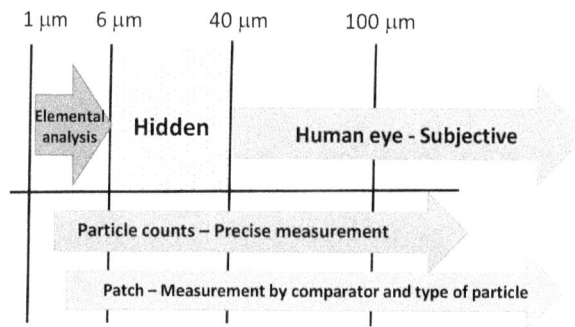

FIGURE 3-30. Visibility of particles.

Humans are very limited in determining contamination with the naked eye, and when the particles are big enough, we would need the proper background to be able to see them. Thus, for small particles, the human eye is useless without a microscope and a filter patch designed for this kind of activity.

There is a *hidden zone* of certain sizes of particles, the ones that cause most of the wear in systems and would be invisible without a particle count test. Figure 3-30 illustrates the justification for always having particle counts.

Wear Metals and Contaminant Sources

See Table 3-3 for details.

TABLE 3-3 Wear Metals and Contamination Sources

Wear Metals	System	Component Source	Caused by
Iron	Engines	Liners, oil pump, valve guides	Water, dirt, glycol, low TBN, high TAN, fuel, sulphation, nitration, oxidation, overheating
	Hydraulics	Cylinder, reservoir, hydraulic tubing, gear pumps, axial pumps, steel body gear pumps, fan pump	Water, dirt, low viscosity, rust
	Transmissions	Clutch discs	Water, clutch slippage, wrong fluid
	Axles, final drives, gearboxes	Planetary carriers	Water, dirt, brake discs
Chromium	Engines	Piston rings	Dirt, water, break-in period
	Hydraulics	Valve spools, pump bearings	Dirt, water, cylinder peeling, valve spools, pilot controls
	Transmissions	Bearings	Dirt, water, bearings
	Axles, final drives, gearboxes	Bearings	Dirt, water, bearings
Copper	Engines	Oil cooler, camshaft thrust washer, rocker arm bushings, connecting rod bushing, bearings	Cooler passivation caused by CI4 Plus or newer oils, dirt, water, fuel, low TBN, high TAN, coolant leak from cooler
	Hydraulics	Hydraulic pump	Dirt, water, aeration, foaming, cavitation, high temperature, mixing, oxidation, high TAN component failure
	Transmissions	Clutch plates, bearing cage and lube orifice (Funk transmission), thrust washers	Clutch slippage, overload, contamination, wrong fluid, loose lube orifice, high temperature
	Axles, final drives, gearboxes	Planetary thrust washers, diff. lock plates, spider gear thrust washers, bushings	Dirt, water, high TAN, wrong oil, overload, wheel spinning, failing thrust washers, defective thrust washers
Aluminum	Engines	Piston, dirt ingestion	Dirt, piston wear, compressor wear
	Hydraulics	Charge pump	Dirt, cavitation, failing pump
	Transmissions	Clutch piston (−), charge pump	Dirt, failing part
	Axles, final drives, gearboxes	None	Dirt

(continued on next page)

TABLE 3-3 Wear Metals and Contamination Sources (*continued*)

Wear Metals	System	Component Source	Caused by
Tin/Lead	Engines	Bearings, camshaft thrust washers	Dirt, water, glycol, fuel, high TAN, oil sulphation, oil oxidation
	Hydraulics	Hydraulic pump, valve plate, rotary group sleeves, swash cradle bearing	Dirt, water, wrong oil, high TAN, age wear, low viscosity, high temperature, high load
	Transmissions	Thrust washers	Dirt, water, wrong oil, high TAN, age wear, low viscosity, high temperature, high load
	Axles, final drives, gearboxes	Thrust washers	Dirt, water, wrong oil, high TAN, age wear, low viscosity, high temperature, high load
Nickel	Engines	Distribution gears	Normal wear, dirt, lubrication film loss, low viscosity
	Hydraulics	Carburized pump bearings	Dirt, water
	Transmissions	Carburized bearings, gears	Dirt, water, high iron content
	Axles, final drives, gearboxes	Carburized bearings, gears	Dirt, water, high iron content
Silver	Engines	Not typical from construction equipment. Solder from certain oil coolers	Leaching, high TAN
	Hydraulics	Not typical from construction equipment	NA
	Transmissions	Not typical from construction equipment	NA
	Axles, final drives, gearboxes	Not typical from construction equipment	NA
Titanium	None	None	Not a typical metal on construction/forestry equipment. However, it could be paint contamination or dust from mines. As of 2012, titanium is part of new additive packages in some CJ4 and CK4 oils.

Measuring and Counting

Gravimetric measurement requires a highly sensitive scale that can weigh parts to an accuracy of ±1 mg or better. This method is practical for small parts of any material to detect gross contamination and does not require any other sophisticated equipment besides the scale.

Parts must be weighed prior to being contaminated, and then, after contamination, they are cleaned, oven dried, and weighed again. The difference between the initial weight and the post-cleaning weight indicates any residual contamination left on or in the part. If there is no difference in these two weights, the part is classified as clean.

This is a good gross measurement method when extremely high purity is not required. A small laboratory helps to conduct these tests, and large parts cannot be tested in this way. The other methods to measure include using a particle counter or a microscope. In turn, a particle counter can be a lab instrument or a portable on-site instrument.

The microscope method depends on a patch and solvent to collect the particles from a given volume sample and analysis by direct observation or by means of a comparator catalog. There are also microscopes available with automated counters (see Figure 3-31).

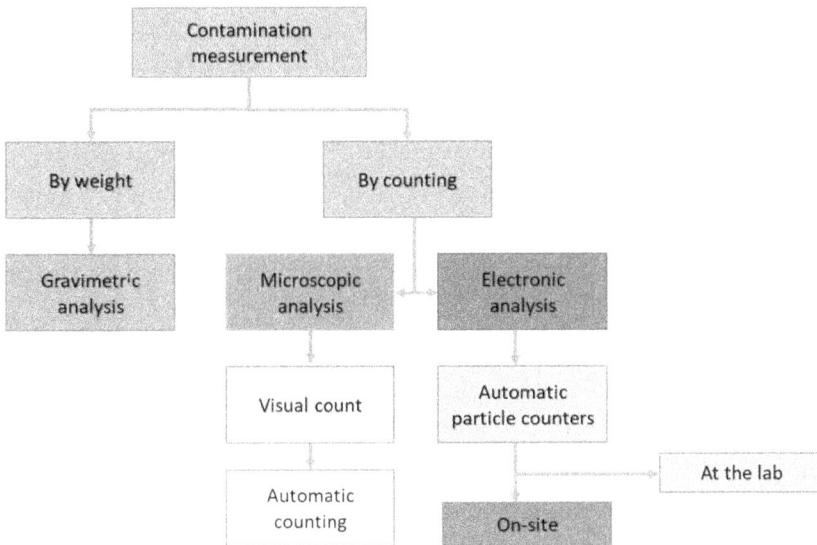

FIGURE 3-31 Measuring contamination.

Particle counts are a very important measurement, especially in hydraulic fluids. However, interpretation requires some understanding of this measurement's behavior. To start with, the type of fluid determines the 4-µm channel reading. In addition, the type of system and the pressure determine the general level of cleanliness needed. Another factor is whether the reservoir is sealed, semisealed, or open. Finally, whether the system is independent of other systems (e.g., combined hydraulics-hydrostatics) makes a difference in the way we interpret particle counts (see Table 3-4). A more detailed discussion of the interpretation of particle counts is provided in Chapter 6.

TABLE 3-4 Particle Counts by System

System	Normal	Abnormal	Critical
Secled Reservoir Hydraulics with AW Fluid	17/15/12–20/17/14	21/18/15–22/19/16	≥X/19/17
Secled Reservoir Hydraulics with Engine Oil or UTHF	19/15/12–22/17/14	23/18/15–24/19/16	≥X/20/17
Unsealed Reservoirs with AW46	18/16/13–20/18/15	21/19/16–25/20/17	≥X/21/18
Unsealed Reservoirs with Engine oil or UTHF	20/16/13–23/18/15	24/19/16–27/21/18	≥X/22/19
Sealed Hydrostatics with AW Fluid	17/15/12–20/17/14	21/18/15–22/19/16	≥X/19/17
Sealed Hydrostatics with Engine oil or UTHF	19/15/12–22/17/14	23/18/15–24/19/16	≥X/20/17

Particle Counts Can Hide Something Else

Particle counts are a great tool to measure fluid contamination, but they can also be a source of frustration when the numbers do not behave as the theory states. Chapter 6 explains more about particle counts and cleanliness codes.

The tradeoff with particle counters, when measuring certain fluids, is that they can suffer from air, water, or fluid mixing. Labs usually follow preparation steps to ensure that the sample is free from air: they vacuum the sample, and they also shake the sample, but water in the sample still could be identified as solid particles. In addition, fluid mixtures of incompatible additives tend to form gels that do not dissolve. These gels tend to give an inflated final code on the particle counter that could mislead the user.

With in-line particle counters or those on a filter caddy, water contamination tends to display codes in which the three numbers of the code are very close to each other. Air contamination tends to show identical numbers in every channel with such counters. A healthy reading should have minimum separations of two of the numbers, such as 19/16/13. International Organization for Standardization (ISO) codes also vary by type of oil. A pure, nonmixed antiwear (AW) hydraulic fluid is a straightforward measurement easily capable of clean readings on the order of 19/16/13. However, engine oil (used in hydraulics) or a universal tractor fluid will always show higher 4-μm particle counts caused by the additive packages typical of those lubricants, and normal readings for them are going to be on the order of 21/17/14 (see Figure 3-32).

Condition	Readings
Mixes	21/19/16
Water	21/21/19
Air	20/20/20

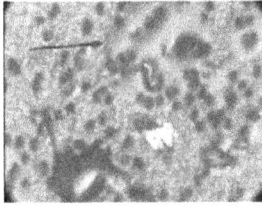

FIGURE 3-32 Particle count variations.

Cleanliness versus Life of Components: British Hydrodynamics

The cleanliness of a fluid or oil has an undeniable influence on the life of a component, yet most equipment users do not take its value seriously. Table 3-5 provides a compelling case for fluid cleanliness requirements. The statistics from a large fleet inspired this figure, which shows values for hydraulics and engines, bearings, friction bearings, and gearboxes. (The four sets of numbers in each box correspond to the larger matrix within the larger chart.) In the left column is the current cleanliness code, and on top is the desired/achievable cleanliness code. If a fleet manager decides to run his or her fleet cleaner, he or she must act to achieve the desired cleanliness. First, he or she needs to service the machines regularly and test and document particle counts to establish a baseline. If there are gels from incompatible mixtures, that needs to be corrected first. Extreme cleanliness levels are not feasible without bypass filtration, and to move to a higher cleanliness target, its adoption needs to be considered. The table shows component life extension if cleanliness goals are achieved. The better cleanliness numbers are not easy to achieve. They require commitment, specific actions, and controls.

TABLE 3-5 Life Extension Table

	12/10/7		13/11/8		14/12/9		15/13/10		16/14/11		17/15/12		18/16/13		19/17/14		20/18/15		21/19/16	
24/22/19	>10	>10	>10	7	>10	6	>10	5	8	4	7	3.5	6	3	4	2.5	3	2	2	1.6
	>10	8.5	10	5.5	8	5	7	4	5.5	3.5	4.5	3	3.5	2.5	3	2	2.3	1.7	1.8	1.3
23/21/18	>10	10	>10	7	>10	5	9	4	7	3.5	5	3	4	2.5	3	2	2	1.7	1.5	1.5
	10	8	9	5.5	7	4	5	3.5	4.5	3	3.5	2.5	3	2	2.2	1.6	1.8	1.4	1.5	1.3
22/20/17	>10	9	>10	7	9	5	7	4	5	3	4	2.5	3	2	2	1.7	1.6	1.5	1.3	1.2
	10	7	8	5.5	6	4	5	3	3.5	2.5	3	2	2.3	1.7	1.8	1.4	1.5	1.3	1.2	1.05
21/19/16	>10	8	9	6	7	4	5	3	4	2.5	3	2	2	1.7	1.6	1.5	1.3	1.2		
	9	6	7	4.5	5	3.5	3.5	2.5	3	2	2.2	1.7	1.8	1.5	1.5	1.3	1.2	1.1		
20/18/15	>10	6	7	4.6	5	3	4	2.5	3	2	2	1.7	1.6	1.5	1.3	1.2				
	8	5	5.5	3.7	3.5	2.5	3	2	2.3	1.7	1.8	1.5	1.5	1.3	1.2	1.1				
19/17/14	8	5	6	3	4	2.5	3	2	2	1.7	1.6	1.5	1.3	1.2						
	6	3.5	4	2.5	3	2	2.3	1.7	1.8	1.5	1.5	1.3	1.2	1.1						
18/16/13	6	4	4	3.5	3	2	2	1.7	1.6	1.5	1.3	1.2								
	4.5	3.5	3.7	3	2.3	1.8	1.8	1.5	1.5	1.3	1.2	1.1								
17/15/12	4	2.5	3	2	2	1.7	1.6	1.5	1.3	1.2										
	3	2.2	2.3	1.8	1.8	1.5	1.5	1.4	1.2	1.1										
16/14/11	3	2	2	1.8	1.6	1.6	1.3	1.3												
	2.3	1.8	1.9	1.5	1.6	1.4	1.3	1.2												
15/13/10	2.5	1.8	1.8	1.5	1.4	1.2														
	2	1.6	1.6	1.3	1.2	1.1														

Legend (each cell of the table is a 2×2 block):

Hydraulics, diesel engines	Ball and roller bearings
Friction bearings and turbines	Gearboxes and others

Particle Counts for Engines

Particle counts on engine oils are feasible, but not with the traditional methods using laser particle counters. Laser technology does not work with dark oils. Other test methods can be used, however, such as measurements from a pore blockage apparatus, although not all labs have this equipment. Another method to measure particle counts in engines is with a patch and microscope kit. The procedure dilutes the sample with solvent before passing it through the filter patch. Practical particle count measurements are achieved by comparing patches with a picture chart. Large contaminants can damage bearings, as shown in Figure 3-33.

FIGURE 3-33   Engine bearing failure caused by dirt.

Soot in Engines

Soot and carbon sediment deposits in engines are the result of a complex process. Figure 3-34 helps in understanding the process.

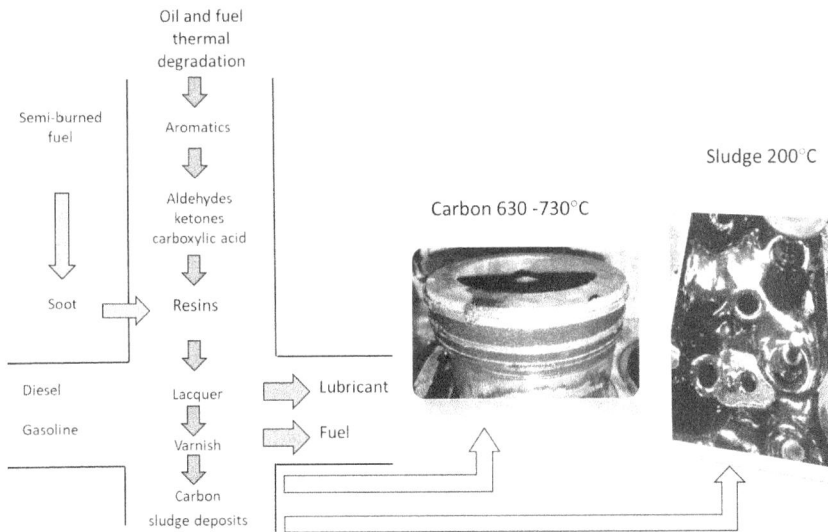

FIGURE 3-34   Carbon and soot.

The soot from semiburned fuel encounters the resins from oil breakdown and tries to cling to them. The use of detergents and dispersants helps keep soot and resin particles from sticking to surfaces and each other. Diesel and gasoline engines suffer from the same phenomenon with a slight difference in the deposits that form on the piston skirt. In diesel engines this deposit is called *lacquer*, whereas in gasoline engines it is called *varnish*.

The resin by-products end up forming soft or hard deposits depending on the operating temperature of the parts in the engine. Soft deposits, as we know, form in the oil sump and valve covers, where temperatures are in the 200°C range. Lacquer or varnish forms in the middle temperature range of 230°C, and hard deposit form on piston heads and valves where temperatures reach 630–730°C.

Cleaning Soot in Engines

Cleaning soot from engines is not an easy task without disassembling the engine. If the soot has not coagulated and has not caused turbocharger or bearing failure, it could benefit from a flushing oil to dissolve it to some degree. Be aware that with most flushing oils the engine cannot do any work or undergo a load of any type. Furthermore, if the cause of soot generation is not established, the soot will recur. Among soot generators are light loads, long idling periods, high altitudes, and air filter restrictions.

Micropatch and Microscope: Creating the Patch

Portable oil labs are on the market at a cost-effective price, provided that the user is willing to assign someone the task of running the tests and collecting the data. Experience shows that most users are not interested in using this tool, but it is important to discuss its merits in this chapter.

Portable labs are not everyone's tool, but they could possibly be so someday. Some users have been unsure of the suitability and value of this tool, but helping validate some of the challenges is discussed here. In addition, these kits contain a quality microscope that allows technicians more insight into the particles within contamination. The tool allows users to identify what kind of contaminants are present (e.g., iron, copper, dirt) and their quantity (as in identifying the ISO codes).

The kits typically come in a briefcase with all the elements needed to run the tests. The key elements are sterile patches, a vacuum pump/funnel, and a 100× microscope. Users need to procure the solvent, which could be mineral spirits or toluene.

Once the sample is diluted, it needs to pass through the filter, which is located at the bottom of the funnel. A hand pump creates a vacuum to draw the sample fluid through the filter. Once the funnel is empty, users just need to move the patch to the microscope and use the picture comparator to assign a value.

The picture on the far right of Figure 3-35 shows the completed patch. The lines on the patch are there to provide a point on which to focus when the patch is placed under the microscope. As you can see, the patch is dry and appears to be clean. Once it is placed under the microscope, however, it becomes apparent that the patch is not clean. It is important to be able to see particles on the patch. They can help in monitoring and maintaining the health of the fleet and in identifying the root cause after a failure.

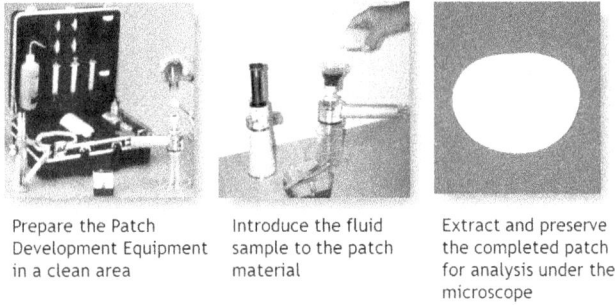

Prepare the Patch Development Equipment in a clean area

Introduce the fluid sample to the patch material

Extract and preserve the completed patch for analysis under the microscope

FIGURE 3-35 Patch and microscope kit.

ISO Code Visibility: Quantitative Code Catalog

Particle count concentrations are ready for comparison with a contamination catalog commonly available from providers such as Pall Corp. and HYDAC. Once the filter patch is ready, as explained earlier, the results can be compared with the patch picture catalog to establish the closest image. Even untrained eyes can perceive the relative levels of contamination and, to some extent, the potential for internal damage to a system. The vertical black line is the ink line used to assist in focusing the microscope on the surface of the patch (see Figures 3-36 and 3-37).

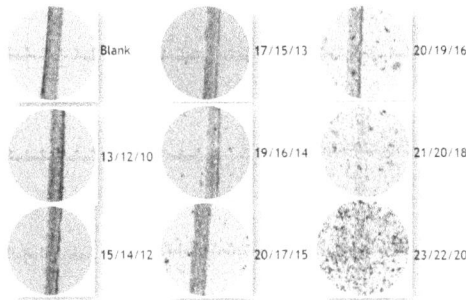

FIGURE 3-36 Patch particle count codes.

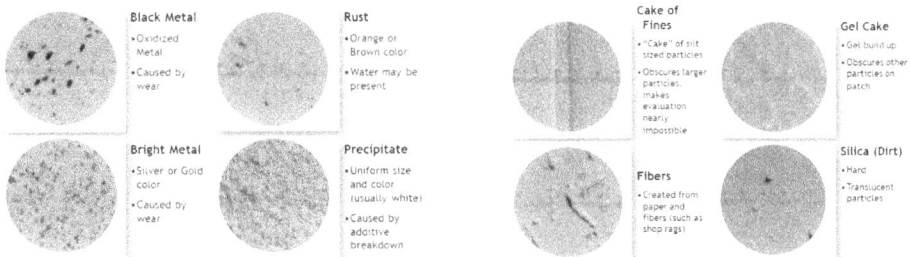

FIGURE 3-37 Patch comparator.

Oil Pump or Dirt Pump?

For most users, fighting contamination is like fighting a ghost. Just as ghosts cannot be seen, neither can contamination unless oil analysis and regular sampling are in place. Without testing, failures come as a surprise, when they could have been anticipated.

Figure 3-38 shows how any hydraulic system suffers from continued contaminant-related wear during a year of operation. In this case, a 50-gallon system is running continuously at a cleanliness level of 22/18/15, circulating dirt particles over and over again. In a year's time, 6800 pounds of dirt will have moved through the pump. During this period, the user cannot see the slow wear that this seemingly small amount of contamination is causing to the system in the absence of oil analysis and trending.

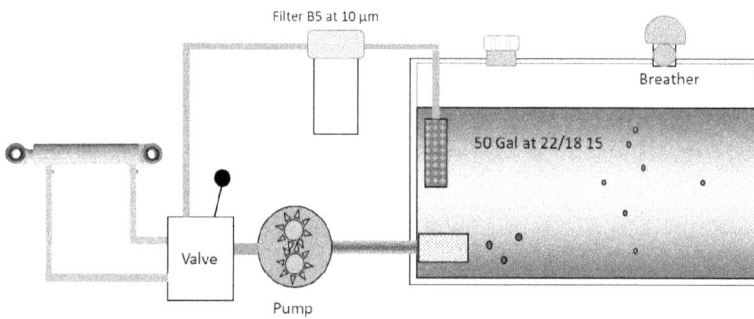

FIGURE 3-38 Oil pump or dirt pump?

Component Tolerances

The importance of particle size becomes evident when given the tolerances built into some of the components used in construction equipment. Gear pumps and vane pumps have tolerances as tight as 0.5 μm between the pump body and gears or vanes, respectively. An axial pump rotary group valve also has very tight tolerances at 0.5 μm. Some valves have a much larger tolerance of 5 μm, which is still extremely small. Awareness of these tolerances is important when assessing the impact of contamination. None of these particle sizes are even close to being visible to the human eye, not even when present in large quantities (see Figure 3-39).

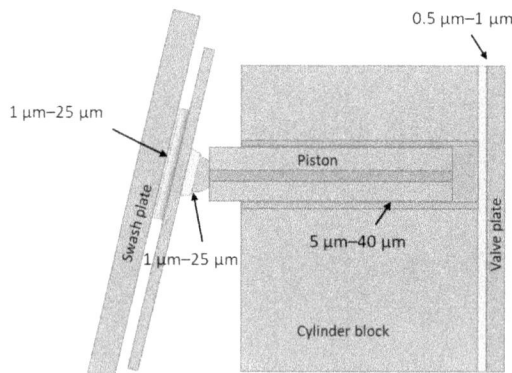

FIGURE 3-39 Axial pump part tolerances.

Internally Generated Contamination

A high count of hard external contaminants, such as silica or aluminum oxide (alumina), can fatigue the surfaces of close-tolerance moving parts, which could cause surface pitting and microscopic fragments in a chain reaction. Elemental analysis and particle counts typically show the presence of these imported particles. Aside from imported contaminants, there are resident and internally generated contaminants. Three-body abrasion is the most common source (see Figure 3-40). For example, silica dust particles can get trapped between the surfaces and break particles out of the surface. These wear particles circulate through the system, further increasing particle counts and wear.

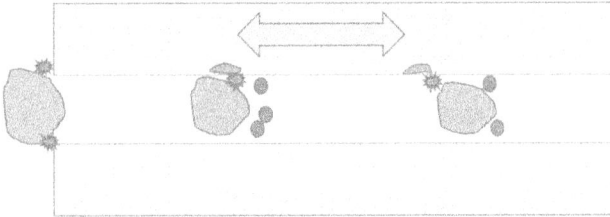

FIGURE 3-40   Three-body abrasion.

Lubricating Oil Film

The lubricating oil film in bearings can be less than 1 µm in thickness. Bearings have a self- defense mechanism when it comes to viscosity loss. The rolling elements put pressure on the oil film, causing it to increase its viscosity to the point of temporarily turning the oil film into a solid. However, if hard contaminants larger than 1-µm tolerance get into the lubricating film, they will cause surface fatigue. If the lubrication film is reduced due to low viscosity, the damaging effect is accelerated (see Figure 3-41).

FIGURE 3-41   Bearing contamination.

Impact of Water on Bearings

Figure 3-42 shows the impact of water on bearing life. Rolling-element bearing design is such that they can last forever if the conditions in which they work are ideal and the load they carry does not exceed their design capacity. Bearings appear to exceed their projected life by 250% when water is not present in the oil. However, eliminating 100% of water contamination is not feasible. At the least, there are ways to control water at a level that is attainable without having to incur large investments to accomplish this goal. To attain 100% of the expected life of a rolling-element bearing, water must be maintained at less than 100 ppm. At 500 ppm of water in the oil, the potential life of the bearing is reduced by 50%. This level is feasible for most users—and achieving 50% of the projected life of a bearing is still an improvement for most and a worthwhile goal.

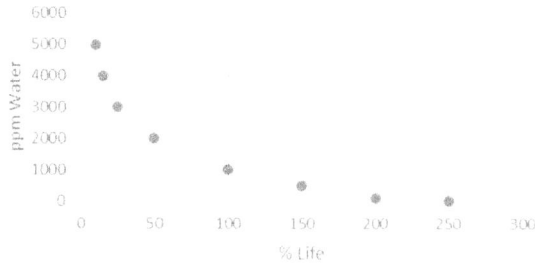

FIGURE 3-42 Water and bearing life.

CONCLUSION

Now that we have reviewed the basic concepts of contamination, let's move to Chapter 4, where we will discuss the means to identify types of contamination in hydraulic systems and the methods to achieve a safe level of cleanliness.

CHAPTER 4

HYDRAULIC CLEANING PROCEDURES

WHERE DOES THE DEBRIS COLLECT AFTER A FAILURE?

To answer this one correctly, first you need to know the type of system, whether it is hydraulic or hydrostatic, and second, you need to know where the pump failed, whether in the slipper side or in the pressure side. Let us assume that it is a hydraulic system for this discussion.

The debris will collect in every component of the hydraulic system (see Figure 4-1). The difference, as we shall see in this chapter, is in the types of pumps and whether they have failed on the slipper side. However, in the worst-case scenario, the only option for a good repair is to remove and clean every single component. The oil cooler is a very difficult part to clean; typically, in massive failures, the cooler requires replacement.

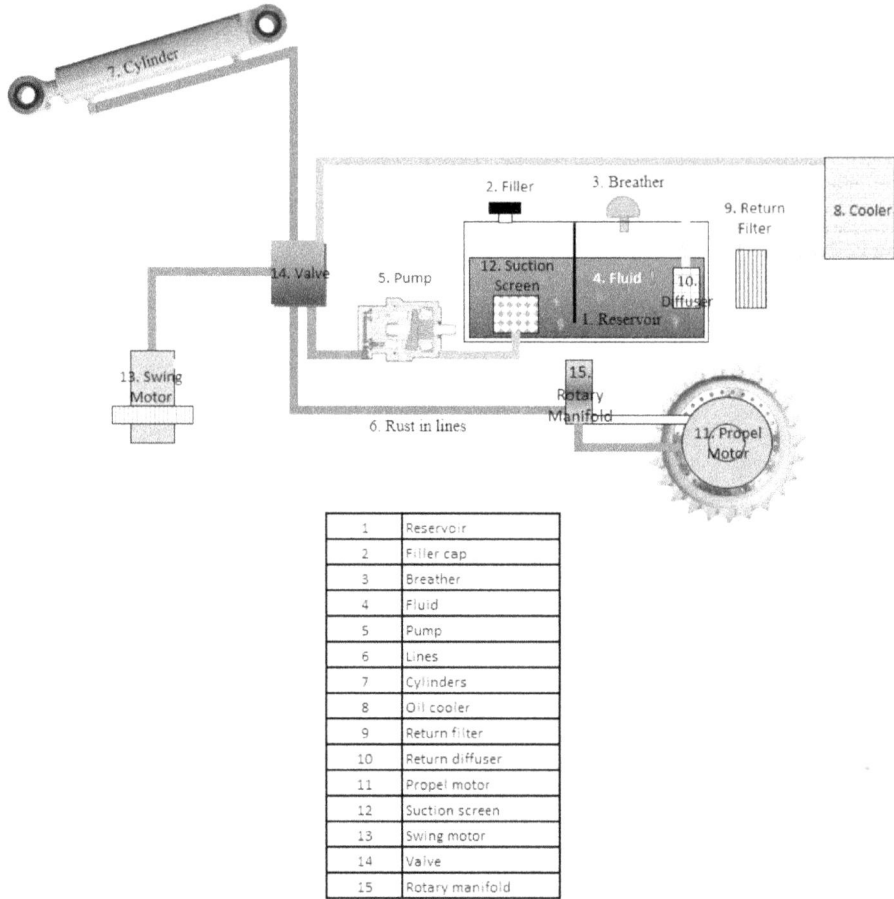

1	Reservoir
2	Filler cap
3	Breather
4	Fluid
5	Pump
6	Lines
7	Cylinders
8	Oil cooler
9	Return filter
10	Return diffuser
11	Propel motor
12	Suction screen
13	Swing motor
14	Valve
15	Rotary manifold

FIGURE 4-1 Debris collection points.

INTELLIGENT FILTER CADDIES

Filter caddies are great tools that help in the process of cleaning hydraulic systems after major repairs, and they also help in attaining the required particle levels before releasing the machine back to the job site. Filter caddies polish the fluid, ensuring that the machine complies with the cleanliness requirements after a major repair (see Figure 4-2).

Many filter caddies have on-board sensors that provide information about particle count and percentage of water saturation, while at the same time removing solid contaminants and water from the fluid. Knowing the cleanliness level of a machine before returning it to work gives peace of mind.

Good filter caddies are equipped with a variable-speed hydraulic pump, particle counter, and water saturation instrument. In addition, they offer particulate filters of different beta ratings and a water-absorbing filter. Ideally, filter caddies connect to the machine via threaded connectors, but they also offer conventional wands. Threaded wands are the cleanest way to use filter caddies, as shown in Figure 4-3.

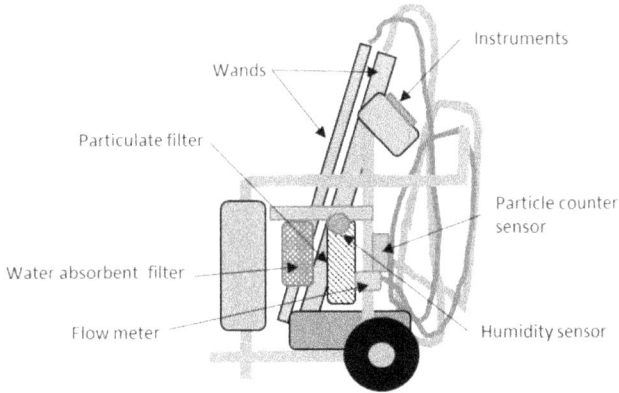

FIGURE 4-2. Intelligent filter caddy.

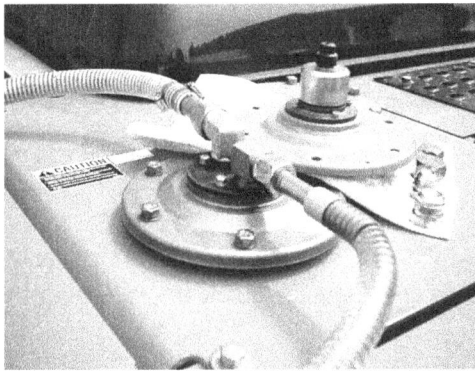

FIGURE 4-3 Adapter for filter caddy wands.

The simplest way to connect a filter caddy to a machine is by inserting the wands into the reservoir, trying to keep them apart for better cleaning results. However, in some machines, the opening is small and may not allow the X configuration. In this case, an offset arrangement also works, making sure that both wand tips are always below the surface of the fluid (see Figure 4-4).

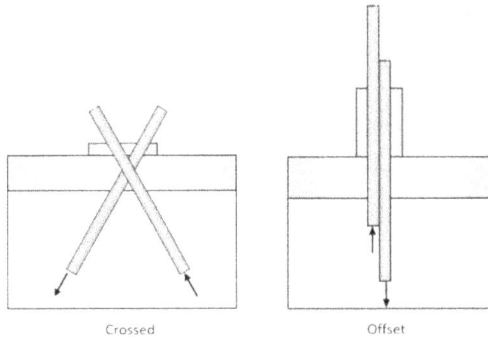

Crossed   Offset

FIGURE 4-4 Wands in X configuration and offset.

For small openings and where threaded connectors are difficult and inserting two wands is impossible, a concentric wand offers an alternative, as shown in Figure 4-5.

FIGURE 4-5 Concentric wands.

To connect an intelligent filter caddy with particle counter and water saturation instrumentation, some rules apply. These rules are essential for a successful cleaning procedure:

1. Do not connect the filter caddy to a machine that has suffered a catastrophic failure and is still waiting for repairs and internal cleaning by other means. Otherwise, the filter caddy pump and particle counter will suffer.

2. The fluid needs to be warm to avoid filter caddy cavitation and to facilitate fluid circulation. Start with a low flow until the caddy pump is able to work at full speed without cavitation.

3. Let the filter caddy purge all air through its purging valves so that particle count readings are reliable.

4. The filter caddy needs dedicated filters for the type of fluid in machine.

5. The machine needs to cycle the functions to bring dirty fluid back to the reservoir where the filter caddy can do its job. However, this generates bubbles and can affect particle count readings.

6. Allow the system to rest for a couple of minutes between cycles before accepting particle count readings as good ones.

7. Run the caddy filter until particle counts exceed cleanliness specifications for the machine in question by at least two numbers for each channel of the particle count code.

8. Always send a sample to the lab after cleaning to confirm that the machine meets the expected specifications.

9. When the filter caddy has finished cleaning, disconnect the suction wand, and operate the caddy filter slowly until it empties its fluid content.

Some techniques exist to optimize the volume of fluid required for a total flush. This approach involves modifying the return filter cover and replacing the filter with a blocked dummy filter. In this way, the waste fluid is dumped outside the system and does not mix with the new fluid. This produces significant savings in the amount of fluid required for total flushing. The cycling of the machine functions makes the fluid flow out of the system through the dummy filter and adapter. The function needs to operate smoothly and easily to avoid excessive fluid backpressure while exiting the system (see Figure 4-6).

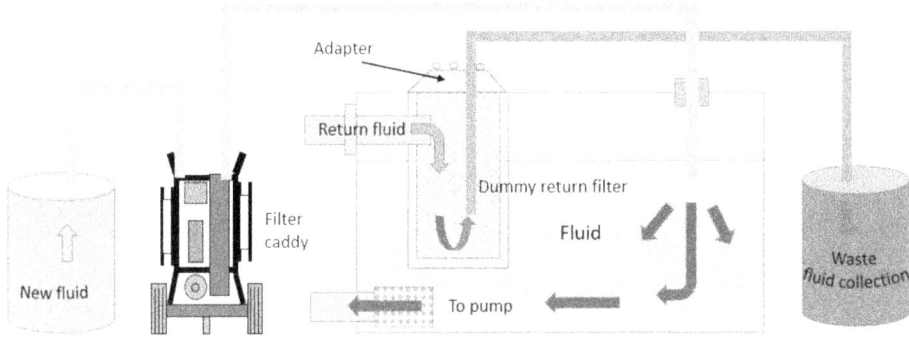

FIGURE 4-6   Flushing with a dummy return filter.

HOW MUCH OF THE OLD FLUID
REMAINS IN THE SYSTEM AFTER FLUSHING?

The answer to this question depends on the type of machine. Experience shows that the amount of fluid left in the system after reservoir flushing can vary from 15% to 60%. With excavators, the ratio is close to 50%, whereas that for a backhoe loader can be 40% (see Table 4-1).

TABLE 4-1   Reservoir Volume as Percent of Total System Volume

System	Reservoir as Percent of System	Number of Flushes to 95% New Fluid
Excavators	45%–63%	5–8
Four-wheel-drive loaders	50%	5
Backhoe loaders	63%	3
Crawler hydrostatics	75%	1.5
Crawler hydraulics	50%	5
Wheel harvesters	74%–79%	2.5

IDENTIFYING TYPE OF CONTAMINATION AND CLEANING PROCEDURES IN HYDRAULIC SYSTEMS: CASE DISCUSSION

The cleaning procedures chosen for hydraulic systems depend on many factors. There are many ways a system gets contaminated, as you will see in the following cases. The cleaning procedures and tools necessary to clean a system depend on the type and degree of contamination, and the latter depends on how long the machine has operated under the specific conditions.

The following contamination types are feasible but not limited to the cases discussed:

- The machine shows dirty fluid and high particle counts.
- The hydraulic system is contaminated with water.
- The fluid is oxidized or has high total acid number (TAN).
- The hydraulic system shows mixed fluids.
- A component is wearing out slowly, and the fluid is showing growing metal particle counts.
- A component has failed and has contaminated the whole system (machine is not operable).
- The system is contaminated with varnish.
- The system shows blackened fluid.

The Machine Shows Dirty Fluid and High Particle Counts

A machine with dirty fluid that still shows integrity with regard to signature, TAN and oxidation is good, so this machine makes a good candidate for fluid polishing with an intelligent filter caddy. However, it is necessary to find the reason why the machine was contaminated. The most common causes are leaking cylinders, field repairs, and poor service procedures (see Figure 4-7).

METALS							ADDITIVES						
Iron (Fe)	Chromium Cr)	Lead (Pb)	Copper (Cu)	Tin (Sn)	Aluminum (Al)	Nickel (Ni)	Magnesium (Mg)	Calcium (Ca)	Barium Ba)	Phosphorus (P)	Zinc (Zn)	Molybdenum (Mo)	Boron (B)
52	3	4	45	<1	10	<1	10	74	1	350	290	<1	<5

CONTAMINANTS						AR	PHYSICAL PROPERTIES			INFORMATION			
Aluminum (Al)	Silicon (Si)	Sodium (Na)	Potassium (K)	Water (%) KF	PQ Index	Particle Counts	Viscosity @ 40C) (cSt)	Oxidation	TAN	Fluid Changed	Fluid Type	Total Hours	Fluid Hours
10	43	9	5	0.05		21/18/15	42	15	1.03	No	AW46	2800	2800

HYDRAULIC OIL ANALYSIS	Excavator with dirt and high particle counts. Fluid signature represents listed type.

FIGURE 4-7 Hydraulics with dirt.

The Hydraulic System Is Contaminated with Water

When a machine shows contamination with water, it is necessary to know whether the water contamination has exceeded the saturation point of the fluid. This is simple to do by checking the color of the fluid or whether the caddy filter shows excessive water saturation:

- Water saturation exceeds 100% and the fluid is milky.
- Water saturation is between 100% and 85%.

For the first situation, there is no practical way to save a fluid at a reasonable cost. Although the technology exists to dehydrate a highly contaminated fluid, the cost does not justify trying it. A total flushing procedure requires total removal of fluid from the system. This means replacing the fluid that is in the cylinders, lines, cooler, and motors—not an easy task. There are two ways to accomplish this: multiple flushing and disassembly. Multiple flushing requires a filter caddy feeding the reservoir of the machine while the machine's pump removes the fluid at the same rate. The machine needs to cycle its functions to bring the stagnant fluid back to the reservoir, where it finds its way out of the system.

For the second situation, a filter caddy with a water-absorbing filter can bring the water numbers down. To capture free water, the fluid needs to be lukewarm only because saturation will be less and thus it is easier to capture free water. The filter caddy may need to operate at a slower speed to avoid cavitation.

We can represent both approaches with the help of this logic tree, as in Figure 4-8.

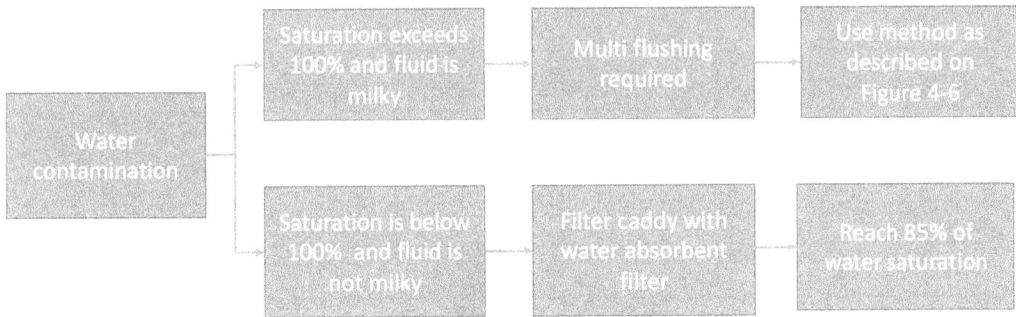

FIGURE 4-8 Water contamination.

Multiple flushing can be very expensive because some machines have numerous cylinders and hydraulic motors. Depending on the type of machine, the percentage of fluid remaining in the system can be difficult to determine, and the machine could hold up to 60% of its fluid volume outside the reservoir. A case with water saturation below 100% may show an oil analysis with water by the Karl Fischer method on the order of 1200 ppm, which is a very salvable fluid using a filter caddy with a water-absorbing filter. The particle counts may show atypical readings, and this is a confirmation of excessive water in the fluid (see Figure 4-9).

METALS							ADDITIVES						
Iron (Fe)	Chromium Cr	Lead (Pb)	Copper (Cu)	Tin (Sn)	Aluminum (Al)	Nickel (Ni)	Magnesium (Mg)	Calcium (Ca)	Barium Ba	Phosphorus (P)	Zinc (Zn)	Molybdenum (Mo)	Boron (B)
3	<1	<1	27	<1	<1	<1	10	1840	1	1167	1100	<1	<5

CONTAMINANTS					WEAR		PHYSICAL PROPERTIES			INFORMATION			
Aluminum (Al)	Silicon (Si)	Sodium (Na)	Potassium (K)	Water (%) KF	PQ Index	Particle Counts	Viscosity (cSt @ 40C)	Oxidation	TAN	Fluid Changed	Fluid Type	Total Hours	Fluid Hours
<1	9	2	<5	0.12	NA	21/21/20	34.4	NA	1.03	Yes	Donax TD	3602	3602

HYDRAULIC OIL ANALYSIS	Excavator with water and atypical particle counts

FIGURE 4-9   High water contamination.

The Fluid Is Oxidized or Has a High TAN

A machine with an oxidized fluid or a fluid with a high TAN has no other remedy than to do a total flush. The fluid is beyond recovery, and the system requires a total flush. Some users opt for updating just the reservoir volume so that the TAN and oxidation creep are kept to an acceptable level. However, this approach will shorten the interval, which today is usually around 3,000 hours for most fluids, if not more (see Figure 4-10). It is important to establish the reasons for such short fluid life, which includes checking high-temperature records.

METALS							ADDITIVES						
Iron (Fe)	Chromium Cr	Lead (Pb)	Copper (Cu)	Tin (Sn)	Aluminum (Al)	Nickel (Ni)	Magnesium (Mg)	Calcium (Ca)	Barium Ba	Phosphorus (P)	Zinc (Zn)	Molybdenum (Mo)	Boron (B)
12	3	4	22	<1	6	<1	7	82	1	380	275	<1	<5

CONTAMINANTS					WEAR		PHYSICAL PROPERTIES			INFORMATION			
Aluminum (Al)	Silicon (Si)	Sodium (Na)	Potassium (K)	Water (%) KF	PQ Index	Particle Counts	Viscosity (cSt @ 40C)	Oxidation	TAN	Fluid Changed	Fluid Type	Total Hours	Fluid Hours
6	15	5	<5	0.06	NA	20/17/14	43.5	27	1.7	No	AW46	1580	1580

HYDRAULIC OIL ANALYSIS	Tree harvester with high TAN. Fluid signature represents listed type.

FIGURE 4-10   Hydraulics with a high TAN and oxidation.

The Hydraulic System Shows Mixed Fluid

Fluid mixtures are perhaps the most common problem in the field, and many machines run their entire lives with mixed fluids. In some cases, the mixes are benign and do not cause major harm to the hydraulic pump. In other cases, the mixes cause reduced pump life and high operating costs. The lubricant compatibility chart in Figure 2-29 is a good tool to assess the risk of continuing to run a fluid mixture.

Cleaning a machine with mixed fluids and adopting a new standardized fluid is the best approach (see Figure 4-11). To save on the amount of new fluid required for such a flushing, the procedure described in the section titled "The Hydraulic System Is Contaminated with Water" works equally well with mixed fluids. Please also refer to Chapter 2 for fluid mixing permissibility.

METALS										ADDITIVES				
Iron (Fe)	Chromium Cr	Lead (Pb)	Copper (Cu)	Tin (Sn)	Aluminum (Al)	Nickel (Ni)	Magnesium (M)	Calcium (Ca)	Barium Ba)	Phosphorus (P)	Zinc (Zn)	Molybdenum (Mo)	Boron (B)	
15	3	4	32	<1	3	<1	5	150	1	450	280	<1	<5	
9	1	2	14	<1	2	<1	0	12	1	560	0	<1	<5	

CONTAMINANTS					WEAR		PHYSICAL PROPERTIES			INFORMATION			
Aluminum (Al)	Silicon (Si)	Sodium (Na)	Potassium (K)	Water (%) KF	PQ Index	Particle Counts	Viscosity (cSt @ 40C)	Oxidation	TAN	Fluid Changed	Fluid Type	Total Hours	Fluid Hours
3	9	5	<5	0.07	NA	22/18/16	43.5	NA	1.5	No	AW46	2850	2850
2	6	<5	<5	0.06	NA	19/16/13	45.3	12	1.1	No	ZF46	1850	1850

HYDRAULIC OIL ANALYSIS	Excavator with mixed fluids. Originally equipped with zinc free fluid

FIGURE 4-11 Hydraulics with mixed fluids.

In this example, the machine started with a zinc-free fluid but at one point switched over to a zinc-based antiwear (AW) product. The machine appears to suffer from the mix, as indicated by the atypical signature, high particle count, and presence of copper.

A Component Is Wearing Out Slowly, and the Fluid Is Showing Growing Metal Readings

One of the beauties of oil analysis is the ability to predict a catastrophic failure based on the trending behaviors of critical metals. The most valuable metal for component progressive wear assessment is copper. Copper is not a time-dependent element, it is seen in hydraulic pumps and motors as part of bronze alloys, which usually enjoy dynamic lubrication. When these components are about to fail, they usually send increased readings of copper and small partner metal readings, namely lead and tin. Iron is typically not involved in the affairs of this component.

If the copper readings are not the result of dirt, water, mixing, fluid oxidation, or a high TAN, you can suspect that the component is undergoing some kind of erosion or pitting, which, if allowed to continue unattended, can result in very expensive repairs. This case supports the need for more frequent fluid sampling and trending analysis to realize the advantages of advance warning.

Depending on the age of the fluid and its physical conditions, the fluid could have life remaining, provided that a filter caddy is used to clean it. The fluid cleaning can be done initially in external tanks (see Figures 4-12 and 4-13).

METALS							ADDITIVES						
Iron (Fe)	Chromium (Cr)	Lead (Pb)	Copper (Cu)	Tin (Sn)	Aluminum (Al)	Nickel (Ni)	Magnesium (Mg)	Calcium (Ca)	Barium (Ba)	Phosphorus (P)	Zinc (Zn)	Molybdenum (Mo)	Boron (B)
18	2	6	45	5	3	<1	13	22	0	225	18	0	<1
12	2	5	37	4	6	<1	15	25	0	228	14	0	<1
22	3	3	22	2	7	<1	14	22	0	233	12	0	<1
10	1	1	8	0	3	<1	15	23	0	245	4	0	<1

CONTAMINANTS					WEAR		PHYSICAL PROPERTIES			INFORMATION			
Aluminum (Al)	Silicon (Si)	Sodium (Na)	Potassium (K)	Water (%) KF	PQ Index	Particle Counts	Viscosity (cSt @ 40C)	Oxidation	TAN	Fluid Changed	Fluid Type	Total Hours	Fluid Hours
1	9	<5	<5	0.06	NA	22/17/15	53.5	16	0.09	NO	ZF46	5000	2000
2	8	5	<5	0.05	NA	21/17/15	54.8	11	0.07	NO	ZF46	4000	1000
4	12	<5	<5	0.06	NA	18/17/14	46.1	15	0.08	YES	ZF46	3000	3000
3	9	<5	<5	0.07	NA	18/16/13	46.2	10	0.07	NO	ZF46	1000	1000

HYDRAULIC OIL ANALYSIS	LARGE EXCAVATOR. Copper trending up

FIGURE 4-12 Component wearing out.

FIGURE 4-13 Failure in progress.

A Component Has Failed and Has Contaminated the Whole System (Machine Is Not Operable)

When a component fails catastrophically, the machine no longer works. Very possibly, shrapnel has circulated thought the whole system, contaminating practically every component in the machine (see Figure 4-14).

FIGURE 4-14. Slipper-side failure (*left*); massive contamination (*right*).

There is no chance of saving the fluid, and the machine needs to go through a methodical cleaning that involves disassembling every major component and replacing the cooler. However, there are going to be differences in the degree of contamination, depending on where the component failed (see Figure 4-15).

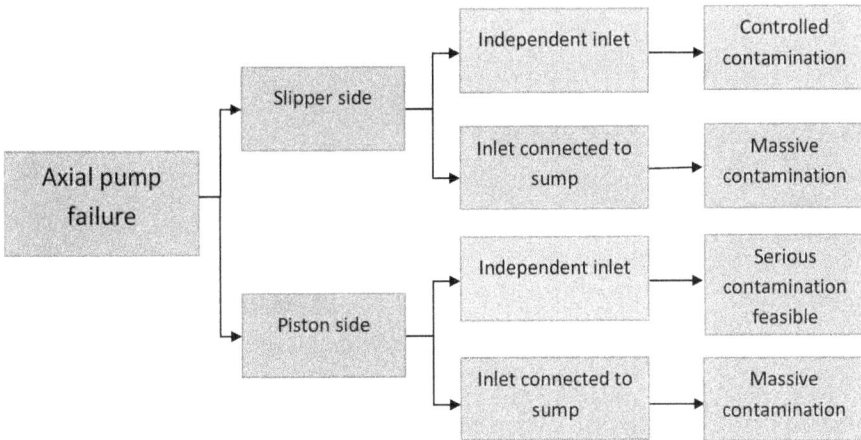

FIGURE 4-15 Axial pump: failure localization versus contamination.

An axial piston pump has two defined areas: the slipper side, which converts the axial force to rotational force, and the delivery side, which pumps fluid through a metal-to-metal seal joint. The different pump designs, between independent inlet and inlet in sump, make a great difference in the way they contaminate the system when they fail. If a failure occurs on the slipper side, most of the debris goes to the sump and then to the reservoir through a case return filter. In contrast, if the piston fails on the delivery side, the debris generated will flow with the fluid to other components. We can safely conclude that slipper-side failures are much easier to control unless the pump has an inlet-in-sump design, as shown in Figure 4-16.

FIGURE 4-16   Inlet-in-sump design.

Cleaning Projectiles

Whenever a component fails and causes massive system contamination, the cleaning of lines and hoses with sponge projectiles is mandatory. These projectiles come in different sizes to fit all diameters found in a machine and can clean lines and hoses very quickly with pressurized air. Other than the air-propelled sponge projectiles, there is no effective procedure to clean lines and hoses, and not using this technology is risky for new hydraulic pumps or motors (see Figure 4-17).

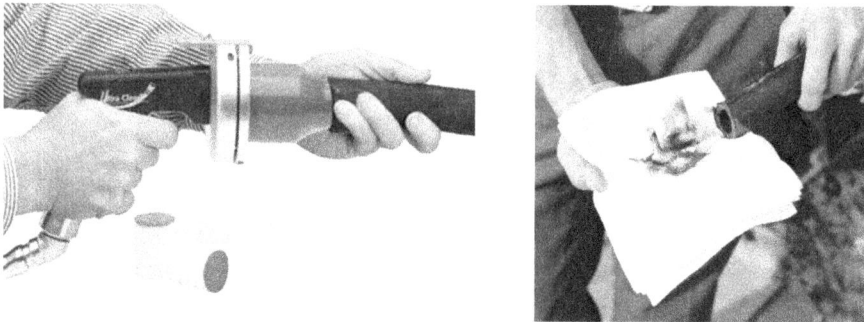

FIGURE 4-17   Ultra Clean tools (*left*); Ultra Clean projectiles (*right*).

The System Is Contaminated with Varnish

Varnish can cause valve malfunction and loss of machine control. Varnish creation is a complex process that is linked to something else happening in the system (see Figure 4-18). Before trying to clean a machine suffering from varnish, it is important to determine the cause of varnish formation. Although cleaning varnish sometimes requires total disassembly of the components, on other occasions it is feasible to dilute the varnish and collect the particles. The scenario illustrated in Figure 4-19 is feasible in a machine with varnish issues.

FIGURE 4-18 Varnish.

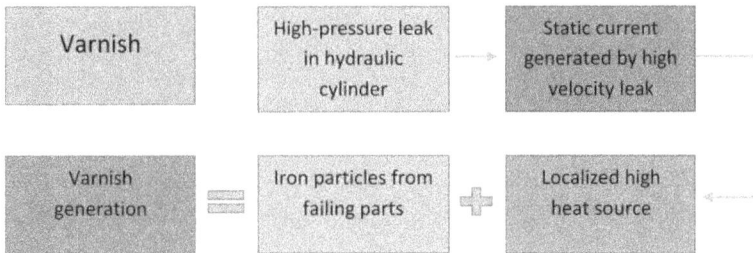

FIGURE 4-19 Varnish-generation sequence.

Cleaning Varnish

Once the cause of varnish deposits is determined and fixed, the cleaning of varnish can start. If the main control valve has no symptoms of malfunction, flushing the machine's fluid is mandatory as the first step. The fresh fluid will act as a solvent to start dislodging the varnish. Installation of a beta 2000 bypass filter is the second step. The investment in the filter will be much cheaper than having to use hundreds of labor hours cleaning the machine. The cleaning process will be slow but steady and could take several weeks before all the varnish dislodges.

The System Shows Blackened Fluid

Another phenomenon in high-pressure hydraulic systems is the blackening of hydraulic fluid. Although not common, this needs discussion to avoid confusing it with varnish or fluid oxidation conditions. Blackening of the fluid is more common in new machines or machines recently reassembled after a repair. The fluid blackening is the result of microdieseling caused by air left in the hydraulic cylinders. A cylinder with air acts as diesel engine in that the heat of compressing the air burns the fluid vapor inside the cylinder causing blackening (see Figure 4-20).

Fluid blackening is not detrimental to a hydraulic system. However, if the blackening causes high particle count readings that mask real readings, then filter caddying the machine should fix the problem.

FIGURE 4-20 Blackened fluid.

CONCLUSION

Proper hydraulic system flushing and cleaning require knowledge of contamination and hydraulic systems. The cases discussed in this chapter are the most typical, but many more types exist, depending on the system, the environment, and the application. To test that we have achieved the cleanliness required for a given hydraulic system and verify what the intelligent filter caddy is telling us, we must take samples properly for lab analysis. This is the topic of Chapter 5.

CHAPTER 5

OIL SAMPLING

Oil analysis is at the center of condition-based maintenance tools. Keep in mind that we will be measuring in parts per million and particles per milliliter and looking for relatively small changes. The only way to get meaningful results is for the sample to be representative of the oil in the machine at full operating temperature while it is running. Wear metals, water, and particulate contamination have a tendency to settle out after shutdown. Fuel dilution tends to rise with idling and low-speed operation. Sludge and oil-degradation products tend to settle out as oil temperature drops.

All the changes that occur after shutdown and as a result of abnormal operation will skew your laboratory results. Therefore, proper sampling technique is critical to obtaining meaningful information on which to base your maintenance decisions. "Garbage in, garbage out" was the cliché in the early days of computing—and it applies equally to oil analysis.

Unfortunately, fluid sampling is not always in line with a maintenance technician's priorities. Maintenance people are generally focused on crucial repairs and keeping up with orders for scheduled service. Their priority is the fleet's uptime and performance as required by the job-site supervisor. Thus, taking a sample to obtain a report that will not show up for several days and that they cannot interpret seems relatively unimportant to them.

This is not a book on management or administration, but technicians need training, or perhaps someone else with a different set of priorities needs to pull the samples. The lab needs to produce data that correlate with the condition of the machine, not with the sampling process. It is beyond difficult to identify trends if the numbers are bouncing around because of erratic sampling techniques. Poorly taken samples can produce misleading results for water, fuel dilution, and other contaminants, and those are just the contaminants generated within the machine. Poorly trained or careless individuals can introduce contaminants or metals from the environment, providing inaccurate machine data that could lead to unnecessary repairs and expenditures. A poorly collected sample also hides the relevant data that otherwise could be used to do wear or contamination trending analysis or to take actions at the right time.

The sample must be representative of the oil in the machine while it is operating for meaningful analysis. The best samples come from a sampling valve located just upstream from the filter. Pulling a sample with a vacuum pump also yields excellent results. A midstream drain sample's integrity is questionable in a clean shop and should not be an option in the field.

SAMPLING METHODS

There are certain rules to obtain a good sample:

- The sample kits need to be clean to prevent contaminants entering from the environment. This means that all the kit's components stay in zip-lock bags or equivalent protection from dust.
- In a dusty environment, you can assemble the components while still in the bag, and the tubing can be punched through the bag to minimize exposure to dust. Remember, we are measuring in parts per million, so if the technician puts the lid to the sample bottle in his or her pocket while pulling the sample, the pennies in his pocket can raise the copper reading.
- Be aware of the challenges of getting a clean sample that is representative of the oil that needs to be sent to the lab. This implies cleaning the area before attempting to pull a sample. (See Figure 5-1.)
- The use of sampling intervals that are as close as possible to a fix-hour schedule is ideal. Certain elements are time dependent, and taking samples at irregular intervals will probably skew calculations and trending.
- Ideally, you take samples from components at full operating temperature while they are running or immediately following shutdown. Idling a cold engine to obtain a warm sample is likely to increase fuel dilution.
- Properly record the information on the machine's component and the hours on the machine and the fluid. This is the most neglected task by technicians, who often fail to record the information completely. The amount of time on the fluid is very important because wear, contamination, and additive concentrations correlate with hours. Further, viscosity trending is a very valuable tool that requires knowing the time on the oil. A machine's hour-meter reading is equally important because as machines age, there are certain indicators that show up, such as silt in the hydraulics or soot in the engine, that need to be accounted for in making decisions on maintenance.
- Send the sample to the lab on the same day as the sample collection day.

FIGURE 5-1 Final drive challenge.

SAMPLING VALVE METHOD

Sampling valves are the "gold standard" for collecting representative samples (see Figure 5-2). They allow samples to be obtained while the fluid is circulating through the system. Here is how to obtain a sample properly (without including safety rules):

1. Place the sample bottle in a zip-lock bag and seal it while in a clean environment.
2. With the fluid at full operating temperature, locate the sampling valve and clean around it before removing the dust cover.
3. Insert the probe and remove enough fluid to purge the valve (see Figure 5-3).
4. Discard the probe, tubing, and purged oil in an environmentally responsible way (they are contaminated and not to be reused for sampling).
5. If oil flow is too low to obtain a representative sample, increase the engine speed.
6. Remove the lid from the sample bottle while both lid and bottle remain in the sealed zip-lock bag.
7. Some companies have "clean" bottles with a plastic sheet between the bottle and lid. If this is the case, hold the plastic sheet against the bottle so that it remains on the bottle.
8. Hold the plastic bag tight across the bottle opening and push the open end of the cap probe hose through the plastic bag (and plastic sheet, if applicable) about 0.5 inch into the bottle. The bag creates a protective covering for the sample so that external contaminants do not get inside the bottle.
9. Press and hold the sample probe to the valve to release oil from the system and fill the sample bottle about three-quarters full.
10. Once a representative sample has been obtained, remove the tube and return the lid to the container, leaving the lid and bottle inside the plastic bag with the protective plastic sheet attached underneath the lid (if applicable).
11. Reinstall the valve dust cap to the sample valve and properly dispose of the cap probe, which you will never use again.

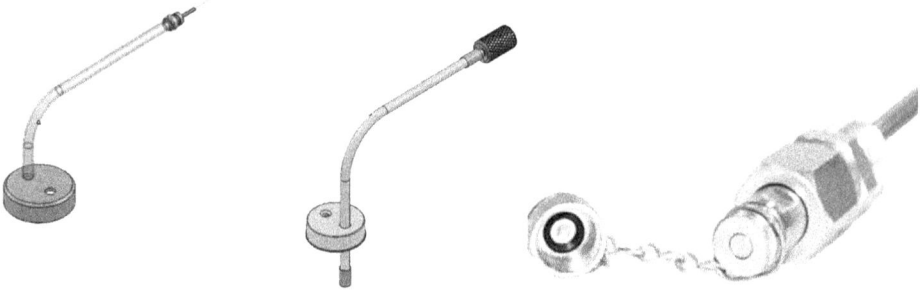

FIGURE 5-2 Needle probe (*left*); cap probe (*center*); needle sampling valve (*right*).

FIGURE 5-3 Push-in sampling valve.

VACUUM PUMP METHOD

Using a vacuum pump to obtain oil samples provides more opportunities for contamination than a sampling valve. Even so, a conscientious technician with a vacuum pump can draw good representative samples. This is the best choice for machinery that is not equipped with sampling ports.

There are two types of vacuum pumps: dual and single function (see Figure 5-4). The dual-function vacuum pump allows drawing from a sump and from a valve, and single-function pumps only allow the use of plastic tubing for drawing the sample.

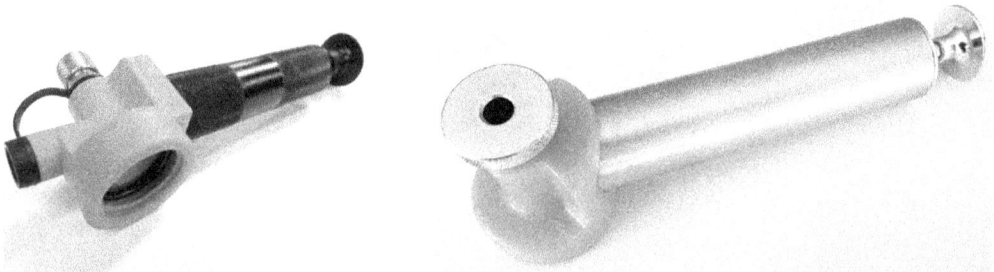

FIGURE 5-4 Dual-function sampling pump (*left*); single-function sampling pump (*right*).

Following the rules set out earlier under "Sampling Valve Method," this method works as follows:

1. The fluid should be at full operating temperature and collected as soon after shutdown as possible. Certainly, wait no more than 20 minutes after shutdown to draw the sample because some of the wear metals and contaminants can settle out.
2. Cut the appropriate length of new tubing from a zip-lock bag.
3. Insert the tubing to the same depth every time.
4. Fill the sample bottle about three-quarters full, taking care not to contaminate the pump.
5. Properly dispose of the used tubing (see Figure 5-5).

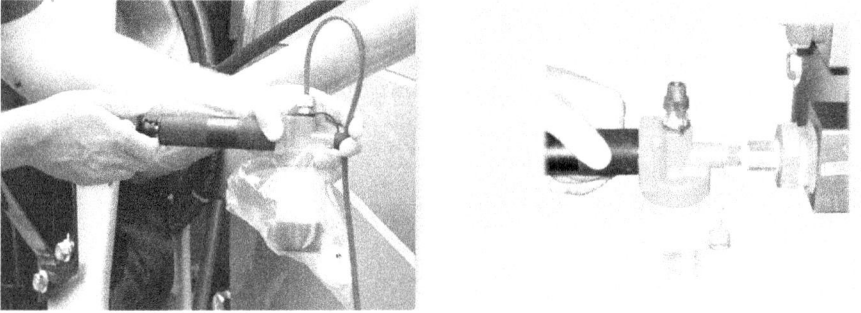

FIGURE 5-5   Sampling by the baggy method (*left*); dual-function pump sampling (*right*).

BOTTLE TYPES

The market offers several options for sampling bottles. The most important characteristic is cleanliness level. There are also bottles in 3- and 5-oz capacitates depending on the amount of fluid required for testing. In addition, there are accordion-type sampling bottles that offer a simplified way of taking samples. The oil needs to be sufficiently warm to flow with the suction. However, these types of bottles may not work properly in winter (see Figure 5-6).

FIGURE 5-6   Standard plastic bottle (*left*); glass bottle (*center*); accordion bottle (*right*).

SUBMITTING SAMPLES

Nobody likes doing paperwork, but the form that goes with the sample is important and needs to be fully complete. We are always on the lookout for abnormal wear patterns, which vary with manufacturer and model. Therefore, analysts need the make and model of the machine listed on the form to reference "normal." Then the component (engine, transmission, etc.) to which the sample belongs must be noted. Most labs will show previous analyses on the report following the current one, making it much easier to identify trends. Serial numbers work best because they are unique to the individual components.

Double-check that the hours on the component and the hours on the fluid are listed on the form, as mentioned in the preceding section. The name of the oil product, viscosity, and amount added during the interval need to be noted on the form. We cannot measure the amount of change if we do not know the starting point (see Figure 5-7).

SAMPLE INFORMATION FORM (SIF)							
NAME	MODEL	SERIAL NUMBER/ ID NUMBER	HOURS/MILES	HOURS/MILES ON OIL	OIL TYPE	VISCOSITY	BRAND
ADRESS							
	ENGINE	HYD/ HYDROST.	TRANSMISSION	AXLE	FINAL DRIVE	GEARBOX	OTHER

FIGURE 5-7 Sample information form.

It seems obvious that contact information has to be on the submission form, but laboratories get samples every day for which they have no idea of where to send the report.

Finally, if there are any abnormalities with a machine, a note should be included with the sample. A conscientious lab technician will sometimes take the extra step if he or she knows what to look for. Knowing that there was a coolant loss, that the unit was running hot, or that the fluid level was increasing on the dipstick may lead the technician to further investigation.

THE LABORATORY

With a good representative sample, we are ready for the lab work. The results will give us an indication of the condition of the oil and the hardware.

SAMPLING POINTS

Points *A* and *B* in Figure 5-8 are the best places for fluid sampling, while point *E* represents a diluted sample. Points *C* and *D* are not good options because they have already passed filtration and will not provide a real representation of the fluid's quality. Point *E* is not the optimal place for sampling. Nevertheless, it is readily available when the machine does not have sampling valves. Make sure that you take your sample from a recently active system.

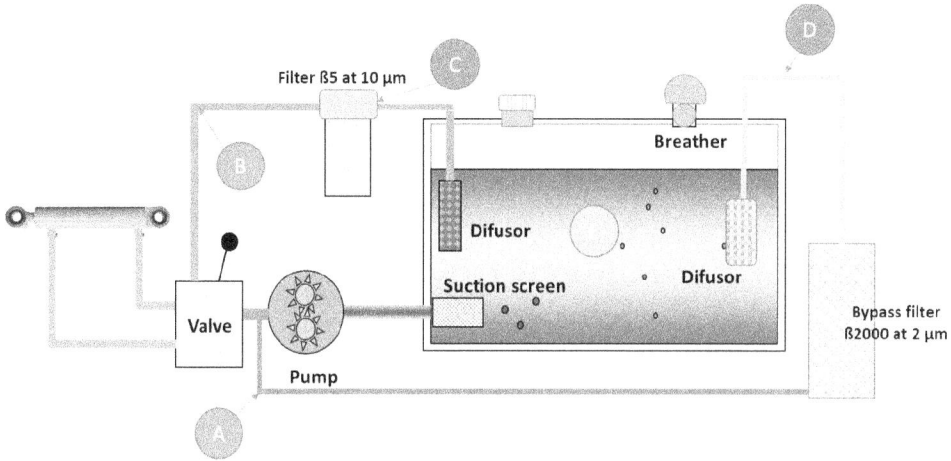

FIGURE 5-8   Sampling points.

SAMPLING VALVE INSTALLATION

Install sampling valves based on a complete understanding of flow and contamination. For example, in a T-connection, some of the particles in the fluid will directly pass and avoid the sampling point as a result of simple kinetic energy. Whenever feasible, install sampling valves in angled lines where the turbulence offers better points of extraction (See Figure 5-9).

FIGURE 5-9   Improper sampling valve location (*left*); ideal sampling location (*right*).

SAMPLING VALVE DEAD ENDS

Sampling valves in equipment make a great contribution to the sampling culture of these times. This does not come without tradeoffs, however. When taking a sample, some line flushing is required because the lines are dead ends without active circulation. Thus, when taking a sample, some bleeding of the line's fluid is necessary, for which an additional sample bottle is always necessary and kept handy (see Figure 5-10).

FIGURE 5-10 Transmission and engine sample valves.

CONCLUSION

Proper sample collection ensures the next step in obtaining reliable results from the lab that represent the actual health of the equipment. Chapter 6 will review the recommended tests for lubricants.

CHAPTER 6

LUBRICANT TESTING

OIL AND FLUID TESTING

Lubricant testing involves a much more intensive spectrum of tests than that required for mobile equipment maintenance. For this reason, we will only cover the testing that directly affects fluid analysis activities for mobile equipment. Some tests are common to all types of lubricants, such as wear metals, viscosity, and total acid number (TAN) tests, whereas others are specific to the type of component/lubricant, such as particle counts, total base number (TBN), fuel dilution, and glycol content. Other tests are similar but with different testing procedures, such as water content by Fourier-transform infrared (FTIR) spectroscopy or water content by Karl Fischer coulometric titration.

Elemental analysis provides direct information on metal generation and also provides indirect information such as glycol contamination and dirt. The presence of sodium and potassium suggests the presence of glycol and is determined by direct measurement, while the presence of silicon and aluminum in certain ratios confirms indirect information revealing dirt as the source. We will cover these tests one by one with a short explanation of how to analyze the results in each case.

In many cases, there is more than one procedure to test a characteristic. What is important is that we adopt the most appropriate test procedure for the equipment and that we use the tests consistently for optimal trending. The following Table 6-1 of suggested tests is provided only for information purposes because the labs have the last word on which tests to use for a given measurement.

TABLE 6-1 Mobile Equipment Tests and Standards

Test parameter	Hydraulics	Engines	Powertrains	Test methods
Wear metals	X	X	X	ASTM D5185 (ICP), D6595 (RDE)
Viscosity	40°C	100°C	40°C	ASTM D445/D7279
Water	ASTM D6304	FTIR/crackle	ASTM D6304/ E203 if >1000 ppm	FTIR ASTM D6304/Crackle/ E203
Soot	—	X	—	Shimadzu ATR/IR/ASTM D7844
Glycol	—	By Na and K	By Na and K	ASTM D6595
Fuel dilution	—	X	—	D3524/D3525 GC/D38280.2
Particle counts	X	—	Optional	ISO 11500/ASTM D7596
Base number	—	X	—	ASTM D4739/D2896
Acid number	X	X	X	ASTM D664/D974
Oxidation	X	X	X	ASTM D5846/D943/D8048/ FTIR
Nitration	X	X		ASTM D943/FTIR
Sulfation	X	—		ASTM D7415/FTIR
PQ index	—	—	X	ASTM D8184

Wear Metals (ICP AES Spectrometer ASTM D5185, D6595 [RDE])

Any lab needs a minimum of equipment to do the basic testing on oils or fluids for service technicians to assess the conditions of a machine. The two methods used for elemental analysis are the inductively coupled plasma (ICP) spectrometer and the rotating disk emission (RDE) spectrometer, or arc spark. Both instruments use a high-energy source to excite atoms within a sample. The atoms give off energy in the form of light. The wavelength of the light emitted is specific to each atom, and the amount of light energy can be converted to a concentration of each element tested (see Figure 6-1).

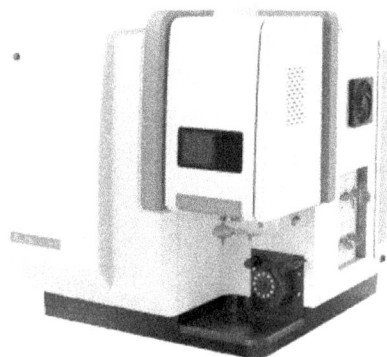

FIGURE 6-1 PerkinElmer Avio 500 ICP-OES ICP AES spectrometer.

The RDE spectrometer uses a high-voltage electrode above a rotating disk to vaporize the sample, whereas the ICP spectrometer uses high-temperature (8500°C+) (15,332°F) argon plasma to vaporize the sample. The RDE can vaporize particles up to 10 μm in size, whereas the ICP spectrometer can vaporize particles up to 5 μm in size due to sample preparation. If a sample is first run in the ICP spectrometer, followed by the RDE spectrometer, certain element concentration measurements can be higher in the RDE device because the ICP device cannot handle particles above 5 μm. Additional testing is needed for wear particle analysis of particles >10 μm in size.

The wavelength and detection limits of the ICP unit for the different elements are listed in Table 6-2. The lab is able to detect the elements that are essential to the mobile equipment industry, and it is up to users and the lab to establish the standard deviations for those elements. In this way, users can make good decisions based on these readings by establishing what is abnormal for each machine in their fleets.

TABLE 6-2 Element Wavelengths

Element	Wavelength (nm)	Detection Limit (ng/mL)	Source
Aluminum (Al)	386.152	2.0	Dirt, engine pistons, torque converters
Barium (Ba)	233.527	1.1	Oil additive
Boron (B)	249.773	3.5	Coolant additive, engine oil additive
Calcium (Ca)	317.933	2.8	Sulfonate additive
Chromium (Cr)	283.563	1.2	Top piston ring, bearings, needle bearings, spools
Copper (Cu)	324.754	1.2	Oil coolers, bushings, overlay engine bearings, hydraulic pumps
Iron (Fe)	259.940	1.3	Engine liners, hydraulic cylinders, oil pumps, valve guides
Magnesium (Mg)	285.213	2.4	Engine oil sulfonate additive
Molybdenum (Mo)	202.03	9.6	Friction modifier
Nickel (Ni)	221.647	10.0	Gears, carburized bearings
Phosphorus (P)	213.618	12.0	Extreme-pressure additive, friction modifier, antiwear agent
Potassium (K)	766.450	0.06	Coolant additive, fertilizers, soaps
Silicon (Si)	251.611	32.0	Dirt, foam inhibitor, sealant
Silver (Ag)	328.068	2.6	Solder
Sodium (Na)	589.592	6.8	Dirt, saltwater, coolant
Tin (Sn)	189.926	32.0	Overlay in Babbitt bearings
Titanium (Ti)	334.941	0.6	Contaminant, engine oil additive, paints
Vanadium (V)	292.402	2.0	Alloys, marine fuels
Zinc (Zn)	213.856	1.8	Antiwear additive, filter wrapping mesh leaching

Viscosity Tests (ASTM D445/D727)

Gravity viscometers and evacuation-time viscometers are used to perform viscosity tests on lubricants. Both are approved methods and yield the same results.

Keep in mind that viscosity varies dramatically with temperature, so the test temperature is critical to any measure of viscosity. Engine oil viscosity testing usually occurs at 100°C. However, if the engine oil is used as a hydraulic fluid, then 40°C is the preferred test temperature. To obtain the viscosity index (VI), the viscosity at both temperatures is needed. VI could be an important measurement when judging the type of base oil used for hydraulic fluids. In the case of engine oils and tractor fluids, VI helps estimate the extent to which the viscosity improver has sheared down.

Testing of viscosity for hydraulic and transmission fluids is normally performed at 40°C, but testing at 100°C is also acceptable. It is a matter of preference. Some manufacturers prefer to test everything at 100°C. The support for such a preference is in the need to see how far the viscosity falls at 100°C testing, which is especially important in components without hydrodynamic lubrication, such as in gearboxes and differentials (see Figure 6-2).

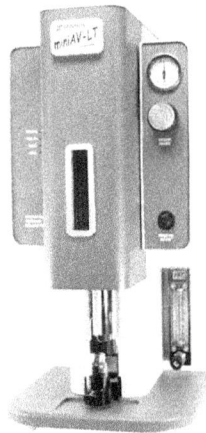

FIGURE 6-2 Cannon Mini AV-LT Viscometer.

Hydraulic Fluid Viscosity Decline

Hydraulic fluids, engine oils (in hydraulic applications), and tractor fluids shear down at different rates depending on the fluid and the component in which the fluid operates. Rust and oxidation (R&O) hydraulic fluids and antiwear (AW) hydraulic fluids are typically very shear stable because they are made without viscosity improvers, which are the weakest link for shearing in a multigrade oil. These hydraulic fluids shear down very slowly, which is why they are preferred for hydraulic applications. In 4,000 hours, they only lose about 15% of their initial viscosity.

In contrast, the viscosity improvers that give us multigrade capabilities in multiseasonal hydraulic fluids, most tractor fluids, and engine oils shear out dramatically in the first 500 hours of hydraulic use. In about 4,000 hours, these fluids can lose up to 35% of their original viscosity. Initially, viscosity drops rapidly in fluids that contain viscosity improvers and then tapers off.

After 1,000 hours or so, the rate of viscosity loss flattens out, and the fluids start to behave more like straight viscosity hydraulic fluids. Figure 6-3 is very useful in practicing oil analysis because it provides a guideline on the expected viscosity at specific hours of use of a given fluid measured at 40°C and helps us determine the hours on the oil when they are not given.

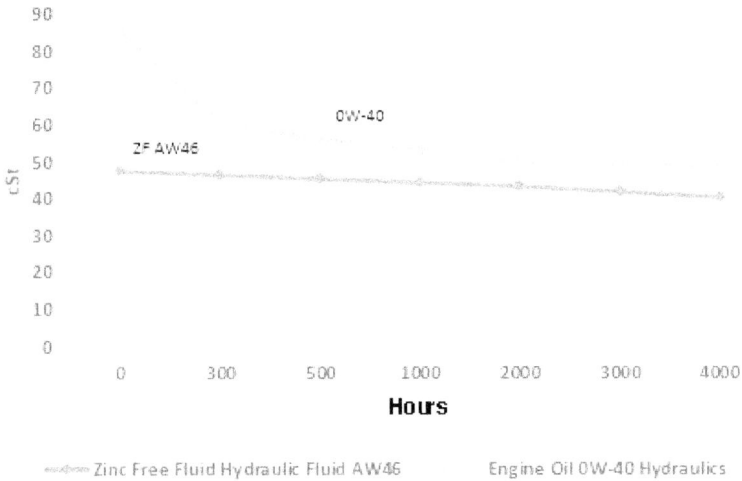

FIGURE 6-3   Hydraulic fluid viscosity decline at 40°C.

Viscosity Variation by Oil Viscosity Grade

Engines produce a very distinct and consistent oil viscosity pattern that we can use to our benefit. For example, some engines shear oil in a visible manner, causing oil viscosity to fall for about 125 hours, and at that point, things reverse and viscosity starts recovering.

The reality is that the oil still shears down, but by-products of combustion (i.e., soot, nitration, and sulfation) take over and combine with oxidation to thicken the oil for the remainder of the drain interval. If allowed to continue, the oil will thicken beyond its viscosity when it was new, ultimately to the point of failure. This viscosity pattern will vary based on the oil, sump capacity, and combustion specifics of the engine.

This viscosity information helps when we are assessing whether the engine behavior is within the expected parameters. If the viscosity does not recover, this means that potentially there is fuel dilution. If the viscosity starts climbing faster than expected, this could indicate soot generation, nitration, or rapid oxidation.

We also need to be aware of fresh top-offs because the addition of new oil changes the viscosity pattern somewhat. If we do not have the information from the oil in use, we still can figure out the level of fuel dilution using Figure 6-4. Such figures can be precise if we know the behavior of the engine when we try to establish the level of fuel dilution.

Some tier III or IV engines may behave differently, and the application is going to make a difference as well. A constant-speed engine with a regular load is unlikely to show viscosity decline by fuel dilution. However, a machine used intermittently with waiting periods may show dilution right away.

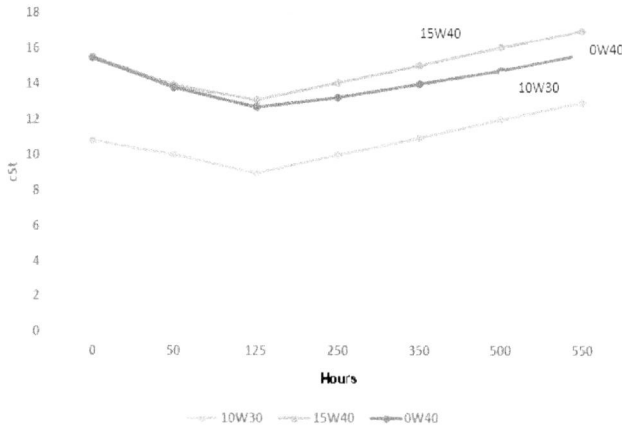

FIGURE 6-4 Diesel engine oil viscosity variatio non-Tier III or IV.

Viscosity Decline in Gear Oils

Gear oils are very viscosity stable. Still, they shear down a little bit (see Figure 6-5).

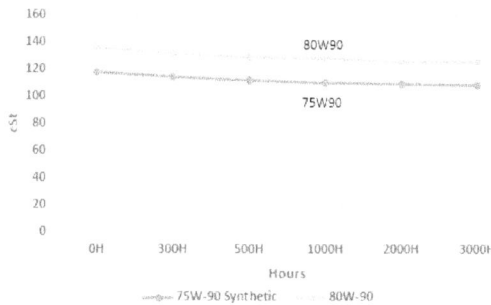

FIGURE 6-5 Gear oil viscosity decline.

Gear oils are also very unconventional when it comes to viscosity protocol and ranges. You can find several gear oils that claim to belong to a given SAE viscosity, but actual measurements indicate that they are mislabeled. Table 6-3 helps in estimating the hours of use for a given oil/fluid viscosity grade when the user does not provide it.

TABLE 6-3 Viscosity (cSt) of Various Gear Oils versus Hours at 40°C

Gear oil	0 h	300 h	500 h	1000 h	2000 h	3000 h	Decline
75W-90 synthetic	118	114	112	110	109	108	−8%
80W-90	136	132	129	128	126	126	−7%
80W-90LS	165	163	162	160	160	159	−4%
75W-90HD	148	145	144	143	142	141	−5%
75W-140 synthetic	184	177	172	168	167	166	−10%
80W-140	271	260	252	247	245	243	−10%
85W-140	335	339	329	322	319	317	−5%

Viscosity Loss in Tractor Fluids

Tractor fluids shear down but with a different pattern depending on the application. Usually, an application that involves a torque converter is going to shear the fluid more than a direct-drive-transmission axle (see Figure 6-6).

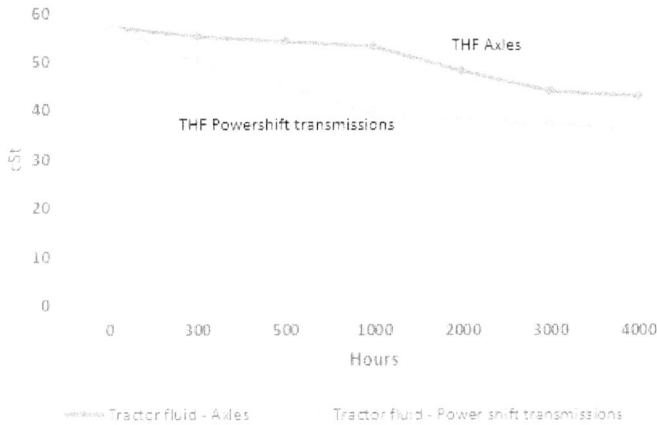

FIGURE 6-6   Tractor fluid viscosity decline.

Water Content Tests (Karl Fischer ASTM D6304/E203)

Precise water content analysis is fundamental in the preservation of high-pressure hydraulic systems and power shift transmissions, especially those with torque converters. There are several methods of measuring water, but Karl Fischer coulometric titration (ASTM D6304) is the most precise and consistent test (see Figure 6-7). Not being able to see the water defeats the purpose of oil analysis. Figure 6-8 lists several of the water analysis tests that are available. Whereas the infrared test is the most common, it fails to provide information below 1600 ppm, which is way beyond the interest point in hydraulics. However, the Karl Fischer titration covers the whole spectrum from very little water to dissolved and free water.

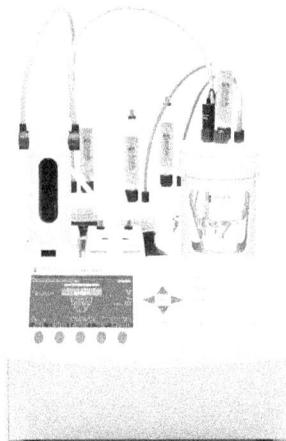

FIGURE 6-7   Hanna Instruments H1903 Karl Fischer Titrator.

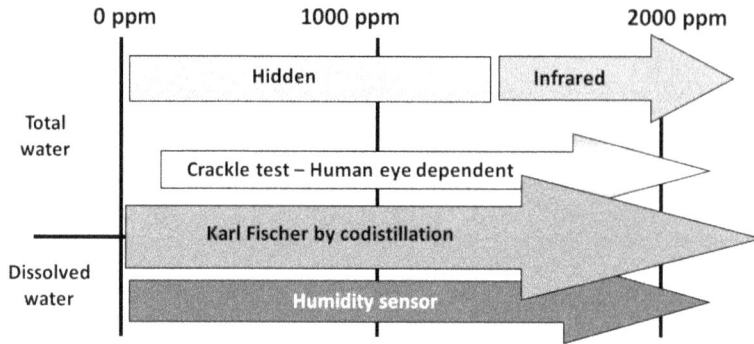

FIGURE 6-8 Visibility of water.

Alternative Methods of Measuring Water Content (ASTM D2412)

Fourier-Transform Infrared (FTIR) Spectroscopy

One method commonly used is FTIR spectroscopy. Although the method is acceptable for engines, it is not acceptable for hydraulics and powertrains. The reasons are simple. Hydraulics use high pressures and high temperatures. Under these conditions, any amount of water >1000 ppm is detrimental to the hydraulic pump. By comparison, conventional FTIR spectroscopy only captures water at 1600 ppm or higher, way beyond the point of interest. With powertrains, especially those using gear oils with high doses of phosphorous, bad reactions take place when water is >1000 ppm. With transmissions using tractor fluids, the tolerance for water is better, unless the transmission has a torque converter. Water can be very destructive to the fins of a torque converter (see Figure 6-9).

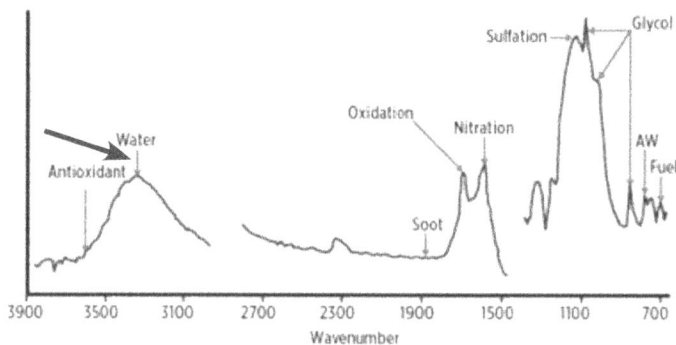

FIGURE 6-9 Detecting water by FTIR spectroscopy.

Crackle Test

Aside from FTIR spectroscopy, discussed previously, a practical and cheap method of testing for water is with the crackle test using a hotplate. The method is no more than a heated frying pan used to test oil for water content. Its name comes from severely contaminated samples that crackle like a French frier (see Figure 6-10).

Figure 6-11 illustrates the capabilities of the test, and although it is highly dependent on the human eye, some technicians claim that it is very precise when done correctly. The problem is that with massive testing volumes, the efficiency decreases as the operator tires, and then the repeatability of the test becomes questionable.

FIGURE 6-10 Crackle test apparatus.

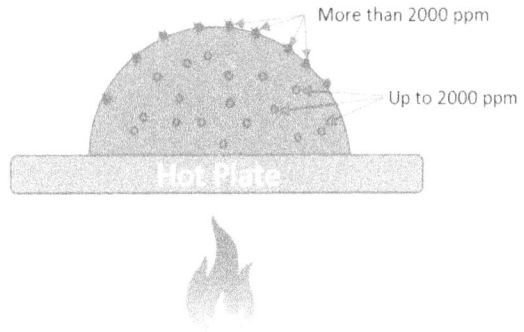

FIGURE 6-11 Crackle test.

Soot: Shimadzu Attenuated Total Reflection (ATR) Method (IR/ASTM D77844)

With the attenuated total reflection (ATR) method of measuring soot, the sample is held in tight contact with the surface of a highly refractive prism that transmits infrared light. The light reflected through the minute surface of the sample makes the spectrum visible.

Soot is a combustion by-product that is typical from diesel engines and needs to be monitored. Soot can alter viscosity, consume dispersant additive, buffer AW additives, and cripple lubrication. There is more than one method of measuring soot, but the Shimadzu ATR method is the preferred test because of its repeatability (see Figure 6-12). The section "Soot in Engines" in Chapter 3 covers this subject in detail.

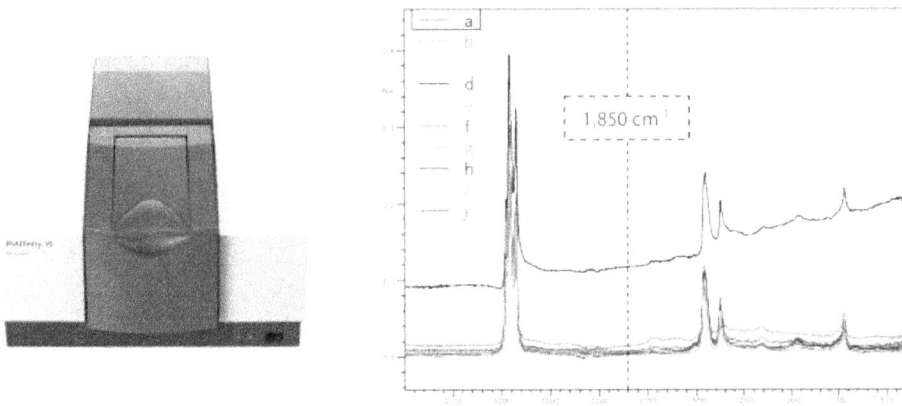

FIGURE 6-12 Shimadzu IRTracer (*left*); Shimadzu soot record (*right*).

Glycol by Sodium and Potassium Readings

Glycol is easily identifiable in oil analysis through the presence of its constituents, namely sodium, potassium, and in some cases, silicon from silicates. Boron also can be present, but it is not a consistent indicator of a coolant leak.

The challenges in determining whether the readings belong to coolant are somewhat similar to those involving dirt because the constituents also could come from different sources and mean very different things. There are no typical ratios, as we saw with dirt, because the formulations for coolant vary widely. In addition, sodium can arise from four different sources and potassium can be from at least three.

Other Methods for Glycol Detection (FTIR ASTM D6595 or ASTM D7889)

Glycol has a strong absorbing band in the wave frequency around 3450 cm^{-1} corresponding to the O–H functional group, as well as a unique band for ethylene glycol at 1070–1030 cm–1 corresponding to the C–O functional group. Interferences from water and oil additives can cause errors in measurement, so a significant amount of signal processing is required to ensure reliable and consistent results (see Figure 6-13).

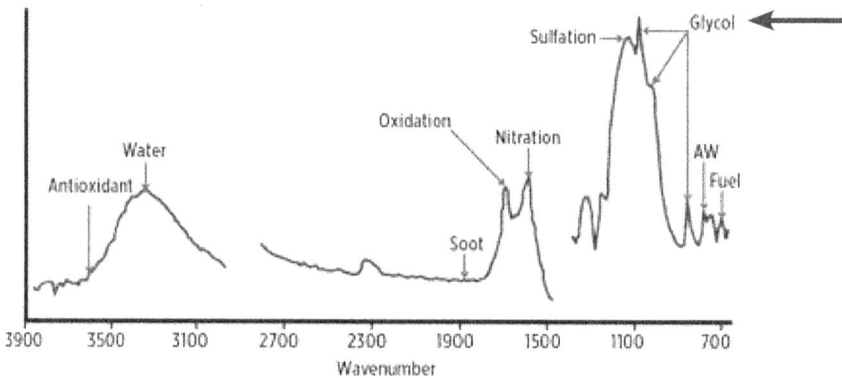

FIGURE 6-13   FTIR spectroscopy glycol detection.

Engine Oil Fuel Dilution

Fuel contamination in diesel engines is a common phenomenon with high-pressure rail or high-pressure unit injectors. Fuel dilution needs to be in check to make sure that the oil viscosity never goes below levels that could cause bearing failure. The most practical method of measuring fuel dilution is to compare hours of use with oil viscosity. As discussed earlier, engine emission Tier types and their applications have a direct impact on fuel dilution. Laboratories test fuel dilution using various methods. With current ultra-low-sulfur fuels, the use of chromatography is necessary to detect fuel because the new fuel species resemble lubricants and hide behind the results when the old method is used.

Chromatography (D3524, D3525, D7993.4-6)

The most widely accepted direct method for analyzing fuel dilution in lubricants is gas chromatography (GC) according to ASTM methods D3524, D3525, and, more recently, D7593.4-6. These methods involve injecting a portion of an oil sample into a gas chromatograph. The chromatograph vaporizes the sample and passes it through an analytical column that separates the sample into its component hydrocarbons in order of boiling point. By integrating the area of fuel peaks, as detected by a flame ionization detector (FID), quantification is feasible. The great advantage of this method is that it can detect biodiesel, which other methods have difficulty recognizing (see Figure 6-14).

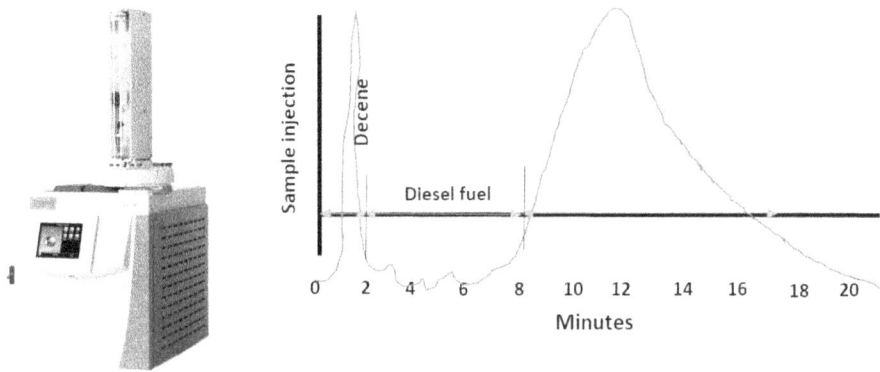

FIGURE 6-14 TRACE 1310 gas chromatographer (*left*); gas chromatography results (*right*).

Setaflash (ASTD3828)

The Setaflash flash point detector is a quick and easy way to detect fuel dilution, but it has the inconvenience of being blind to biodiesel. The procedure uses the flash point to indirectly measure the amount of fuel dilution. The flash point is the lowest temperature at which an ignition source causes the vapors of the sample to ignite under set conditions. In the presence of lighter hydrocarbon fuel components in a lubricant, the flash point temperature will decrease (see Figure 6-15).

FIGURE 6-15 Setaflash tester.

Oil Viscosity Trending

As explained earlier in this chapter under "Viscosity Tests," engines have a characteristic way of shearing down oil, depending on the engine type and application. Once the user learns his or her engine's shear-down pattern, it is easy to use viscosity and hours of use to establish fuel dilution in a practical way (see Figure 6-16).

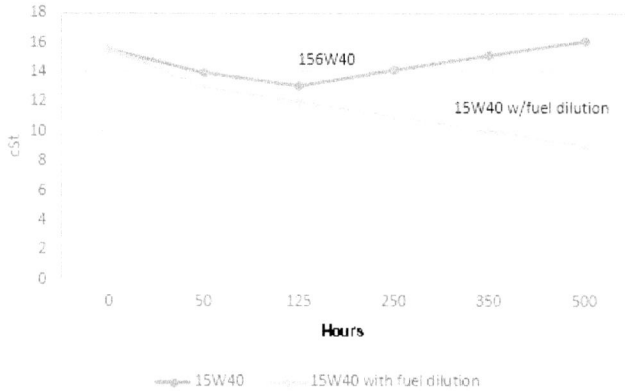

FIGURE 6-16   Engine with fuel dilution.

Tier III, interim IV, and final Tier IV engines could experience fuel dilution in certain applications and under certain operating modes. Some manufacturers are fine with a fuel dilution of up to 7% these days, provided that the viscosity never falls below 8.5 cSt, which is a risk when using SAE 10W30 oil at fuel dilutions approaching 7%.

Particle Counts (ISO 11500)

A particle counter is an instrument that detects and counts particles. By its very nature, a particle counter is a *single particle* counter, meaning that it detects and counts particles one at a time. The particle counting relies on light scattering, light obscuration, or direct imaging. A high-energy light source illuminates the particle as it passes through the detection chamber. The particle passes through the light source (typically a laser or halogen light), and a photo detector captures the passing of the particle.

If direct imaging is used, a halogen light illuminates the particles from the back within a cell while a high-definition, high-magnification camera records the passing particles. A recorded video and computer software measure particle attributes. When using light blocking (obscuration), the detection occurs by the loss of light. The amplitude of the light scattered or blocked is measured, and the particle count goes into standardized counting bins. The image to the right in Figure 6-17 shows a light-scattering particle counter diagram.

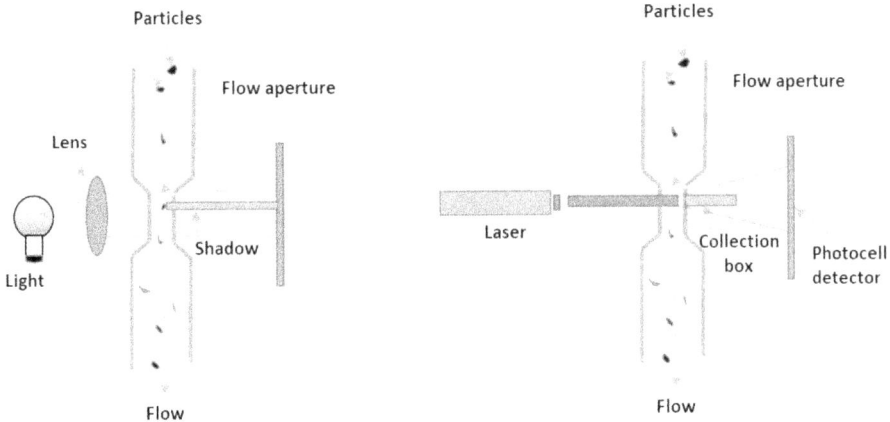

FIGURE 6-17 Light blockage (*left*); light scattering (*right*).

LaserNet Fines (ASTMD7596)

A relatively new and exciting technology, LaserNet Fines, offers the advantages of particle counts and ferrography together. This technology allows particle classification from 4 μm and particle profile determination starting at 20 μm. The parts >20 μm and <100 μm fall into 5 major categories: cutting, sliding, fatigue, nonmetallic, and water/air. The instrument can separate water and air from the counts and provide an accurate count of particles.

LaserNet Fines offers a great advantage for hydraulic systems where mixing causes gels. By separating gels from the total counts, accurate particle counts can be provided. For readers interested in ferrography, this technology provides valuable information on particle morphology (see Figure 6-18).

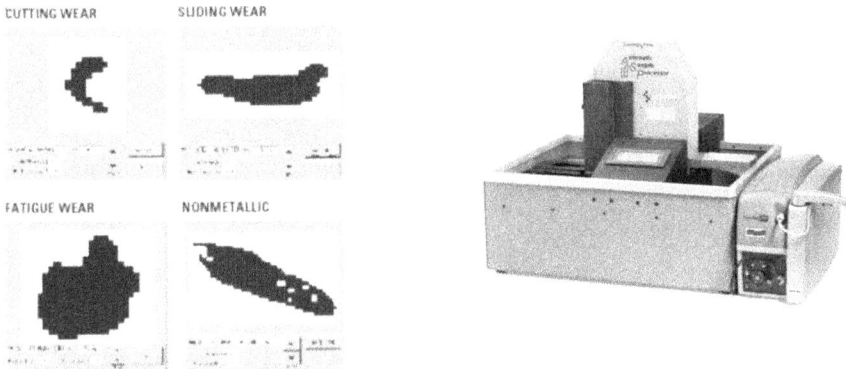

FIGURE 6-18 LaserNet 200 Fines apparatus.

LaserNet Fines constraints are that the technology is expensive, there are not many apparatuses available, and interpretation of the ferrrography requires expertise, a commodity not too abundant at labs or users' locations.

Pore Blockage Particle Counter

The pore blockage method is a widely used approach in which automatic particle counts are obtained. In this method, a volume of fluid is passed through a mesh screen with a clearly defined pore size, commonly 10 μm. Two instruments apply this method.

One instrument measures flow decay across a membrane as it starts plugging while pressure is held constant, first with particles >10 μm in size and later by smaller particles as the larger particles partially plug the screen. The second instrument measures the rise in differential pressure across a screen as it starts plugging with particles while the flow rate is constant. Both instruments use a software algorithm that turns the time-dependent flow decay or pressure rise into an ISO cleanliness rating according to ISO 4406:99.

While pore block particle counters do not suffer from the same problems as optical particle counters with respect to false-positive results caused by air, water, dark fluid, and so on, they do not have the same dynamic range as optical particle counters. In addition, because the particle size distribution uses estimation, the counters depend on the accuracy of the algorithm to accurately report ISO fluid cleanliness codes according to ISO 4406:99.

Nevertheless, these counters accurately report the aggregate concentration of particulates in the oil, and in certain situations, particularly dark fluids such as diesel engine oils and other heavily contaminated oils, pore block particle counting does offer advantages. This method is ideal for hydraulic fluid mixes that tend to form gels and inflate particle count results because of gels (see Figure 6-19).

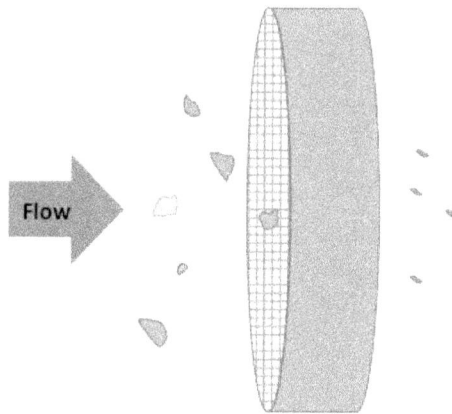

FIGURE 6-19 Pore blockage.

Particle Cleanliness (ISO Code 4406 1999)

The International Organization for Standardization (ISO) created the cleanliness code 4406:1999 to quantify particulate contamination levels per milliliter of fluid at three sizes: 4, 6, and 14 μm. This is *not* a test method but rather a user-friendly way of displaying contamination. This ISO code comes in three numbers that could be, for example, 19/17/14. Each number represents a contaminant level code for the correlating particle size. The code includes all particles of the specified size and larger. It is important to note that each time a code increases by 1, the quantity of particles in that range *doubles*.

The example in Figure 6-20 shows a 100-mL sample of oil that contains particles larger than 4, 6, and 14 μm in the amounts shown on the left. The number of 4-μm and larger particles in this sample was 3 million, which correlates with a code of 22. The number of 6-μm and larger particles was 200,000, which correlates with a code of 18. The number of 14-μm and larger particles was 7,000, which correlates with a code 13. Therefore, the ISO cleanliness code for this sample is a 22/18/13. High-pressure hydraulic systems and hydrostatics are required to operate in a 20/16/13 cleanliness code at a minimum. Gains in component longevity are feasible when the systems operate constantly at higher levels of cleanliness.

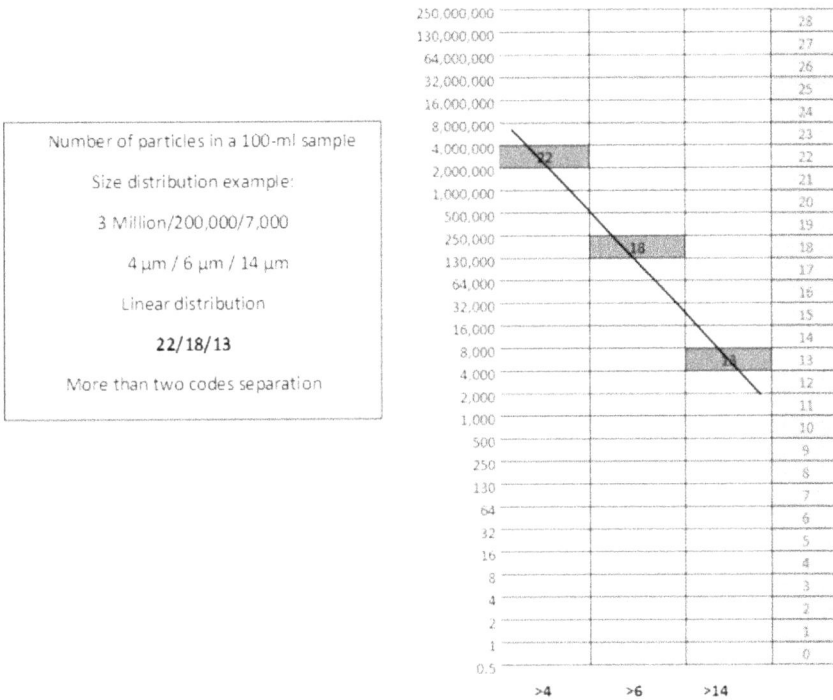

FIGURE 6-20 Particle count chart.

The old ISO code 4402 from 1987 was used to measure 5- and 15-μm particles in a sample. With improvements in technology over the years, particles deemed as 5 μm in the past were closer to 6.4 μm with the newer measurement techniques, and the 15-μm particles were actually in the range of 13.8 μm.

By the same token, the 2-μm particles in the old standard now appear as 4.3 μm. Therefore, a new standard (ISO 11171) came to the world, and a new 4-μm channel was added. For those still using old testing equipment, the 2-μm readings can be seen as equivalent to the current 4-μm size.

The two notations are still related and correspond. Some users of old equipment may use the old notation and only provide 5- and 15-μm readings, such as 18/13 (see Figure 6-21).

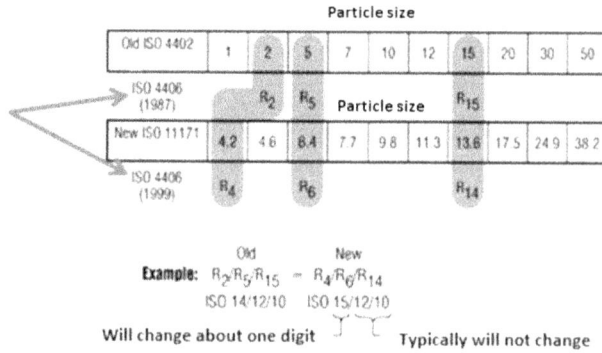

FIGURE 6-21 Particle count correlations.

National Aerospace Standard (NAS) 1638 Cleanliness Code: Update to AS4059

Although not common in the mobile equipment industry, the NAS 1638 measurement is still around, but now under update AS4059. There are also particle counters capable of expressing measurements in different standards. NAS 1638 was the forerunner of other contamination coding. The concept of the code appears in Table 6-4 and is based on a fixed particle size distribution of the contaminant over a size range of >5 to >100 μm.

TABLE 6-4 NAS 1638 Contamination System

Class	Maximum Particles/100 mL in Specified Size Ranges (μm)				
	5–15	15–25	25–50	50–100	>100
0	125	22	4	1	0
0	250	44	8	2	0
1	500	89	16	3	1
2	1,000	178	32	6	1
3	2,000	356	63	11	2
4	4,000	712	126	22	4
5	8,000	1,425	253	45	8
6	16,000	2,850	506	90	16
7	32,000	5,700	1,012	180	32
8	64,000	11,400	2,025	360	64
9	128,000	22,800	4,050	720	128
10	256,000	45,600	8,100	1,440	256
11	512,000	91,200	16,200	2,880	512
12	1,024,000	182,400	32,400	5,700	1,024

The aircraft component cleanliness of 1960 inspired this particle distribution method. From this basic distribution, a series of 14 classes is defined, covering very clean to very dirty levels. The interval between each class is double the contamination level of the previous class. This principle is a feature of many of the classification systems that have developed since.

Particle Counts by System in Mobile Equipment

There are different interpretations as to what the correct cleanliness levels of different systems should be. Table 6-5 shows typical readings of different systems using different fluids in mobile equipment of all types. The "normal" level may require extensive use of a filter caddy or, as an option, a permanent bypass filtration installation. Nevertheless, it is going to depend on application as well. There are also considerations regarding the type of system, type of fluid, and type of reservoir before determining what the system cleanliness should be.

TABLE 6-5 Particle Counts by System and Type of Fluid

System	Normal	Abnormal	Critical
Sealed reservoir hydraulics with AW fluid	17/15/12–20/17/14	21/18/15–22/19/16	≥X/20/17
Sealed reservoir hydraulics with engine oil or UTHF	19/15/12–22/17/14	23/18/15–24/19/16	≥X/20/17
Unsealed reservoirs with AW fluid	18/16/13–23/18/15	21/19/16–25/20/17	≥X/21/18
Unsealed reservoirs with engine oil or UTHF	20/16/13–23/18/15	24/19/16–27/21/18	≥X/22/19
Sealed hydrostatics with AW fluid	17/15/12–20/17/14	21/18/15–22/19/16	≥X/20/17
Sealed Hydrostatics with engine oil or UTHF	19/15/12–22/17/14	23/18/15–24/19/16	≥X/20/17

Particle count specifications are complex for any product line. This is so because machines use all types of systems and reservoirs. If fluid mixing occurs, this adds to the challenge. With these variations, the table ends up with a large list of specifications.

When there is fluid mixing, the 4-μm channel loses its importance because any high 4-μm code could be the measurement of coarse additives. Sealed reservoirs, as in the case of some hydrostatic crawlers, promote the best cleanliness and lowest particle counts. In addition, filtration in hydrostatics is generally superior to that of hydraulic systems.

Total Base Number (TBN) Test (ASTM D2896 and D4739)

The *base reserve*, also known as the *total base number*, is a measure of the level of alkalinity in an oil. It is best determined by measuring the amount of acid required to neutralize the base. The resulting number is equivalent to the alkalinity of that amount of potassium hydroxide measured in milligrams per 1 gram of oil (mg OH/g). There is more than one ASTM procedure for testing TBN, and each test yields different results. When evaluating the change in TBN, make sure to use the same standard consistently (see Figure 6-22).

The TBN in an engine oil tends to go down as the oil's acid-neutralizing power wears out. When the TBN reaches a given point, we generally agree that it is time to change the oil. In the past, this magic number was 5.

The need to neutralize acids with high TBNs diminishes with fuel sulfur content below 15 ppm. TBN levels in new oils have trended down since ultra-low-sulfur diesel fuel has become mandatory. Even so, extended engine oil drain intervals of up to 500 hours are common without the penalty of corrosive wear.

FIGURE 6-22 Metrohm Tritosampler Light (1T/2T) TBN/TAN Tritrator.

Labs can determine the TBN based on ASTM D4739. The differences are in the chemicals used for titration. D4739 uses hydrochloric acid, whereas D2896 uses perchloric acid. Typically, used oils are tested by D4739. ASTM D2896 is used in testing new oils and is often the test method used for marketing because it produces a higher number. What is important is to stick to one method to avoid inconsistent trending.

Coulometric Titration

Coulometric titration methods are an option so long as the oil is not dark. For these methods, a visible color change indicates the endpoint with an indicator, which reacts to a change in pH. TBN measurement can be challenging due to the dark nature of crankcase oils, especially when soot may be present, but the test works for new oils. Table 6-6 lists accepted values.

TABLE 6-6 Accepted Values for TBN

Test oil	TBN limits with CJ4 or CK4 oils	TBN with CI4 oils
Diesel fuel with <15 ppm sulfur	3.0–3.5	3
Diesel fuel with <500 ppm sulfur	4.0–4.5	4
Diesel fuel with >500 ppm sulfur	4.5–5.0	5

TAN (ASTM D664/D974)

An oil's total acid number (TAN) increases with use. Ideally, we should change an oil when its TAN and TBN intersect because, theoretically, the oil has not enough alkaline reserve left to prevent corrosion from the increasingly acidic oil. Hydraulic and powertrain fluids could have the suggested normal values for TAN listed in Table 6-7.

TABLE 6-7   Typical TAN Values

Fluid	TAN
Hydraulic fluids AW	<1
Automatic transmission fluid (ATF)	<1.5
Tractor fluids	<5
Engine oils as hydraulic fluids	<4

We have used these parameters independently; TBN for engines and TAN for hydraulics and powertrains, but given the low TBNs in modern engines' oils today, the use of both measurements has become the norm. With regard to TANs for engine oils, the suggestions listed in Table 6-8 apply based on fuel sulfur content.

TABLE 6-8   TAN Limits for Engines Oils

Fluid	TAN limits with CJ or CK oils	TAN with CI4 oils
Diesel fuel with <15 ppm sulfur	3.0–3.5	3
Diesel fuel with <500 ppm sulfur	4.0–4.5	4
Diesel fuel with >500 ppm sulfur	4.5–5.0	5

TANs for gear oils can have different values depending on the type of gear oil (see Table 6-9).

TABLE 6-9   TAN Limits for Gear Oils

Fluid	TAN
Standard gear oils	<3
LS gear oils	3.5–4.9

Oxidation (ASTM 943, D5846, D8048 FTIR [Fast Fourier Transform Infrared] Spectroscopy)

Labs also carry an FTIR spectrometer to collect signatures of high-wavelength contaminants such as fuel, glycol, soot, oxidation, and sulfation. FTIR spectroscopy is a versatile tool used to detect common contaminants, lube degradation by-products, and additives within lubricating oils. It has become a widely used technique for quickly assessing multiple lubricant characteristics (see Figure 6-23).

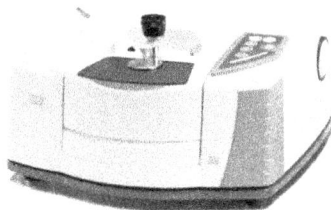

FIGURE 6-23   Thermo Scientific Nicolet iS20 FTIR Spectrometer.

Ideally, labs will use a FTIR apparatus to measure many different parameters, as detailed in Table 6-10. However, not all measurements enjoy a definition from which we can derive an action in maintenance. The results, though, can suggest the need for further analysis when the FTIR spectrometer indicates something of interest in a sample. For more definition, we depend on other tests that are more precise or simpler to do. For example, the FTIR spectrometer indication of small amounts of glycol is not very reliable, but it is easily identifiable via readings of sodium, potassium, silicon, and sometimes boron.

TABLE 6-10 Wavelengths of Various Substances

Substance	Wavelength	Weakness
Water	3400	Up to 1600 ppm only
Glycol	3400	Better with Na and K
Oxidation	1750	Needs new oil reference
Soot	1750	Better with ATR
Nitration	1630	Needs new oil reference
Sulfation	1150	Needs new oil reference
Glycol	1080	Better with Na and K
Glycol	1040	Better with Na and K
ZDDP	980	Better with ICP
Glycol	880	Better with Na and K
Diesel	800	Weak, better with GC
Gasoline	750	Better with flash point

However, FTIR spectrometry is a great tool for a general view of the lubricant condition, especially oxidation, nitration, and sulfation. Still, the lab needs the new oil signature to establish the degradation level because a change in the values is more meaningful than the values themselves. In the world of mobile equipment, measuring oxidation only has value in large hydraulic systems that work at high pressures and temperatures, such as excavators. It also has value in axles and transmissions from high-speed, high-load equipment, such as articulated dump trucks. It has little practical value for medium-sized engines (see Figure 6-24).

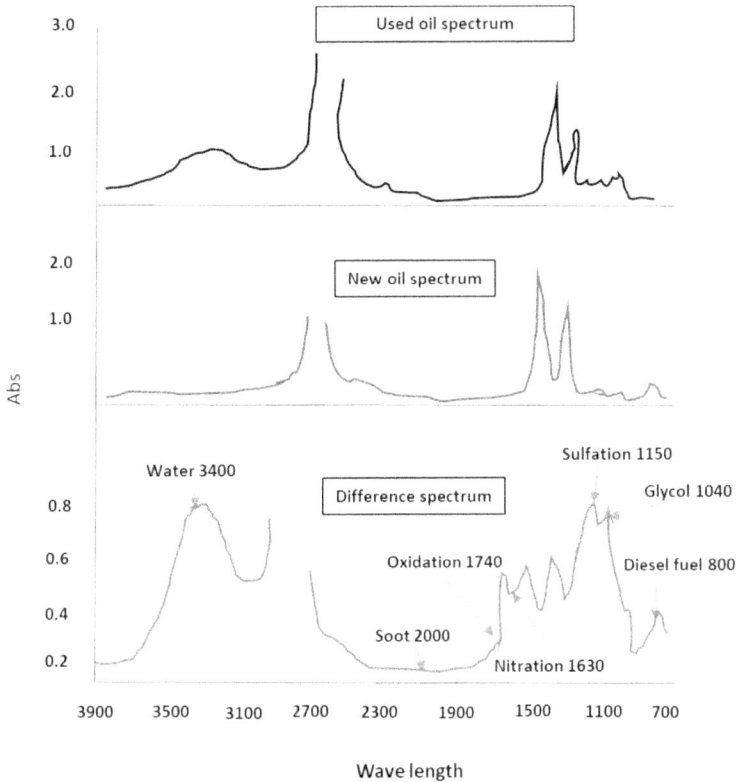

FIGURE 6-24 Wavelengths for new and used oils.

Nitration (ASTM D943)

FTIR spectroscopy also can be used to get readings on nitration. However, nitration is seldom a concern with modern diesel engines, and it is more of an issue with natural gas engines. Nevertheless, nitration still can take place in old diesel engines, and it contributes to an increase in oil viscosity. Nitration in older engines is usually due to cold operating temperatures and blowby past piston rings.

Nitration is mainly a by-product of combustion as nitrogen oxides. Current engine technologies focus a great deal on reducing nitrogen oxides by retarding injection and improving engine breathing through better valve design and turbocharging. The goal is to achieve stoichiometry and completely consuming the injected fuel. It is noteworthy that biodiesel fuels increase nitrogen oxides. The FTIR spectrometer is a good tool to measure nitration, as Figure 6-24 shows. The lab needs a sample of the new oil to establish the degree of nitration.

Sulfation (ASTM D7415)

Sulfation is the direct result of combustion of high-sulfur fuels. Water is a by-product in engine combustion, which reacts with sulfur in the injected fuel and drains slowly into the oil sump during operation. Once the combustion by-products end up in the oil sump, they cause corrosion that shows up as high metal readings, especially iron, copper, and lead. The exhaust gas also suffers from the impact of sulfur corrosion by affecting variable-geometry (VG) turbocharger operating mechanisms and poisoning the diesel oxidation catalyst (DOC), if present.

With the implementation of ultra-low-sulfur diesel (ULSD) fuels at 15 ppm, sulfation issues have practically disappeared. However, in many markets, the use of high-sulfur fuels is still an issue, and a sulfation test is great to have because sulfation increases oil viscosity in addition to corrosion.

For a successful test, the reference values of the new oil are necessary. Otherwise, the results could be inconclusive or misleading. Figure 7-24 earlier shows how an oil changes its spectrum profile as it ages. This is the information used by the lab to determine the remaining life of a lubricant.

TABLE 6-11 Physical Property Values

Property	Normal	Abnormal	Critical
Oxidation	<25	25–30	>30
Nitration (in engines)	<25	25–30	>30
Sulfation (in engines)	<25	25–40	>40

PQ Index (ASTM D8184)

The *particle quantification (PQ) index* measures the total concentration of ferrous metal in an oil sample. Whereas inductively coupled plasma (ICP) analysis measures only small particles, PQ index completes the picture by measuring the concentrations of all iron particles, including those larger than 6 or 7 μm. This information is especially useful for powertrain components, which tend to generate larger wear particles than engines (see Figure 6-25).

FIGURE 6-25 Optimus Instruments FG-K17000 Analex PQL PQ Index Tester.

The advantage of the PQ index test is that it provides a number relating to the concentration of large ferrous particles that otherwise would not be visible by ICP analysis. However, this technology still needs to show the value of the concentration because collecting a representative sample that includes larger ferrous particles is a challenge. The second issue is the interpretation of what is meaningful coming from a component. Solid tables need to be developed based on a history of sampling on many different machines.

The test procedure detects the presence of any ferrous metal in a sample by the level of distortion in the magnetic field. PQ index results are displayed as a number, which is an arbitrary unit of measurement that correlates well with more expensive direct-reading ferrography for large particle testing.

The key to interpreting wear patterns in gearboxes is in the ratio of large iron particles to small iron particles. For example, if the PQ index increases dramatically while the ICP iron measurement in parts per million remains consistent or goes down, we know that larger ferrous particles are growing in number.

Based on the PQ index, some labs have developed a sense for the numbers coming out of this test and provide a table with numbers that are considered normal and abnormal. Every manufacturer still needs to develop a taste for this test and make good use of it. Plenty of data and field experience are required to correlate these numbers to prefailure warnings.

Because of the variety of components used by manufacturers and applications, it would not be surprising that several tables emerge to cover the different situations that the diversity of components produce (see Table 6-12).

TABLE 6-12 PQ Index Table

Fluid	Max	Abnormal
Hydraulics	10	50
Engines	10	50
Power shift transmissions	50	100
Manual transmissions	300	750
On-highway differentials	500	1250
Final drives	500	1250
Off-highway final drives	500	1250
Tandems	600	1500

GREASE TESTING

A mobile equipment fleet manager is seldom going to require grease testing unless the grease he or she uses is suspected of causing wear or not meeting specifications. Users can test new grease in inventory to check whether it meets the stated specifications. However, if a user wants to test used greases coming from a machine for wear or contaminants, this is going to be a challenge. The reason is that the amount of grease required for testing is around 1 oz, and it will be difficult to find a machine that provides this amount from one spot. However, new grease can only be tested to see if it meets certain specifications. Table 6-13 shows a comprehensive set of tests for grease. Highlighted in bold are those tests that a fleet manager may consider having done, if the need arises.

TABLE 6-13 ASTM Grease Testing

Lab Test	Standard
Dropping Point	ASTM D566, IP396
Cone Penetration	ASTM D217, IP50
Oxidation Stability	IP142, ASTM D952
Copper Corrosion	IP112
Corrosion Preventative Properties	IP220, ASTM D1743
Oil Separation	ASTM D1742
Wet Worked Stability	ASTM D7342
Low Temperature Torque	ASTM D1478
Fretting Wear Protection	ASTM D4170
Evaporation Loss	ASTM D972
Anti-Wear Properties	Four-Ball, ASTM D2596
Wheel Bearing Leakage	ASTM D1263
Compatibility of Mixtures	ASTM D6185
Water Washout Test	ASTM 1264

ASTM Grease Testing Methods

Dropping Point

The dropping point of a grease is the temperature at which it passes from a semisolid to a liquid state. The dropping-point test determines the cohesiveness of the oil and thickener of a grease (see Figure 6-26).

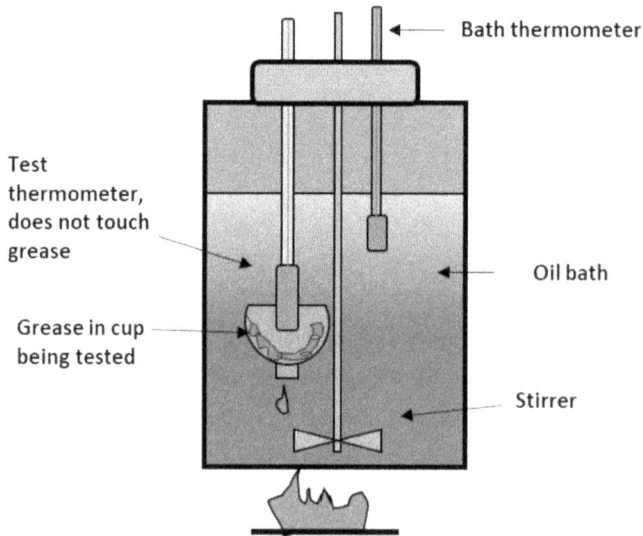

FIGURE 6-26 Grease dropping-point test apparatus.

The dropping point establishes the maximum usable temperature of the grease, which is typically set at from 50°C to 100°C below the experimentally determined dropping point. Furthermore, the dropping point identifies whether the correct grease is in use and helps determine whether the used grease is still usable for the intended purpose.

Cone-Penetration Test

This test (ASTM D217 or IP 50) measures grease consistency. Under prescribed conditions, a standardized cone is dropped into the grease for five seconds. Penetration is measured in tenths of a millimeter and is used to determine the grease's National Lubricating Grease Institute (NLGI) consistency number. The higher the penetration measurement, the thinner is the grease, and the lower is the NLGI number (see Figure 6-27).

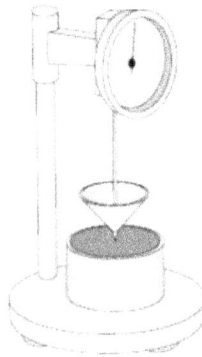

FIGURE 6-27   Grease cone penetration apparatus.

Oil-Separation Test

An oil-separation test monitors the separation of oil from grease at elevated temperatures for a defined period of time. The tendency of oil to separate during storage can be an important characteristic. When the base oil begins to separate from the thickener in a grease, the remaining material may change in consistency and potentially affect the ability of the grease to function as designed (see Figure 6-28).

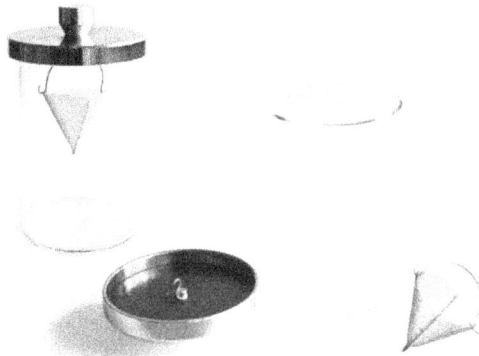

FIGURE 6-28   Grease oil-separation test apparatus.

Antiwear Properties: Four-Ball Weld Test

The four-ball extreme-pressure (EP) test locks three half-inch-diameter steel balls in a test cup filled with the test grease. A fourth ball makes contact with the three lower balls. An 80-kg load is applied, and the top ball is rotated for ten seconds while the three balls in the cup remain stationary. After taking measurements of the size of the wear scar on the three stationary balls, the test is repeated at different loads. The test is typically repeated ten to twelve times at progressively higher loads until the balls weld together.

- The **weld point**, reported in kilograms, is the load at which the four balls weld together. A higher weld point indicates a more effective EP lubricant.
- The **load-wear index** (LWI), also reported in kilograms, quantifies the relative wear protection of a lubricant under load. A higher LWI indicates a more effective extreme pressure lubricant.
- **Last nonseizure load** (LNSL), also reported in kilograms, is the highest load before welding occurs. A higher LNSL indicates a more effective EP lubricant (see Figure 6-29).

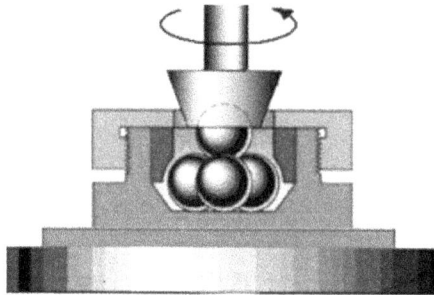

FIGURE 6-29   Grease four-ball test.

Water Washout Test

The water washout test method evaluates the resistance of a lubricating grease to washout by water from a bearing when tested at 38 and 79°C (100 and 175°F) under the prescribed laboratory conditions. It is not equivalent to service evaluation tests (see Figure 6-30).

FIGURE 6-30   Linetronic Technologies Water Washout Grease Apparatus
LT/WW-205600/M (ASTM D1264).

Grease Thief Tests

An alternative to traditional grease testing is by die extrusion and preparation from the Grease Thief approach. The Grease Thief only requires a small amount of grease, making it the preferred method for testing used grease. Some of the test results as shown in Table 6-14 are an indirect match with traditional testing through ASTM standards Table 6-13, The Grease Thief approach provides a reasonable match for the ASTM dropping-point and cone-penetration tests. However, FTIR spectrometry and the RULER (Remaining Useful Life Evaluation Routine) test are not actual tests of grease capabilities, as is the traditional four-ball test (ASTM D2596). The Grease Thief method focuses more on grease oxidation, which for mobile equipment may have limited use. After all, greasing intervals in modern equipment are around 50 hours these days, if not daily. Thus, measuring antioxidant capabilities becomes a futile exercise.

TABLE 6-14   Grease Thief Tests

Grease Thief test	Conditions Measured
Extrusion	Consistency under speed
Colorimetry	Aging, overheating
FTIR spectrometry	Progression of oxidation
RDE spectroscopy	Detection of wear metals
RULER	Antioxidants presence
Ferrous debris monitor (FdM+)	Wear concentration
Analytical ferrography	Shape and quality of particles
Rheometer	Consistency under load

Extrusion

In the Grease Thief extrusion test, the grease flows through an orifice under varying speeds and conditions during extraction. The consistency of the grease is compared with that of the baseline grease to measure changes in consistency and determine thinning or thickening of the grease.

Colorimetry

Grease colorimetry testing uses a patent-pending method that includes a visible spectrometer, a light-path sample cell, and a thin-film grease substrate to gather a visible-light (400–700 nm) spectral graph of the grease sample. This method validates observed appearance changes in greases, trends darkening due to aging or overheating, characterizes dye formulations of new greases, and can even approximate the concentration of certain particulate contaminants.

FTIR Spectrometry

The Grease Thief procedure uses FTIR spectrometry to obtain a fingerprint of the used grease, which the analysis compares to the new grease fingerprint. A two-spectra overlay allows the analyst to see the differences. Oxidation and grease mixing become visible with this methodology.

RULER Test

In the RULER test, antioxidant remaining in the grease becomes visible, thanks to variable voltage applied to the sample. This test is helpful in determining the remaining useful life of the antioxidants in the sample. This test also works to determine the residual antioxidant purging out of a bearing to determine whether the regreasing cycle is set appropriately.

Rheometer Test

The final test in Grease Thief analysis is the rheometer test, which is used to examine the consistency of grease. Using the cone and plate rheometer normal force measurements determines oscillation stress, recoverable compliance, and G1 values.

CONCLUSION

We have discussed oil and grease testing in this chapter. Of the two, oil testing is the more important activity. However, choosing the right oil test can make the difference between poor and outstanding oil analysis management. As we move to the next four chapters, we will see how these tests play a fundamental role in making oil analysis a real proactive maintenance tool.

CHAPTER 7

OIL ANALYSIS BASICS

OIL ANALYSIS INTERPRETATION
KNOWLEDGE FUNDAMENTALS

Oil analysis is at the center of the condition-based maintenance (CBM) tools. However, to excel in interpretation of an oil analysis, several areas of knowledge are required. If the user lacks knowledge in one or more of these areas, his or her ability to interpret an oil analysis cannot reach its potential.

1. The format and tests, although they sound like a minor aspect of oil analysis, are the foundation on which the whole concept rests. A good format allows the user to interpret the data in the least possible time. In contrast, the format is not an issue if a computer is doing the analysis.

2. The tests chosen determine whether the user will be able to use the oil analysis as a proactive maintenance tool. This is also true whether a human or a computer does the analysis. Here the cost-effectiveness comes into play. A cheap report with a few tests will lessen the ability of the user to find issues that otherwise might help to prevent downtime.

3. The lubricant, with its signatures and physical properties, tells us about its proper use and its health while in the machine. The signature is the identification of the product, which helps differentiate it from other products and protect them from misapplication; it also helps assess lubricant aging.

4. Knowing the environment allows the user to understand the possible sources of external contaminants. A machine working in Florida, for instance, will have a greater tendency to show water than one working in Arizona. Conversely, a machine in proximity to the ocean or farmland may show more sodium or potassium.

5. Finally, knowledge of the machine itself and the metals produced by its components is critical for a correct assessment of the oil analysis. For example, the type of reservoir and whether or not the hydraulics and hydrostatics have a common sump will greatly influence the results.

The type of pump used, the number of hydraulic cylinders, and whether it is a closed- or open-loop system all influence the readings. Familiarity with the equipment is essential for proper interpretation of results. Not all suppliers are willing to share wear data generated from their components, and some keep these numbers confidential. However, users can develop their own tables.

Why We Need Oil Analysis

- Machine systems, whether hydraulics, engines, or powertrains, breathe and are not as sealed as one may think. Small particles of dirt still manage to pass even through new seals on hydraulic cylinder rods.
- Users do not fully monitor the proper use of lubricants at the job site.
- Users end up using whatever is cheapest or most convenient for them or for the fleet.
- Lubricants get contaminated gradually and oxidize.
- Machine components get old and produce more wear metals.
- Field repairs usually happen in harsh environments that contribute to contamination.
- Knowing machine conditions and impending repair or maintenance situations permits scheduling to improve uptime.
- Knowing the condition of fluids may allow extended drain intervals, increasing uptime and allowing maintenance personnel to work on tasks that are more important.

What Can We Measure with Oil Analysis?

Through oil analysis, users can measure at least five important parameters that affect equipment life:

- The environment
- The lubricant
- The machine component health
- The effectiveness of maintenance practices
- Failures in progress

Oil analysis is not only good for measuring the level of metal generation, but it also tells much more about the machine, the environment, and how well the maintenance of the machine is going.

Oil analysis can open new windows of opportunity for users. Because some of those factors change with location and application, the results may vary, but the value of oil analysis will not. Conditions in Florida may produce different results from those in Texas, but oil analysis will tell us what those differences are.

Oil Analysis Formats

There are plenty of laboratories for oil testing. Many of them, if not most, are well equipped to do the testing commonly required for advantageous oil analyses. One of the real differences among laboratories is in the tests they perform. Many labs succumb to the desire to be the low-cost provider and thus often disregard critical information. The omitted information may be quite valuable, but the lab takes advantage of the user's lack of knowledge in analyzing the test results.

Skipping key tests to achieve lower cost is a false economy, especially if it hinders the ability of users to understand the root cause of some of the indicators. Keep in mind that many labs around the world have evolved from the initiative of people who have worked in a lab and felt that they could run the business better. Thus, they provide a low-cost service that their customers want, but if those customers had greater understanding, they would want more complete testing. While following an American Society for Testing and Materials (ASTM) procedure to produce a test result is straightforward, machine-specific in-depth analysis is what really unlocks value.

There are plenty of cheap oil analysis reports available on the market. Some reports discredit the value of this powerful tool. However, for amateurs or uninterested users, a cheap oil analysis provides a false sense of security, but at a high cost.

When it comes to choosing a laboratory, what should matter most is that the report provides the information required to perform rigorous proactive maintenance. Given the volume of reports that a fleet produces, a format that is easy to read and interpret becomes a necessity.

The example in Figure 7-1 comes from an oil supplier that provides free oil analyses to customers who bought a specific number of lubricants. The problem is that this report has no value from a CBM point of view. Reports like this one too often result in costly failures and disappointment.

TEST RESULTS					
Si - Silicon, ppm	5	5	3	15	10
Fe - Iron, ppm	8	8	10	22	13
Cu - Copper, ppm	1	1	95 D	5	2
Al - Aluminum, ppm	1	1	1	5 B	2
Cr - Chromium, ppm	1	1	1	5 B	1
Pb - Lead, ppm	2	2	16 B	5	1
Mo - Molybdenum, ppm	1	1	1	4	1
Ni - Nickel, ppm	0	0	<1	4	0
Sn - Tin, ppm	2	3	13	6	2
Ag - Silver, ppm	0	0	0	3	0
Glycol (Antifreeze)	Neg	Neg	Neg	Neg	Neg
Water, Percent	0.00	0.00	0.20	0.90	0.00
Fuel Dilution %, Percent	0	6	0	0	0
Viscosity 100C, Centistokes	14.19	15.23	14.31	15.47	15.87
VIS Grade	40	40	40	40	40
Soot, Percent	<1.0	<1.0	<1.0	1.0	1.0

FIGURE 7-1 Incomplete horizontal report.

This report does not measure sodium or potassium, so it cannot indicate when glycol is leaking into the engine. It does not include the total base number (TBN), which helps determine the optimal time to change the oil. Without additive levels, how can you know whether the wrong oil is in use or whether you should suspect that hydraulic fluid is leaking into the engine?

Look for a supplier whose reports are complete and have a format that is easy to read, perhaps with the help of large fonts and colors. Some laboratories go the extra mile by flagging additives that do not belong to the reported oil.

In the example in Figure 7-2, the lab has flagged additives in the lubricant from a four-wheel-drive (4WD) loader that should not be there, indicating that a bad seal is allowing hydraulic fluid to leak into the axle. If the laboratory only reported wear metals, there would be a progressive increase in wear with each sample submitted but no indication of an anomaly.

Comments	Check for possible source of ABRASIVES entry (such as faulty filter elements, housings, seals, breathers, fill points, etc). Abrasives (Silicon)are at a SIGNIFICANT level; Bushing/Thrust metal is at a MODERATE LEVEL; Aluminum is most likely in the form of alumina/silica (Dirt); Flagged additive levels are different than what should be present for the lubricant that is identified for this unit. (This does not imply that the lubricant does not meet proper API, SAE, or ISO classifications.); Lubricant change acknowledged.

Sample #	Wear Metals (ppm)							Contaminant Metals (ppm)						Multi-Source Metals (ppm)						Additive Metals (ppm)				
	Iron	Chromium	Nickel	Aluminum	Copper	Lead	Tin	Cadmium	Silver	Vanadium	Silicon	Sodium	Potassium	Titanium	Molybdenum	Antimony	Manganese	Lithium	Boron	Magnesium	Calcium	Barium	Phosphorous	Zinc
1	358	1	4	8	76	2	0	0	0	0	20	1	0	0	11	0	2	0	146	33	257	0	1612	136
2	353	1	5	11	89	0	0	0	0	0	47	4	0	0	7	0	3	0	70	21	250	0	2237	165
3	485	2	7	15	148	0	0	0	0	0	73	6	4	0	11	0	4	0	84	32	288	9	2294	188

FIGURE 7-2 Lab report.

Choose your laboratory carefully. Look for a complete battery of tests, provided in a timely manner, in a format that is easy to comprehend. Avoid the cheapest labs, and seek value. The cost of high-quality oil analysis is nothing compared with the cost of downtime and suboptimal machine life.

Horizontal versus Vertical Displays

The industry has both horizontal and vertical displays. A good report display facilitates reading and interpretation, allowing easy diagnosis in a few seconds. What is important is that the test grouping is logical and that the tests facilitate interpretation of the machine's health. When the lab groups the elements by category, it is much easier to read the report and understand the results. For example, if metals are all listed in one group, it is much easier to assess wear. The same applies to additives; by having them all in one group, it is easier to identify the type of oil in use. Some labs walk the extra mile by showing aluminum twice, once under metals and once under contaminants, which lets the user interpret dirt much more easily. Grouping sodium and potassium together makes it much easier to identify coolant contamination. Having all oil physical properties in one box allows a quick assessment of the remaining life of the oil. Finally, also having the data from all the previous reports displayed allows for trending analysis and tracks changes over the life of the machine.

A vertical display is perfectly fine, provided that the groupings are logical and conducive to easy interpretation. Figure 7-3 shows a good vertical display of the elements along with ideal groupings.

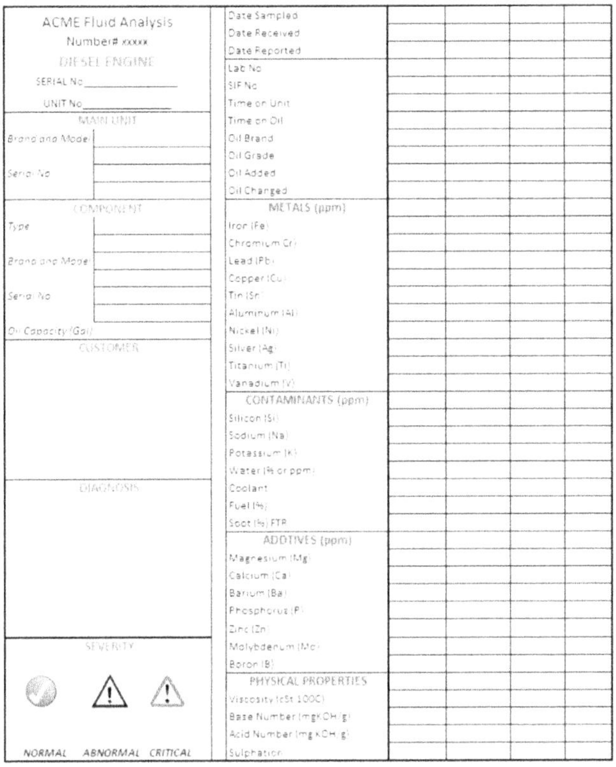
ACME Fluid Analysis
Number# xxxxx
DIESEL ENGINE
SERIAL No
UNIT No
MAIN UNIT
Brand and Model
Serial No
COMPONENT
Type
Brand and Model
Serial No
Oil Capacity (Gal)
CUSTOMER
DIAGNOSIS
SEVERITY
NORMAL ABNORMAL CRITICAL

Date Sampled
Date Received
Date Reported
Lab No
SIF No
Time on Unit
Time on Oil
Oil Brand
Oil Grade
Oil Added
Oil Changed
METALS (ppm)
Iron (Fe)
Chromium (Cr)
Lead (Pb)
Copper (Cu)
Tin (Sn)
Aluminum (Al)
Nickel (Ni)
Silver (Ag)
Titanium (Ti)
Vanadium (V)
CONTAMINANTS (ppm)
Silicon (Si)
Sodium (Na)
Potassium (K)
Water (% or ppm)
Coolant
Fuel (%)
Soot (%) FTP
ADDTIVES (ppm)
Magnesium (Mg)
Calcium (Ca)
Barium (Ba)
Phosphorus (P)
Zinc (Zn)
Molybdenum (Mo)
Boron (B)
PHYSICAL PROPERTIES
Viscosity (cSt 100C)
Base Number (mgKOH/g)
Acid Number (mg KOH/g)
Sulphation

FIGURE 7-3 Vertical display lab report.

Required Information

Users need to understand that they cannot get optimal results if they fail to fill out the form when submitting the sample. For quality results, the sample needs to identify the machine (preferably by serial number), the component from which it was drawn, the hours or miles on the component and on the lubricant (identified by specific product name), and lubricant viscosity. Customer name and address seem obvious, but any lab will tell you that this is sometimes missing.

WHAT SHOULD WE TEST IN ENGINES?

What we measure in engines needs to be related to its significance to the life of the engine. Table 7-1 lists the desirable test objectives for engines and the means to accomplish them in order of importance.

The success of any oil analysis depends heavily on the information provided by the user when submitting the sample. This information, especially hours on the equipment and hours of oil use, is missing most of the time or is imprecise. The same happens with the brand and type of lubricant. This is the equivalent of going to your doctor and keeping mute about the reasons for your visit. The lab needs the hours to evaluate time-dependent metals and to establish whether physical properties correlate with those hours. In addition, the lab needs the brand and viscosity of the oil to establish viscosity trends and oxidation percentage.

TABLE 7-1 Engine Tests

Engines	
What should we test?	**Test Method**
Dirt	Si, Al, Na
Glycol/water	Na, K, Si, crackle
Viscosity increase/decrease	Viscosity at 100°C
Soot	FTIR spectroscopy, viscosity increase
Fuel	Zeta flash, FTIR spectroscopy, chromatography, viscosity
Alkaline reserve/total acid number	Total base number (TBN)/total acid number (TAN), KOH/g
Metals	Pb, Sn, Cr, Fe, Cu by inductively coupled plasma (ICP) spectroscopy
Additive consistency	FTIR spectroscopy
Degradation/aging	Oxidation, FTIR spectroscopy
Sulfation/nitration	FTIR spectroscopy

There needs to be a good return in value for an oil analysis, and to realize this return, users need to understand what they are buying. Knowing what information the lab will provide is fundamental. The lab is usually happy to provide training on the essentials, but some labs may steer users toward the lab's capabilities rather than the users' needs.

The format for the oil analysis report varies depending on the lab, but the elements of the report should be the same across the different suppliers. For learning purposes, we will use a horizontal display for all components in the report, as in Figure 7-4 (the black numbers refer to the subsections in this section).

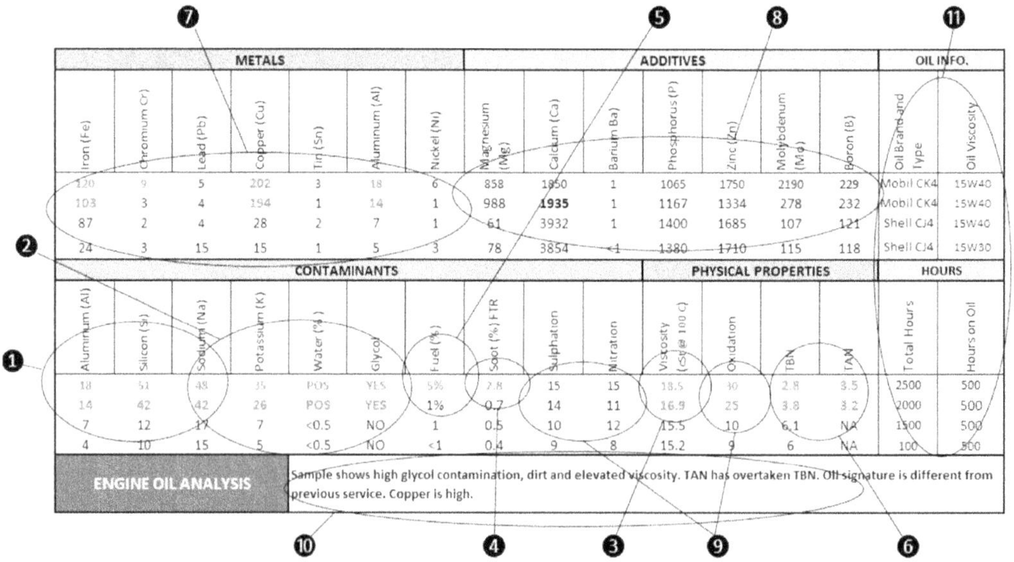

METALS							ADDITIVES							OIL INFO.	
Iron (Fe)	Chromium Cr)	Lead (Pb)	Copper (Cu)	Tin (Sn)	Aluminum (Al)	Nickel (Ni)	Magnesium (Mg)	Calcium (Ca)	Barium Ba)	Phosphorus (P)	Zinc (Zn)	Molybdenum (Mo)	Boron (B)	Oil Brand and Type	Oil Viscosity
120	9	5	202	3	18	6	858	1850	1	1065	1750	2190	229	Mobil CK4	15W40
103	3	4	194	1	14	1	988	**1935**	1	1167	1334	278	232	Mobil CK4	15W40
87	2	4	28	2	7	1	61	3932	1	1400	1685	107	121	Shell CJ4	15W40
24	3	15	15	1	5	3	78	3854	<1	1380	1710	115	118	Shell CJ4	15W30

CONTAMINANTS										PHYSICAL PROPERTIES				HOURS	
Aluminum (Al)	Silicon (Si)	Sodium (Na)	Potassium (K)	Water (%)	Glycol	Fuel (%)	Soot (%) FTR	Sulphation	Nitration	Viscosity (cSt@ 100 C)	Oxidation	TBN	TAN	Total Hours	Hours on Oil
18	51	48	35	POS	YES	5%	2.8	15	15	18.5	30	2.8	3.5	2500	500
14	42	42	26	POS	YES	1%	0.7	14	11	16.9	25	3.8	3.2	2000	500
7	12	17	7	<0.5	NO	1	0.5	10	12	15.5	10	6.1	NA	1500	500
4	10	15	5	<0.5	NO	<1	0.4	9	8	15.2	9	6	NA	100	500

ENGINE OIL ANALYSIS	Sample shows high glycol contamination, dirt and elevated viscosity. TAN has overtaken TBN. Oil signature is different from previous service. Copper is high.

FIGURE 7-4 Horizontal engine lab report.

1. Dirt

Dirt is a combination of several elements, namely silicon, aluminum, and sodium. Silicon is a special case and deserves its own section because it could be four different things. It may be indicating:

1. Foam inhibitor
2. Dirt (which is very harmful)
3. Silicate from a coolant leak (which is also bad)
4. Sealant leaching into the oil from elastomers or sealant (which is benign)

The trick is to identify the source of the silicon. We will start with silicon from foam inhibitor. This type of silicon is made up of massive molecules, at least as compared with oil molecules, suspended in oil that disrupts the oil film of bubble walls. It weakens the bubble, analogous to a crack, making the bubble pop more easily.

Most engine oils and hydraulic fluids contain 2–8 ppm of silicon from foam inhibitor. Gear lubes generally use polymethacrylate foam inhibitors, which work the same way in creating a weakness in bubbles to inhibit foaming. You can think of polymethacrylate as plastic dissolved in the oil, and therefore, it does not show up in the oil analysis. Silicon and aluminum are the two most abundant metals in the Earth's crust. Roughly 90% of the land surface on planet Earth is made of silicon-containing material. Thus, our exposure to silicon is going to be constant in whatever we do. Silicon, in the form of dirt, is among the hardest particles we need to deal with, easily causing damage when those particles are caught between the tolerances of metal surfaces in relative motion. As though this is not bad enough, it comes with aluminum oxide (alumina), which is even harder. A quick scan of the relative hardness of silicon and alumina in Table 7-2 makes it easy to see why the control of dirt is essential to a long machine life.

TABLE 7-2 Material Hardness

Mineral	Symbol	Mohs Hardness
Diamond	C	10
Alumina	Al_2O_3	9
Chromium	Cr	8.5
Silicon	Si	6.5
Steel	—	4–4.5
Iron	Fe	4
Copper	Cu	3
Bronze	—	3
Brass	—	3
Aluminum	Al	2.5–3
Lead	Pb	1.5
Tin	Sn	1.5
Sodium	Na	0.5–0.6

The ratio of silicon to aluminum (in the form of aluminum oxide) in the Earth's crust is between 4 and 1, although the ratio varies, and you are likely to see ratios between 4:1 and 6:1 depending on geography. Thus, say we have a report from an engine with 30 ppm silicon. We know that this brand of oil typically has 6 ppm silicon when it is new, so the change is 24 ppm. We would expect to see 4–8 ppm of aluminum if the silicon is from dirt.

Of course, aluminum could be higher if the piston skirts or other parts are made of aluminum. Because the air intake system is the most likely entry point, we will look for elevated chromium and iron levels to see if dirt is chewing on the rings and cylinders. Another possible source for silicon is silicates from a coolant leak. However, the silicon will not show up all by itself but with other coolant additives. Potassium or sodium is the most likely accompaniment, but it could include molybdenum, boron, and/or phosphorus. Ideally, when also running coolant analysis, we can know what additives are in the coolant and know what elements to look at for changes.

Finally, silicon in used oil can be coming from new gaskets or sealant made of silicone. Silicon levels can remain high in some engines for three or four drain intervals, although they will diminish with each draining. Again, knowledge of the machine is important. If silicon is the only element that is reading high, and there has been a recent repair, it is easy to identify the source of the additional silicon as from the gaskets or sealant used in the repair.

Flow Path for Dirt (Al and Si)

The flowchart in Figure 7-5 is a great tool to determine the origin of silicon based on the metals reading that also show up in the oil analysis. The task is to determine whether the silicon readings represent dirt, coolant, or foam inhibitor from the oil, or perhaps aluminum is coming from an accessory on the engine.

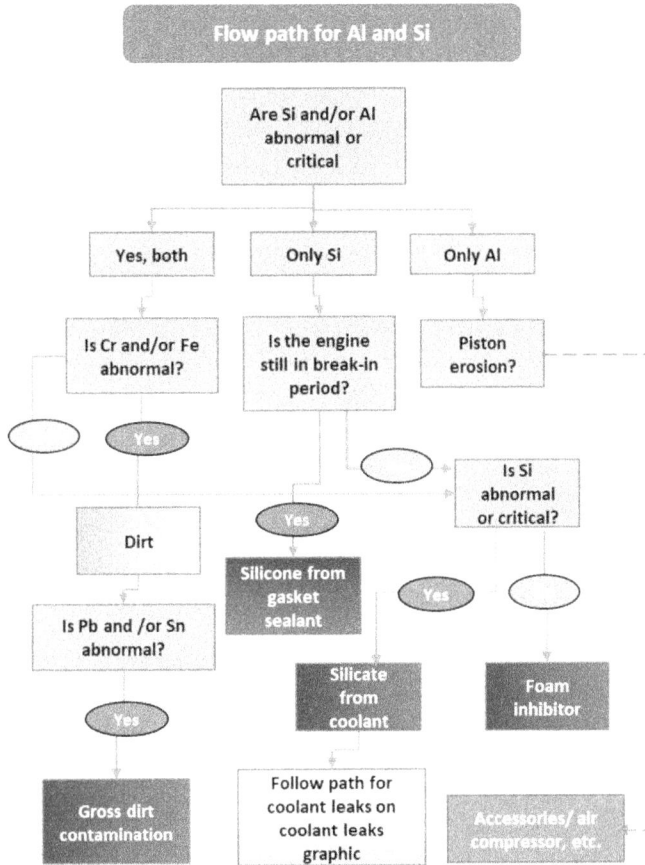

FIGURE 7-5 Flow path of dirt.

2. Glycol

You may be wondering why we bother looking at sodium and potassium as indicators of a coolant leak when most labs test for glycol. It is easier to just look for a positive or negative result for glycol.

The reason is that the glycol test is not very sensitive. Elevated levels of coolant additives indicate coolant entry at much lower concentrations than the glycol test. It is better to catch the leak while it is still small than to wait for it to get bigger and experience the damage that often accompanies a positive glycol result. With regard to water, a lab does the blotter test initially because engine oils usually contain water (much more than a hydraulic fluid would), and the blotter test is in general an accepted test. However, it is human dependent and sometimes fails. Nevertheless, Fourier transform infrared (FTIR) spectroscopy will catch its presence if it is >1600 ppm.

Sodium

Sodium is a contaminant that can alert us to a coolant leak. In contrast, it can be from other sources, and we need to determine whether the increase is from coolant or another source. Sodium can be a detergent additive in oil, although it is not common, and it is more likely to be in gasoline engine oil.

Environmental factors can contribute small amounts of sodium to oil. Dredging equipment working in saltwater and highway trucks driving on salted roads in the winter are likely to pick up a little bit of sodium. Even sweat from the person drawing the oil sample can contribute to sodium content.

Sodium is also more prevalent in certain states or provinces. A machine working in Florida or Ontario is going to show more sodium than a machine working, say, in Illinois or Iowa. Again, we need to know the levels of the various elements present in the oil and the coolant to make a solid determination. If the coolant contains both sodium and potassium, we can expect a coolant leak to increase sodium and potassium at the same ratios as in the coolant.

Potassium

Potassium is another contaminant that indicates a coolant leak. As with sodium, it can come from other sources and requires analysis. Potassium is seen in soil, and it is sometimes greater in soil with fertilizer. Samples from equipment working in farmland often will have some potassium noise.

Another potential source of sodium and/or potassium is detergent, and not the detergent additive in oil. If equipment is washed down with a detergent solution, particularly with a pressure washer, some detergent can find its way past seals and breathers to push up sodium and potassium levels. As with silicon and sodium, we can look at the other elements within our antifreeze, such as sodium and boron, for example, and determine whether the potassium level was from a coolant leak.

Silicon

High silicon readings without a corresponding aluminum counterpart suggest that the reading could come from silicates in coolant.

Flow Path for Coolant Leaks

The flowchart in Figure 7-6 shows the readings of coolant elements and metals that a coolant leak could produce. This will help you to determine whether the readings represent a coolant leak or not. As a bonus, the flowchart helps establish the locations of possible coolant leaks within the engine.

A coolant leak could be misinterpreted based on the presence of its constituents. Potassium alone could be from fertilizer or soap, whereas sodium alone could be from dirt. An indirect way to check for coolant is to keep an eye on metals, namely lead and tin. If silicon is also present, and lead is abnormal, this is confirmation of a coolant leak. If copper is high, this is an indication of an oil cooler leak. However, if iron is high, it is probably because of exhaust gas recirculation valve failure or liner perforation through cavitation, depending on the level of sodium and potassium readings.

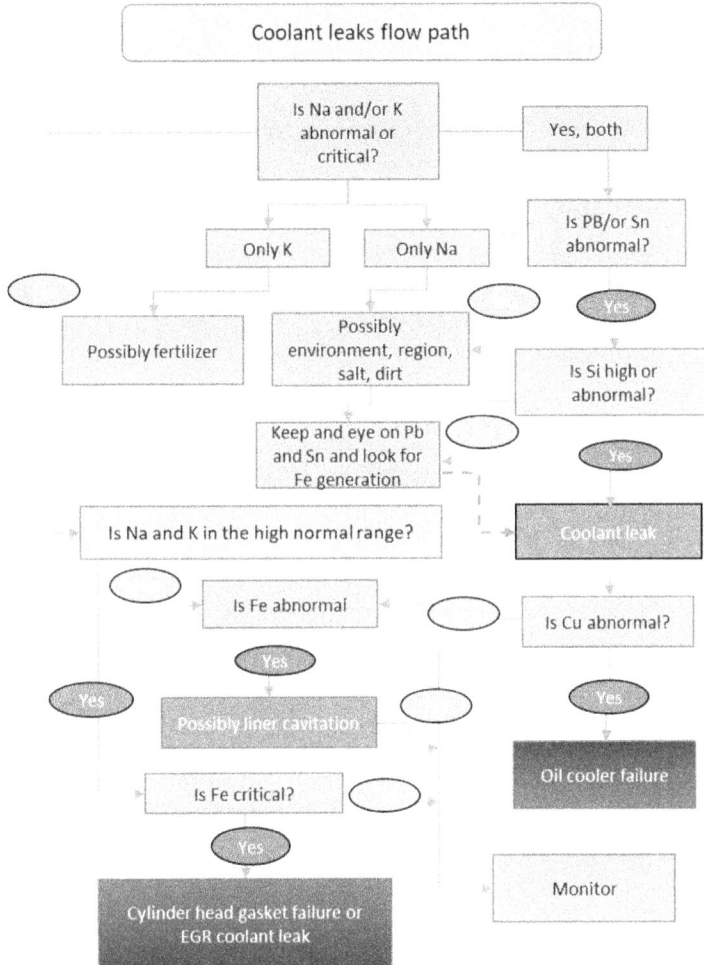

FIGURE 7-6 Coolant leak flowchart.

3. Viscosity

Through viscosity, we can measure fuel dilution and soot indirectly, and we can also see how much oxidation or sulfation has occurred. Although the oxidation test is not popular, an increase in viscosity and a decline in TBN give an indication of its severity. Keep in mind that with Tier emission engines, viscosity is not going to behave as in the past, meaning that the viscosity could go down during the service interval and could reach dangerous levels. You can also have fuel dilution and soot combined, giving a false sense of good health in an engine.

4. Soot

Soot is an important test these days with Tier low-emission engines, and there are two ways to interpret a soot text. One is direct FTIR spectroscopy, and the other is through a viscosity increase.

5. Fuel

Fuel contamination is also a new factor with Tier low-emission engines because these engines tend to suffer from fuel dilution more than pre-Tier-era engines. There are several ways to test for fuel dilution, as discussed in Chapter 6. We can evaluate fuel dilution via various means, but with current use of ultra-low-sulfur diesel (ULSD), chromatography is the best approach. We can also estimate fuel dilution through viscosity, provided that we know the oil viscosity to start with and the registered behavior of the engine in this regard. The engine's application is also a factor because two identical engines working in different applications (e.g., dirt pushing versus loading trucks) are going to have different results. Also, long idling periods worsen the fuel dilution situation.

6. TBN and TAN

The TBN, or alkaline reserve, is an indicator of an oil's ability to neutralize acids. TBN declines throughout the oil drain until the oil no longer effectively controls acids, and wear metals increase. There are limits for TBN, but with the use of CJ4 and CK4 oils, the TBN is much lower, and readings play along well with ULSD. However, TAN can overtake TBN much more easily than in the past, especially with fuels having more than 15 ppm of sulfur. Thus, measuring TAN for engines has become important in this new era.

7. Metals

Some metals by themselves are not that important, but certain metals are critical. For example, lead, tin, and chromium are the ones that require more attention, whereas iron is a truly time-dependent element. Copper means something, especially when an oil cooler fails and contaminates the engine with coolant. However, for the most part, copper is mostly the result of passivation and no longer constitutes a metal of interest for oil analysis.

We need to discuss the metals that are significant in engines, namely lead, tin, chromium, iron, copper, and aluminum. Lead and tin are significant because they are part of the bearing Babbitt overlay material (see Figure 7-7). They are soft and sensitive to acids and gross dirt and coolant contamination. Lead is the first metal to show up with acidity as well as with coolant or gross dirt contamination. When tin shows up along with lead, bearings could have reached a critical point. However, determining bearing condition depends more on lead readings.

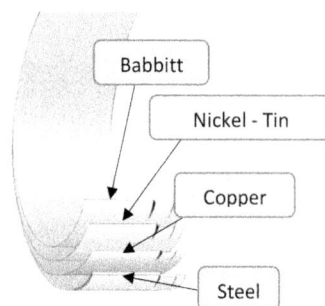

FIGURE 7-7 Bearing cutaway showing metals content.

Coolant is particularly dangerous for bearings because coolant is very good at displacing lubricant and is also a bad lubricant under load. Chromium comes from the top piston rings and shows up in oil analysis when there is dirt ingestion, liner washout with fuel, or overheating. Chromium is a critical metal because it is not abundant, and its presence suggests that serious contamination has occurred in the upper part of the engine.

Sometimes the chromium readings do not convey the seriousness of the issue because in an overheated engine the upper part of the pistons could smear the rings with melted aluminum and mask the readings. However, generally, by the time this occurs, engine operating issues will be very noticeable (see Figure 7-8).

FIGURE 7-8   Melted pistons.

Iron is a time-dependent element that comes mainly from liners. Iron readings go up with time, but the numbers spike up abnormally with dirt, soot, coolant, or water contamination and/ or high TAN or low TBN. Iron by itself is not a serious indicator of wear, but when it comes with chromium and/or coolant or water, it suggests that serious contamination has occurred.

Copper is a difficult metal to make use of for engine wear assessment. However, it could trace consistency in engine oil type use. The reason for this is that oil coolers produce copper as their normal passivation occurs and their readings override those from bearing backup material, or bushings from connecting rods or valve trains. Copper goes up every time a new type of oil chemistry is used until a new passivation occurs. Thus, it makes sense to stick to one type of oil so that the copper readings help in determining whether something other than the cooler passivation is producing copper.

Aluminum, as discussed previously, can be dirt or piston material. However, with technology changes, pistons could come in two sections (see Figure 7-9) using different materials. The upper part of the piston is made of steel, and the skirt is still made of aluminum. This change in technology also brings changes in metal production when analyzing used engine oils.

FIGURE 7-9 Two-section piston.

Minor Metals, Ni, Sn, Ag, V

Several minor metals may show up in engine oil analysis. These metals have limited use (with the exception of tin) in determining the condition of an engine. There are several reasons for this. First, these metals, if used, show up in very small amounts. Second, these metals may not even exist in the equipment in question. Third, if the metals are part of certain alloys, they will appear only as trace metals but not as an indication of wear. Then it becomes difficult to establish a wear condition based on trace metals that could or could not be part of the alloys used in the engine. Instead, the analysis needs to rely on traceable host metals that are the main component in parts that depend on lubrication to perform.

Nickel (Ni) could be present in many gears of the engine valve drive train. It is also a backup material underlayer on engine bearings, but when it shows up, the readings from lead and tin will be the defining criteria to determine wear and not the nickel reading. Tin (Sn) is part of the Babbitt material on bearings. When it shows up with lead, it is bad news coming from engine bearings.

Although it could be part of solder in old diesel engines, silver (Ag) it is not a typical metal found in mobile equipment. If used, it could be in oil coolers. Vanadium (V) is definitely not a metal used in alloys in mobile equipment. However, it is used in marine diesel fuel, and measuring it for these applications makes sense because it could be corrosive. Vanadium can show up in mobile equipment by cross-contamination during service, as happens in real life.

8. Additives

Chapter 2 discussed oil additives and their functions. Their importance from the oil analysis point of view is that they tell us whether the correct formulation was used. They also tell about changes during use and wrong top-offs, and they help in controlling maintenance so that there is consistency in lubricant use to avoid copper generation by continuous passivation.

9. Oxidation

Oxidation provides a good sense of oil degradation/aging, but oxidation testing is no longer popular for mobile equipment engines unless requested as a special test. However, to do the test, the lab needs to compare it with a new oil baseline to be able to establish oxidation. Oil degradation through TBN reduction is another way to assess oxidation indirectly and is often called the "poor man's oxidation test." Oxidation testing for gasoline engines is probably more appropriate because gasoline engines tend to oxidize oil faster.

10. Sulfation and Nitration

Sulfation and nitration tests are not practical anymore, but if needed, they are available for countries with high-sulfur fuels. An increase in the nitration level may be of value in old diesel engines and natural gas engines.

11. Lab Comments

The oil lab always provides a box with comments. In most cases, the person doing the analysis understands the task but may not be familiar with the equipment in question. He or she depends heavily on what the user provides with regard to type of oil and especially hours on the machine and hours of use of the oil. For large fleet owners, this task goes to a team that processes the information and issues recommendations for action, if any. However, the task is monumental. There needs to be an automated way to do things using all the tools available to the user through the lab website tools and through the original equipment manufacturer's (OEM's) proactive maintenance applications and telematics.

12. Oil Type, Brand, and Hours of Operation

These are the most difficult pieces of information lab personnel have to deal with. This begs the question of how good the user wants the report to be. This takes us to the next question: how good the user wants to be at interpreting the data from oil analysis internally. Oil type and brand and hours on equipment and on the oil are pieces of information that the lab requires, but most users fail to provide it accurately, if at all. Technicians in charge of collecting samples generally argue that the information is not available because they reach the machines for sampling after hours. Whether this is a valid excuse is a matter of continuous aggravation.

OIL ANALYSIS FOR HYDRAULICS

The suggested testing shown in Table 7-3 is presented in order of importance to hydraulic systems. A discussion needs to take place between the user and the lab providing the service so that these tests are included in order to obtain the maximum value from the user's investment.

TABLE 7-3 Hydraulics Testing

Hydraulics	
What should we test?	**Means to test**
Particle count	Micron 6/4/14 readings
Dirt	Ratio of Al/Si
Moisture	Karl Fischer titration
Wrong fluid/mixes	Viscosity at 40°C and signature
Metals	Cu, Fe, Cr by ICP spectroscopy
Acidity	TAN, KOH/g
Degradation/aging	Oxidation FTIR spectroscopy
High-temperature records	Machine telematics

The sample report for hydraulics presented in Figure 7-10 includes TAN only and not TBN and introduces particle counts. Another change is the temperature for the hydraulics viscosity test, which is 40°C, although it is also acceptable to do it at 100°C. It is a matter of choice. At 40°C, the scale is larger and allows better appreciation of the changes than the results at 100°C. Another difference in the report is that the water test is by Karl Fisher co-distillation. We will use the same horizontal format as the engines' table for this hydraulics table, but with slight changes to the testing.

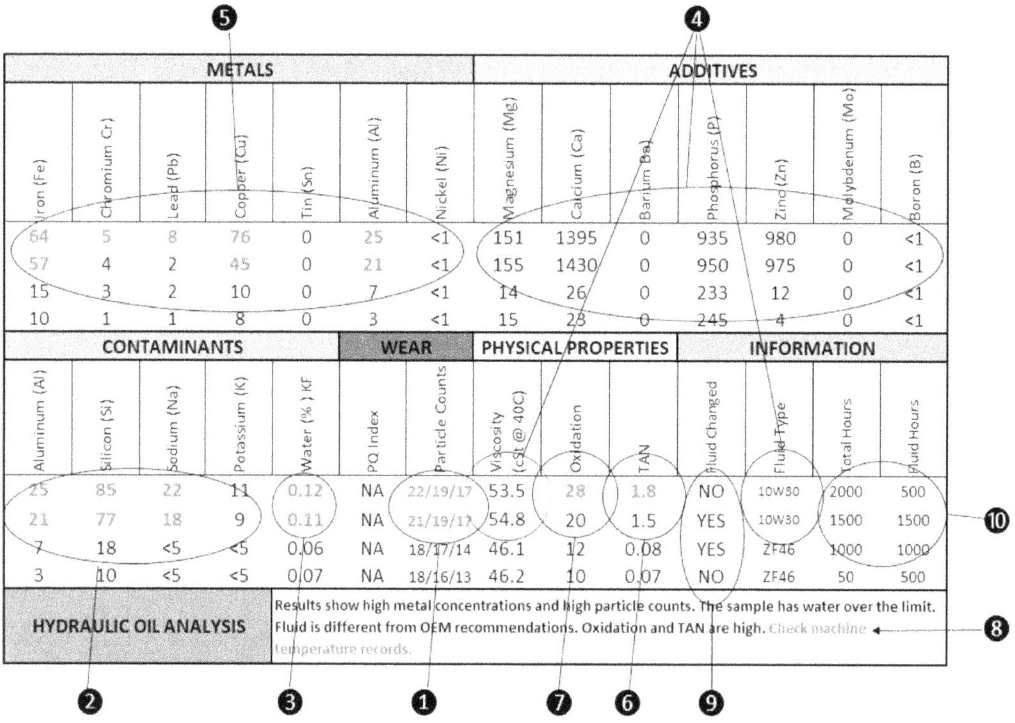

METALS							ADDITIVES						
Iron (Fe)	Chromium Cr)	Lead (Pb)	Copper (Cu)	Tin (Sn)	Aluminum (Al)	Nickel (Ni)	Magnesium (Mg)	Calcium (Ca)	Barium Ba)	Phosphorus (P)	Zinc (Zn)	Molybdenum (Mo)	Boron (B)
64	5	8	76	0	25	<1	151	1395	0	935	980	0	<1
57	4	2	45	0	21	<1	155	1430	0	950	975	0	<1
15	3	2	10	0	7	<1	14	26	0	233	12	0	<1
10	1	1	8	0	3	<1	15	23	0	245	4	0	<1

CONTAMINANTS					WEAR		PHYSICAL PROPERTIES			INFORMATION			
Aluminum (Al)	Silicon (Si)	Sodium (Na)	Potassium (K)	Water (%) KF	PQ Index	Particle Counts	Viscosity (cSt @ 40C)	Oxidation	TAN	Fluid Changed	Fluid Type	Total Hours	Fluid Hours
25	85	22	11	0.12	NA	22/19/17	53.5	28	1.8	NO	10W30	2000	500
21	77	18	9	0.11	NA	21/19/17	54.8	20	1.5	YES	10W30	1500	1500
7	18	<5	<5	0.06	NA	18/17/14	46.1	12	0.08	YES	ZF46	1000	1000
3	10	<5	<5	0.07	NA	18/16/13	46.2	10	0.07	NO	ZF46	50	500

HYDRAULIC OIL ANALYSIS	Results show high metal concentrations and high particle counts. The sample has water over the limit. Fluid is different from OEM recommendations. Oxidation and TAN are high. Check machine temperature records.

FIGURE 7-10 Hydraulics fluid lab report.

1. Particle Counts

In hydraulics, particle counts are important because they give a good idea of how clean the system is, although they do not tell what the particles are unless we use LaserNet Fines, a technology that is able to classify particles by their profile. However, particle counts also can be a source of aggravation because they do not work well with fluid mixes. In mixed-fluid conditions, particle counts show higher numbers produced by gels that form when incompatible fluids mingle, as explained in Chapter 3. In contrast, particle counts can be a source of good information when the system has water, for instance; the particle count readings are atypical and help to confirm water contamination.

2. Dirt

Dirt assessment in hydraulics is similar to the approach used in engines, and the same rules apply. Hydraulics show less dirt than engines, but its detrimental effect could be more serious than in engines because in hydraulics there are high pressures and high flows.

3. Moisture

Water content is very critical in hydraulics. This is why hydraulics need strict specifications for water content, especially if the fluid in question is an antiwear (AW) low-calcium fluid. You do not want water at >1000 ppm in high-pressure systems if you want your axial pump to survive. To catch water at these low levels, Karl Fischer coulometric titration is used. FTIR spectroscopy catches water content only at >1600 ppm, which is too high. It costs more to do the Karl Fischer test, but it can save a pump from failing.

4. Wrong Fluid/Mixes

While an engine is very tolerant of the different types of oils available and generally gets topped off with the first oil available, hydraulics are a different story. The first issue with mixes is high particle counts, the second is foaming, and the third could be loss of AW capabilities. Please refer to the compatibility chart in Figure 2-30 for more details.

5. Metals

Metals in hydraulics boil down to very specific ones that have to do with most axial hydraulic pumps. The major metals found are copper, tin, and lead, whereas iron usually comes from hydraulic cylinders and rust from the reservoir or the hydraulic tool, if any (see Figure 7-11).

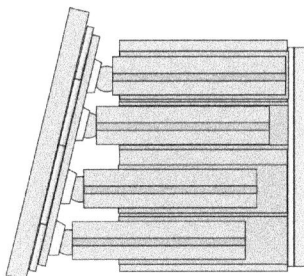

FIGURE 7-11 Axial pump.

6. Acidity

Measuring acidity is important with AW fluids and not so much with calcium-based fluids. The TAN readings from AW products are very low, typically <1.5. When readings go up, the system does not wait long before showing rust or corrosion of yellow metals. With calcium-based fluids, namely engine oil and tractor fluids used in hydraulics, there is a large TBN reserve in those fluids, and measuring TAN is just a formality.

7. Degradation, Aging

Hydraulic fluids can undergo >4,000 hours of use these days, running intermittent high pressures as high as 6,000 lb/ft^2, with temperatures exceeding 100°C. For this reason, the use of more advanced hydraulic fluids made from group III base oils is popular. Therefore, measuring aging through oxidation is not out of the question. Again, the lab needs a sample of the new fluid to establish the degree of degradation. However, any mixing of the fluid while in use turns this test useless. In such cases, the assessment of fluid health needs to rely on TAN.

8. Temperature Records

Temperature records do not come from the lab but rather directly from the machine through telematics. Users need to correlate information from the machine to hydraulic lab results when judging oxidation, high acid numbers, or changes in fluid color (see Table 7-4).

TABLE 7-4 Temperature Record for Hydraulics

Hour meter	<49°C	50–59°C	60–69°C	70–79°C	80–89°C	90–99°C	100–109°C	>110°C	Total >90°C
4302	4	1.2	3.5	7.5	20.2	51.3	4.1	0.7	56.1
4202	3.1	1.8	3.8	12.4	20.9	51.9	5.7	0.5	58.1
4102	2.2	1.1	1.9	12.7	12.6	55.7	13.2	0.4	69.3
4002	0.7	0.4	1.5	26.1	17.2	50	3.7	0.3	54
3902	0.8	0.3	2.6	26	30.2	39.5	0.2	0	39.7
3802	1.3	0.9	2.1	34	31.8	29.6	0	0	29.6
3702	7.8	4	8.7	48.5	21.9	8.7	0	0	8.7

9. Fluid Changed

For hydraulics, this box is important. It helps to establish correlations with oxidation and TAN as well as mixing.

10. Total Hours and Fluid Hours

As with other components, knowing the hours of use is good for time-dependent element analysis, TAN growth, and viscosity shear-down.

OIL ANALYSIS FOR POWER SHIFT TRANSMISSIONS AND GEARBOXES

The report for power shift transmissions and gearboxes looks like the hydraulics report with the addition of the PQ index test. Water content is determined by crackle test and Karl Fischer titration, if needed. Particle counts are optional. With transmissions in general, the story is a little different from hydraulics and engines. Because of frictional elements, we care about dirt, moisture, and TAN to start with, followed by other measurements that could be indicators of fluid degradation and viscosity shear-down (see Table 7-5 and Figure 7-12).

TABLE 7-5 Power Shift Transmission and Gearbox Testing

Power Shift Transmissions and Gearboxes	
What should we test?	**Test method**
Dirt	Ratio of Al/Si
Moisture	Crackle or Karl Fischer
Acidity	TAN
Additive signature	FTIR spectroscopy
Degradation/aging	Oxidation
Viscosity changes	Viscosity at 40°C
Rust/corrosion/wear	Fe, Cu, Cr, Pb, Al by ICP spectroscopy
Particle quantification index	PQ index
Particle counts	4/6/14 Micron readings

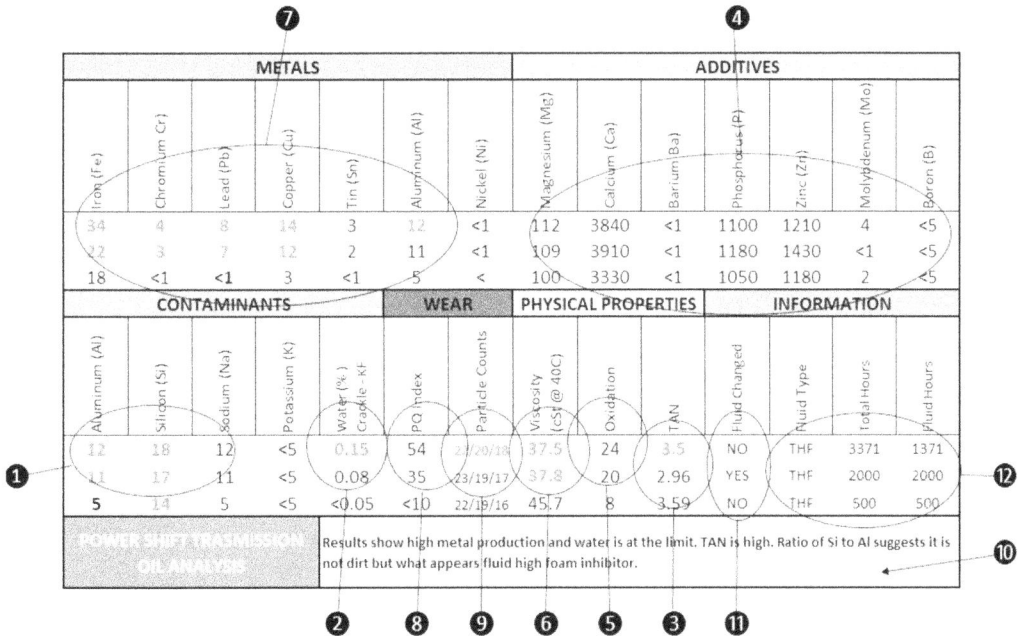

FIGURE 7-12 Power shift transmission lab report.

1. Dirt

Dirt is important because many powertrains are close to the ground and are more prone to dirt and water entry. The interpretation of dirt is similar to that for engines and hydraulics.

2. Moisture

Samples are passed through the crackle test (hot plate), and if positive, a Karl Fischer coulometric titration is run. Water in transmissions is not as critical as in hydraulics, but there are limits because of the types of fluids in play. Tractor fluids usually handle water well up to 1500 ppm. In contrast, automatic transmission fluids (ATFs) need to have water below 1000 ppm.

3. TAN

Transmissions have engaging elements that generate heat and metal wear, which cause the fluid to become acidic more quickly. This oxidation causes the TAN to increase over time. In transmissions, measuring TAN makes more sense than measuring oxidation because the TAN measurement is easier and more reliable.

4. Fluid Signature

With specialty fluids, it is important to follow the original equipment manufacturer's (OEM's) recommendations. If a transmission requires a specialty product such as a synthetic zinc-free ATF, make sure that the signature reported by the lab reflects that product.

5. Oxidation

The oxidation test is no longer routine for transmission fluids, but it still works. Keep in mind that the lab needs the signature of the new fluid for a meaningful evaluation of oxidation.

6. Viscosity Changes

Viscosity at 40°C and/or at 100°C can help us with several indications. It confirms whether the fluid in use is the correct one for the season/application. Viscosity results also confirm the decline in viscosity. Refer to Tables 2-12 and 2-13 for information on typical viscosity loss depending on transmission configuration and type of fluid used. Are the issues with excessive viscosity loss the result of mechanical shearing, or does the viscosity indicate a reasonable drain interval? Is there evidence that the wrong fluid is in use?

7. Metals

With power shift transmissions and gearboxes, the metals of interest are similar to those in hydraulics. In certain transmissions, the clutch discs are made with organic materials like paper or carbon and are not going to produce any traceable elements when they wear. However, power shift transmissions with a torque converter or retarder, such as the ones used in articulated dump trucks, will produce more aluminum (see Figure 7-13).

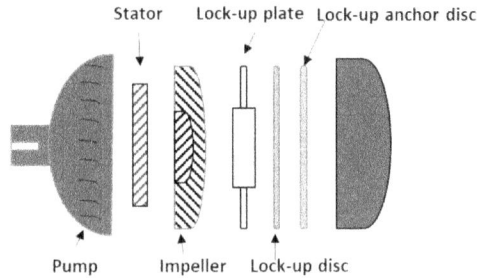

FIGURE 7-13 Torque converter.

8. PQ Index

The PQ index test is becoming popular because it complements elemental analysis. Spectrographic analysis only measures particles up to 6 or 7 μm in diameter, whereas the PQ index indicates both the presence and the concentrations of larger ferrous particles. Each transmission produces specific metal particles and a defined chart of normal and abnormal readings. However, this technology still needs to prove itself useful to this industry because sample collection at present is unable to catch a representative quantity of larger particles (see Figure 7-14).

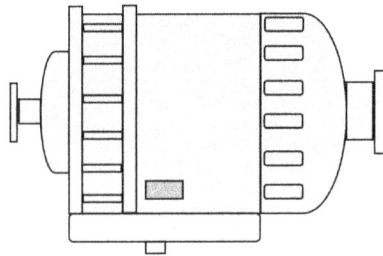

FIGURE 7-14. Power shift transmission.

9. Particle Counts

These are feasible but not typical. Some power shift transmissions produce copious amounts of materials from the clutches that render a particle count test difficult to assess. Metals are the main focal point. Other transmissions may run extremely clean. The choice to do a particle count depends on the relevance of this technology to the transmission.

10. Comment Box

The lab could have important messages about the results, and these are recorded in the comment box. For instance, the report could say that the TAN does not correlate with the hours or that the viscosity has gone higher instead of lower, suggesting an internal leak.

11. Fluid Changed

The same is recorded here as for hydraulics, and this information is valuable for the lab to evaluate time-dependent elements, TAN, and viscosity.

12. Information on Type of Fluid and Hours

Because of the use of specialty fluids, whether they are ATFs or tractor fluids, knowing the type of fluid is essential to understanding whether the machine is running with the proper fluid and whether the information matches the results.

OIL ANALYSIS FOR AXLES AND FINAL DRIVES

With axles and final drives, the reports are similar to power shift transmission reports with two exceptions. First, there is no need for particle counts, and second, it is suggested that viscosity be measured at both 40 and 100°C (see Table 7-6 and Figure 7-15).

TABLE 7-6 Axle and Final Drive Testing

Axles and Final Drives	
What should we test?	**Test method**
Dirt	Ratio of Al/Si
Moisture	Crackle or Karl Fischer test (ASTM D6304)
Friction modifier	TAN, FTIR spectroscopy
Additive signature	FTIR spectroscopy
Degradation/aging	Oxidation, FTIR spectroscopy
Viscosity changes	Viscosity at 40°C and 100°C
Rust/corrosion/wear	Fe, Cu, Cr, Pb, Al by ICP spectroscopy
PQ index	PQ index

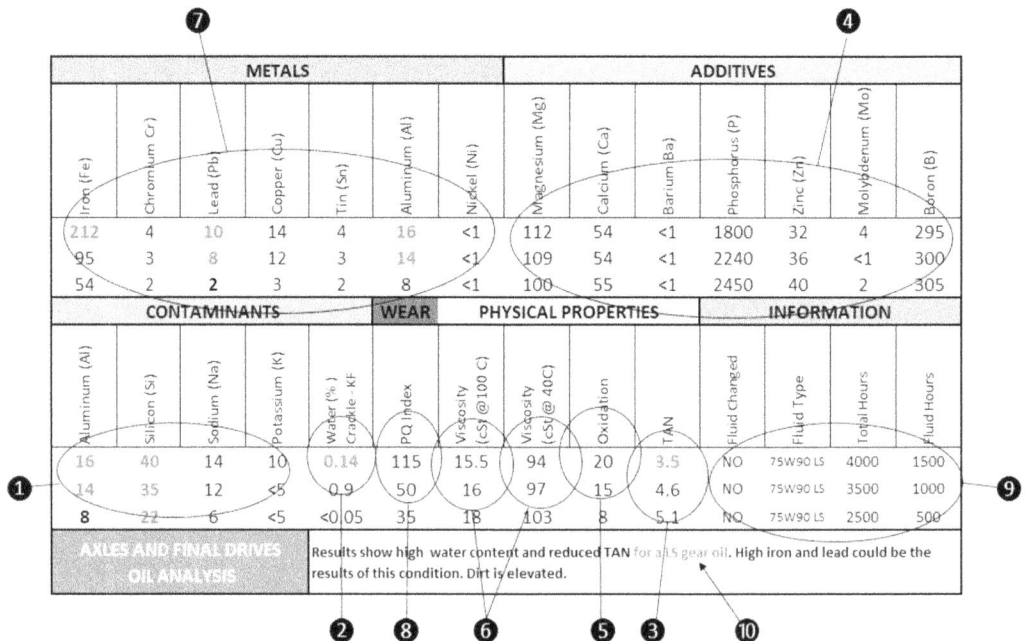

FIGURE 7-15 Axle and final drive report.

1. Dirt

Axles are the closest element to the ground, and this causes more issues with dirt ingestion through breathers and seals when wading. Very typically, dirt and water come together in axles. If you have dirt, very possibly you will find water if you do the right water test.

2. Water

When using gear oils, water needs to be <1,000 ppm because gear oils have plenty of phosphorus, and water and phosphorus do not get along well. If the axles use a tractor fluid or TO-4 oil, they can operate fine with up to 1,500 ppm of moisture. Catching water below 1,000 ppm requires the right test. FTIR spectroscopy alone will not catch it until it is too late. The crackle test could give a hint of the presence of water, and if the test is positive, Karl Fischer titration is in order. However, if the axle uses a gear oil with a high dose of friction modifier, water needs to be <500 ppm to avoid metal generation and spikes in the TAN.

3. Friction Modifier/TAN

Some axles depend on specialty gear oils that carry a heavy load of friction modifier. This mean that they could have a TAN of between 3.5 and 5.5, which is normal. For brakes and limited-slip (LS) differentials to operate without chattering, the right dose of friction modifier is necessary. Oil analysis is the right tool to get this information via the TAN. If the TAN is below specifications, a supplemental additive can restore it. It is normal for LS additives to evaporate while providing protection to bevel gears and disks.

4. Additive Signature

Attention to the additive signature in axles is important. First, is the axle using the right oil or fluid? Second, does the signature represent the product that is supposed to be in the compartment?

With wet brakes operated with hydraulic fluid from a different system, the possibility of an internal leak is feasible (see Figure 7-16). Therefore, the oil additive signature and the viscosity go hand in hand to provide information regarding internal leaks or improper lubricant use. The section "Tractor Fluid and Automatic Transmission Fluid Signatures" in Chapter 2 provides information on these additives.

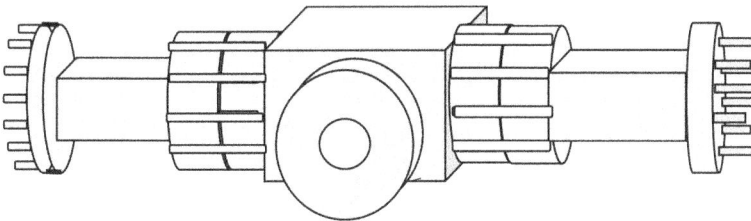

FIGURE 7-16   Wet brake axle.

5. Oxidation

Axles without shared compartments have reduced oil volumes. Final drives also have small compartments. Additionally, axles could have brakes and differential locking disks that generate heat. These circumstances favor more fluid oxidation and wear metals than large, filtered components (see Figure 7-17).

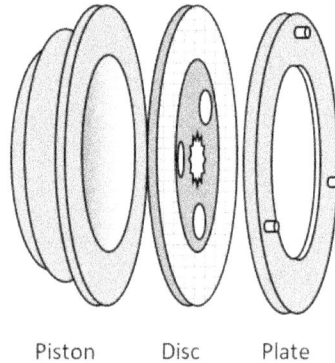

Piston   Disc   Plate

FIGURE 7-17   Wet brake detail.

Metals also catalyze fluid oxidation. Measuring oxidation is not a bad idea, but again, the lab needs the signature of the new oil. The oxidation test is not routine for gear lubricants, but it still works. As with transmissions, an increase in TAN is an easy and reliable way to estimate oxidation indirectly.

6. Viscosity

In axles with gear oil, viscosity is going to be very stable. However, is the machine using oil of the correct viscosity for the season and application? The viscosity decline in gear oils is small but begs the question of seasons. Is the axle using the correct oil viscosity for the current ambient temperatures? With tractor fluids, the viscosity declines, as discussed in Chapter 2. What is important to note is that with wet brakes and hydraulic differential locks, hydraulic fluid leaking into the differential will show up in reports as a difference in the viscosity and oil signature. Axel fluids are perhaps the only case where viscosity testing at both 40 and 100°C is fully justified. Because they are an extreme-pressure and sliding application type, axles deserve to have the two viscosities measured, just to make sure that the viscosity never goes below the established threshold.

7. Metal Generation

For components without filtration, metal generation becomes the judging factor. Many components run clean even without filtration, but it depends on design, the fluid/oil used, running temperatures, and load. Metals become an issue with wet brake piston seals. Gradual accumulation of ferrous silt in the sealing area damages the seals and causes internal leaks. Some axles have carburized bearings that do not produce chromium, but the axle still may contain planetaries with needle bearings with traces of chromium.

Brakes and differential-lock disks may not produce metal particles that are visible to oil analysis if they are made of organic materials. However, the plates these discs work with may produce iron readings. If the discs are made of sintered bronze, then oil analysis results will show copper in the readings (see Figure 7-18).

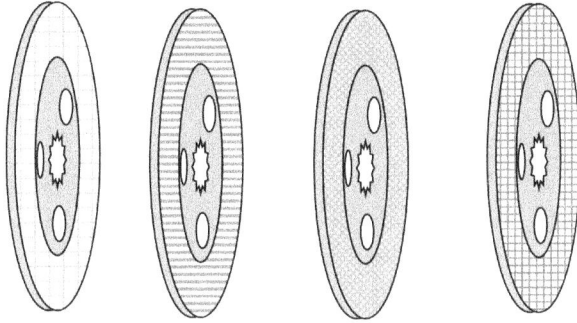

FIGURE 7-18 Wet brake disk types.

Most axles and final drives have magnets to collect gross ferrous material. In many cases, these magnets are made of rare earth metals (e.g., neodymium) and are very powerful. However, interpretation of what they collect could be very misleading. The fact that the ferrous material collected is visible does not mean that the component is about to fail; it is simply an exaggerated visual of the particles that otherwise would be spread out inside the compartment. Do not react to magnets that reflect this kind of ferrous accumulation. Pay attention to large chunks that could bring trouble to gears (see Figure 7-19). Some failures are not going to be visible to oil analysis, like a big chunk coming from a gear (see Figure 7-20).

FIGURE 7-19 Rare earth magnet.

FIGURE 7-20 Gear failure.

8. PQ Index

This test complements ICP elemental analysis, providing information on large ferrous material concentrations that otherwise are ignored by the spectrograph. The fluid test results should be compared with a defined chart for normal and abnormal readings, which is specific to each axle. As indicated earlier, this test still needs to show its value in this industry because collecting a representative sample with the correct ferrous content is a challenge. Refer to the section "PQ Index (ASTM D8184)" in Chapter 6 for PQ index values for different components.

9. Oil Hours Information

Hours on the oil are important for the lab because most axles and final drives are not filtered, and thus iron readings are going to grow exponentially. Knowing the hours allows the lab to establish whether the readings are out of range. When the hours are not available, the viscosity can help in establishing approximate hours on the oil, provided that you know the type of oil and the original viscosity.

10. Comments from the Lab

The comments from the lab sometimes can have important messages regarding the results. For example, a decline in TAN is counterintuitive for most applications and could be confusing for the uninformed. In contrast, if the viscosity and oil signature change because of an internal leak, a comment could indicate this.

Lab Website and Results

Handling the data coming from oil analyses on a large fleet can be a monumental task if the user is not equipped with the tools to handle the inflow. If the data inflow is not managed properly, the investment in this technology can go to waste easily. The typical outcome of badly managed oil lab reports is that only the critical cases get any attention and the abnormal cases are ignored. However, let us not forget that normal cases promote real proactive maintenance by monitoring tendencies and that critical and abnormal results are more in the postmortem class.

Labs usually provide web-based applications to deliver the information and sort the cases with graphics that are a great help for fleet management. The applications are also available for mobile devices for those who want to closely follow small fleets (see Figure 7-21). The value of trends cannot be overstated, and if these are presented in graphics, it is much easier to pinpoint a growing tendency. A visual aid is much easier to distinguish than simple numbers (see Figure 7-22).

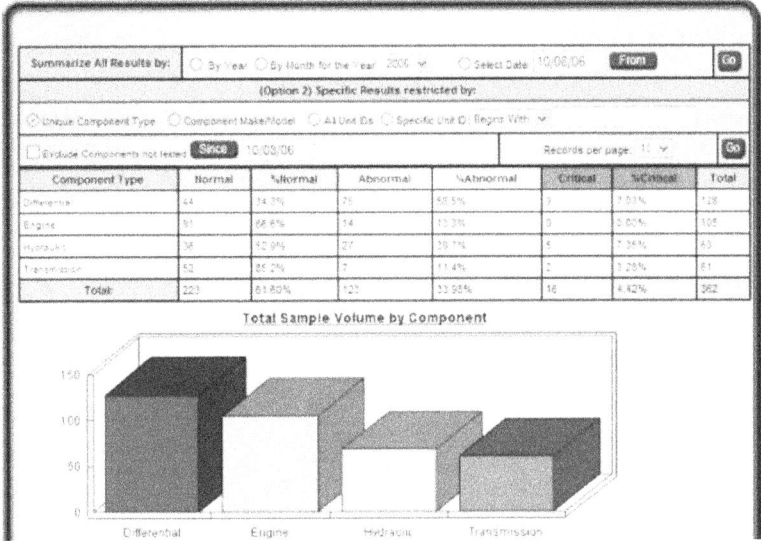

Component Type	Normal	%Normal	Abnormal	%Abnormal	Critical	%Critical	Total
Differential	44	34.3%	75	58.5%	9	7.03%	128
Engine	91	86.6%	14	13.3%	0	0.00%	105
Hydraulic	34	52.9%	27	39.7%	5	7.35%	68
Transmission	52	85.2%	7	11.4%	2	3.28%	61
Total:	223	61.60%	123	33.95%	16	4.42%	362

Total Sample Volume by Component

FIGURE 7-21   Australian Laboratory Services Pty Ltd (ALS) web report.

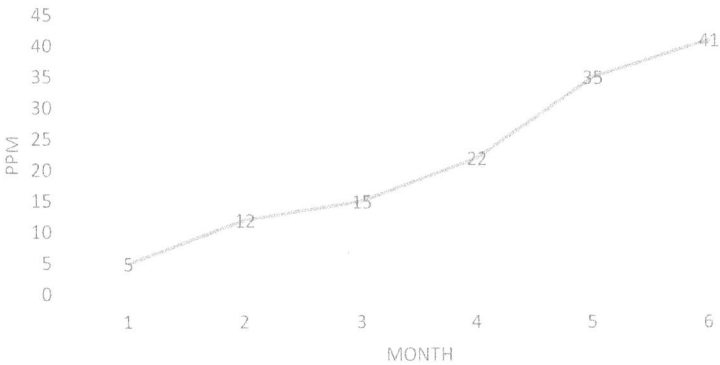

FIGURE 7-22   Copper trending.

If the fleet does not have wear metal tables calculated with standard deviations, as explained in Chapter 8, the whole oil analysis activity loses value to the point of making it a failure.

Some Rules Can Help

Oil analysis interpretation can be hard for people who do not know the equipment well. What the numbers show also can hide many more secrets that need to be brought to the surface. Oil analysis is not about reading metals (which many people interpret as the ultimate job) but rather about looking for the root causes of abnormal readings.

Let us put this into perspective. A fleet with 100 machines produces 2,500 oil analysis reports per year. Imagine if the fleet were made of 500 or perhaps 1,000 pieces of equipment. The handling of such a large amount of data is almost unmanageable. Especially in terms of alerts, the quality and importance of the messages are sometimes lost as a result of poor interpretation of the data. After all, interpretation of the oil analysis usually comes from a trained individual who needs to produce a recommendation every 35 seconds. There is no way a human could squeeze in a message from the lab in this time. Users need to pitch in with additional expertise and make this technology work for them. Here is where a computer program could be used to manage some automatic rules in seconds, providing alerts consistently for the most critical issues.

Elements for a Good Oil Analysis Interpretation

Users need to have some sets of model-specific information immediately at hand (see Figure 7-23).

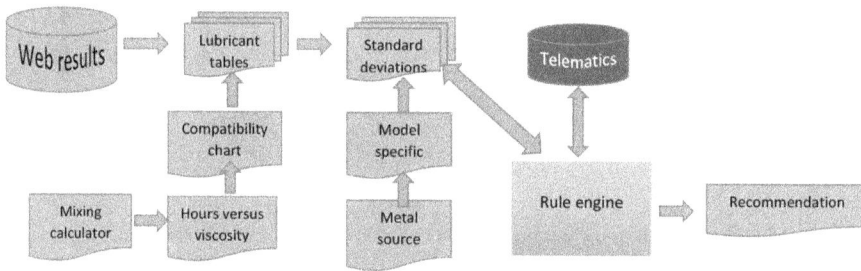

FIGURE 7-23 Data sets for oil analysis.

Logic Flow Path for High Viscosity on a Diesel Engine

Let us use the example of an engine whose oil viscosity appears to be out of range, but we do not yet know what is causing this. From the lab's point of view, very possibly they may say that there is a coolant leak or that there is high soot or a low TBN. It is up to the user to dig into the possibilities and find out the root causes. The logic tree in Figure 7-24 helps us to understand the thought process behind this hypothetical case.

From the example, we can see that telematics and tables combine to make the oil analysis a much stronger diagnostic tool by providing not only the causes for abnormal viscosity situations but also for the possible root causes of the suspected coolant leak. One can argue that there are more possibilities to this example—and there are. For example, we could say that nitration and sulfation are present, but in a well-developed application, there would be other rules to handle that part and solve the question by interacting with the rule of the example.

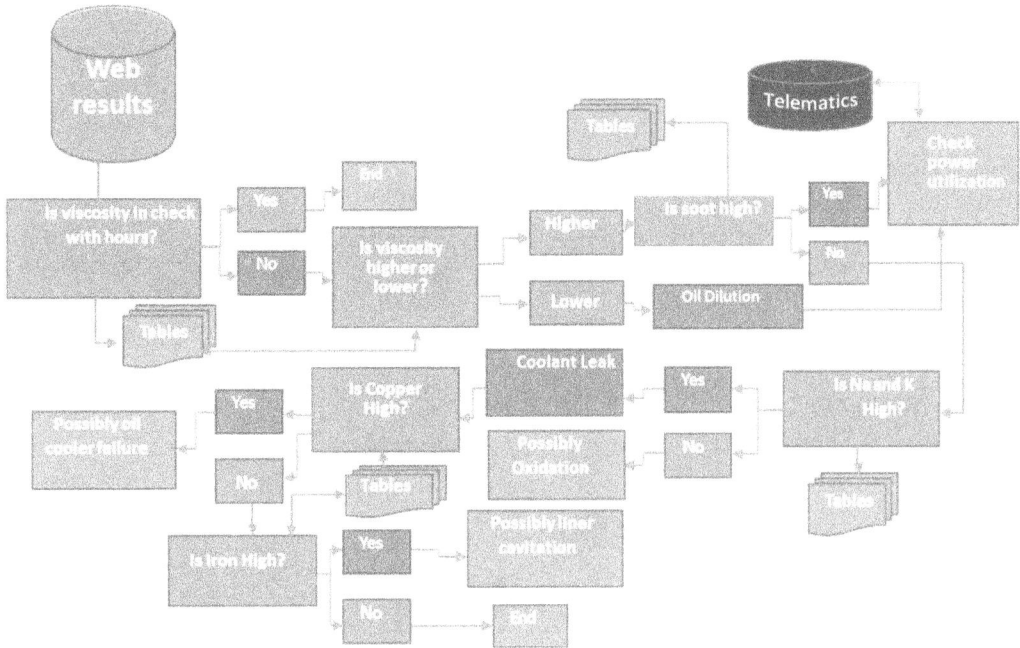

FIGURE 7-24 Engine oil viscosity analysis example.

Logic Flow Path for High Copper in Hydraulics Example

Let us deal with an example where a high-pressure system is showing too much copper. Copper is the most critical metal in most high-pressure hydraulic systems. It is important to understand what is causing this high reading and to try to solve the problem before it is too late. The task is to get to the root cause of the high copper production. The possibilities are many, and the logic tree goes through a process of elimination. However, we need not only specific tables for wear metals but also tables of viscosities versus hours, contaminants limits, and physical properties, and finally, we need telematics regarding hydraulic operating temperatures. When we incorporate all this information, the power of oil analysis reaches its potential, which is amazing (see Figure 7-25).

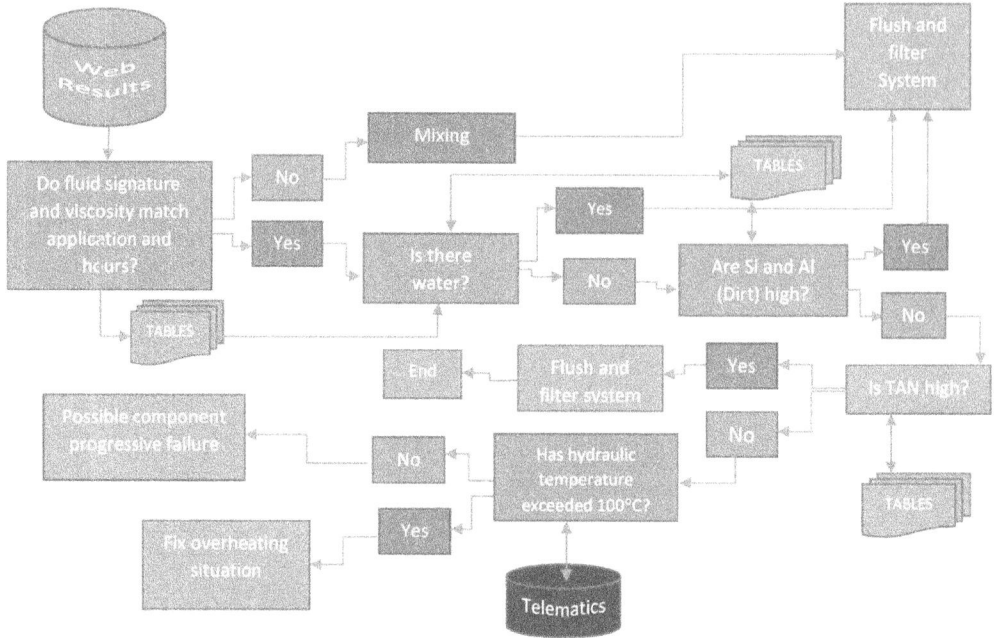

FIGURE 7-25 Copper in hydraulics example.

Maintenance Applications

Controlling the maintenance of a fleet is a tough job, especially if the fleet is made up of multi-brand pieces and types of equipment. Controlling routine maintenance is the basic requirement that every program should strive for. However, integrating telematics and fluid analysis results is a much more complicated task, and just a few systems are capable of doing this.

Most of the main manufacturer players offer scheduled maintenance capability for free, together with their telematics services. However, these services do not integrate into users' accounting and work-order flows, making them difficult to use. Additionally, most fleets are mixed, and users cannot depend on several different applications to handle their maintenance requirements. If we add fluid analysis data to the equation, the task is even more complicated.

Because abnormal fluid results need immediate attention, any fleet management application needs to accommodate the requirement to trigger work orders when oil analysis so dictates. Equally important is that telematics data integrate into the fleet management software in such a way that they trigger a work order when the telematics data pinpoint an issue.

The ultimate application would be able to schedule services based on hour meter readings, take fluid analysis results, and run them through a rule engine and decide whether the readings require action, and finally, cross-check these results with telematics and field inspections to produce the ultimate proactive service when needed (see Figure 7-26).

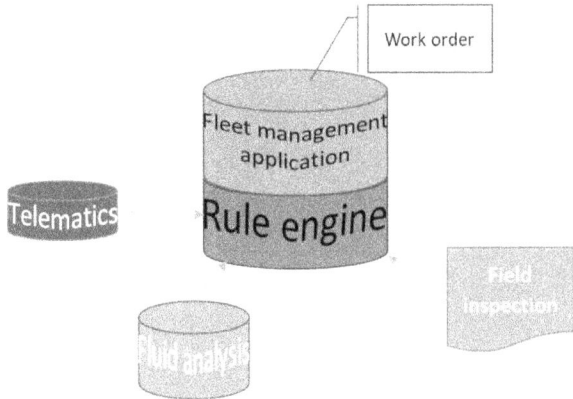

FIGURE 7-26 Fleet management integration.

CONCLUSION

Now that we have examined the basics of interpreting oil analyses, it is time to put that knowledge to use. Let us start with hydraulics in Chapter 8.

CHAPTER 8

WEAR TABLES
AND STANDARD DEVIATIONS

Wear tables are the values of wear metals, contaminants, and physical properties of a given machine that allow comparison of lab results with expected, abnormal, and critical values.

Why Do We Need Wear Tables?

Wear tables allow users to establish whether their machines are wearing normally and whether the maintenance is going well. In the world of oil analysis, having wear tables is essential to interpreting results. Without them, oil analysis is simply a forensic tool without major value. It is in the interest of users to squeeze the power of oil analysis out by correctly producing wear tables that lead to enhanced interpretation of oil analysis results.

How Do We Calculate Wear Tables?

Typically, the lab produces wear tables based on statistical analysis of a large population of a given machine component that represents that population. Users also can produce their own wear tables through standard deviation calculations, as discussed in this chapter.

What Is Standard Deviation?

Standard deviation is a measure of how spread out the numbers are from the mean (average) or median. A low standard deviation means that the values are close to the mean. A high standard deviation means that the numbers are more spread out from the mean.

Types of Standard Deviations

Currently, there are two formulas for standard deviation calculations: population and sample. Population calculations are more difficult to achieve because they involve collecting data from an entire population. Sample calculations can handle the wear tables of a fleet on their own.

WHICH STANDARD DEVIATION IS MORE APPROPRIATE FOR AN EQUIPMENT FLEET?

In equipment fleets, using the sample calculation is more appropriate because it gives a true representation of the machines' behavior in a specific environment with specific maintenance standards.

Population Standard Deviation

$$\sigma = \sqrt{\frac{\sum (x_i - \mu)^2}{N}}$$

Sample Standard Deviation

$$S_x = \sqrt{\frac{\sum (x_i - \bar{x})^2}{n - 1}}$$

where s_x = sample standard deviation
n = size of the population
x_i = each value from the population
$\bar{x}$ = population mean
μ = mean of the data set

Standard deviation is the square root of the variance, and variance is the average of the squared differences from the mean (average) or median. In equipment, we use median values instead of mean values, as explained later in this chapter.

Let us use the example of the heights of a population of seven people (Figure 8-1). To arrive at an understanding of the difference in heights, the variance, we look at how far each person is from the median. The dotted line represents the median value, which is 179 cm.

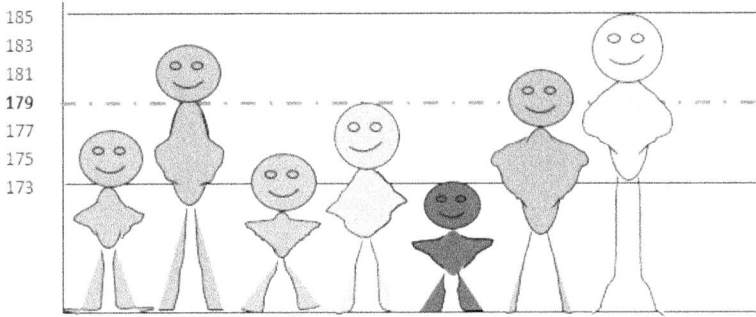

FIGURE 8-1 Standard deviation example.

Person	Height	From Median	(From Median)²
1	177	–2	4
2	175	–4	16
Median → 3	183	4	16
4	**179**	0	0
5	173	6	36
6	181	2	4
7	185	6	36

Sum of the squares: 112

Variance: $112 \div 7 = 16$

Standard deviation: $\sqrt{16} = 4$

The median height of the sample of people is 179 cm. In a large population, we would use Excel's median function to find the median or calculate the average of the two center numbers if there is an even number in the population.

The variance of this small population is 112/7. That is,

$$\text{Variance} = (-2)^2 + (-4)^2 + 4^2 + 0 + (-6)^2 + 2^2 + 6^2 = 112$$
$$112/7 = 16$$

The standard deviation is the square root of the variance of 16, which is 4. Hence, the standard deviation for the height of this sample of seven people is obtained manually as 4.

The example was designed to show the concept of standard deviation, where the figures were calculated from the entire population. There is more variability in samples than in entire populations, and a different formula therefore is used to calculate variance. For samples, variance is calculated as the sum of the squares divided by the sample size minus 1. In our example:

$$\text{Variance} = 112/(7 - 1) = 18.67$$
$$\text{Standard deviation} = \sqrt{18.67} = 4.32$$

Do not be intimidated by the math—Excel does it for you. When using Excel, use the STDEV.S function to calculate the standard deviation of a sample instead of the STDEV.P, which is used to calculate the standard deviation of an entire population.

Parametric Bell Curve

Continuing with our example involving the heights of seven people, 68.27% of the people in the group fall within 1 standard deviation, and 99.99% of the group falls within ±4σ (the sigma symbol is used as an abbreviation for standard deviation). In fact, roughly 68% of any population is going to fall within 1 standard deviation of the mean. A high variance will create a much wider bell curve, but about 68% of the population will be within 1 standard deviation (see Figure 8-2).

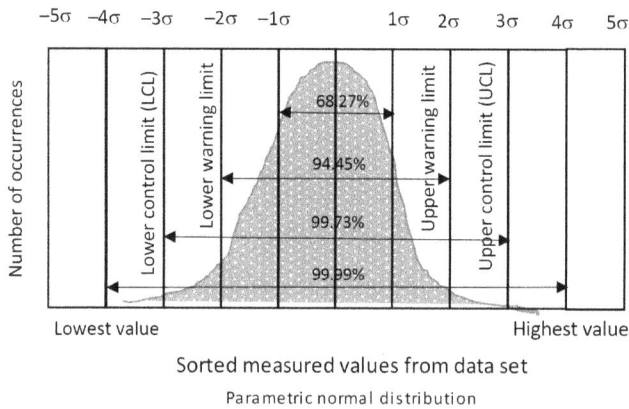

FIGURE 8-2 Parametric normal distribution.

Now let us move from the example of people to data coming from an oil analysis. The values for wear metals in a machine population have a bell-shaped distribution. This type of bell-shaped graphic is called *parametric*.

For our discussion, the normal values fall within ±1σ. However, we only care for values above the mean or median and not for those below the mean or median. And the reason is simple: Why should we worry about a machine that does not produce wear metals?

We establish the normal value by adding 1σ to the median value. We establish the critical values by adding 2σ values to the median. The abnormal values are anything in the middle (see Table 8-1).

A median of 3 with no negative numbers and standard deviation of 16 is not possible. I recommend 68 for median, <84 normal, 85-99 abnormal, and 100 abnormal.

TABLE 8-1 Example

Wear table example using standard deviation of 16	Median	Normal	Abnormal	Critical
16	68	<84	85–99	>100

We can use this calculation in equipment to compare the oil analysis results of a given machine against the calculated results of a given population to see if the machine in question falls within a given tolerance range.

Example Applied to a Fleet

The following example around a small imaginary fleet of five bulldozers helps us understand the concept of standard deviation. These tractors show different figures for iron production in their engines, measured at 500 hours of operation on the oil. Let's proceed to calculate the mean values of the measurements using Excel and then set the standard deviation.

As we can see in Figure 8-3, the median for iron at 500 hours is 45 parts per million (ppm). The standard deviation is 11.45 ppm. Therefore, the value accepted as normal is the median + 1 standard deviation (σ), and the critical value is the median + 2 standard deviations (2σ). Abnormal values fall between these two values.

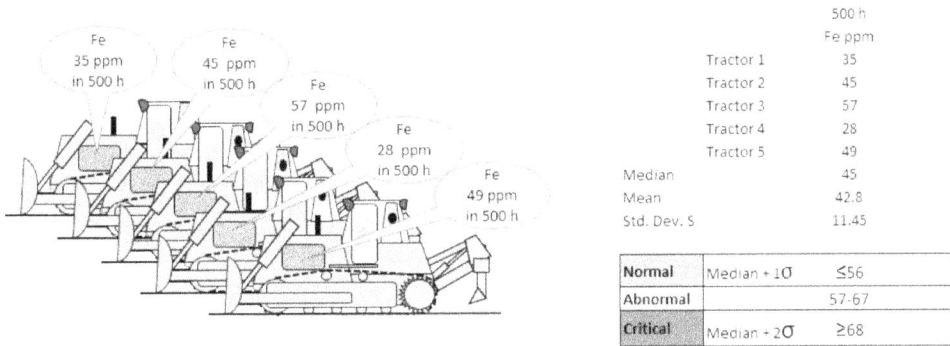

	500 h Fe ppm
Tractor 1	35
Tractor 2	45
Tractor 3	57
Tractor 4	28
Tractor 5	49
Median	45
Mean	42.8
Std. Dev. S	11.45

Normal	Median + 1σ	≤56
Abnormal		57-67
Critical	Median + 2σ	≥68

FIGURE 8-3   Standard deviation applied to a fleet.

Parametric Curve and Cumulative Curve

The parametric graphics produce two-tailed cumulative curves as in the example in Figure 8-4. Wear metals typically produce this kind of distribution and graphic.

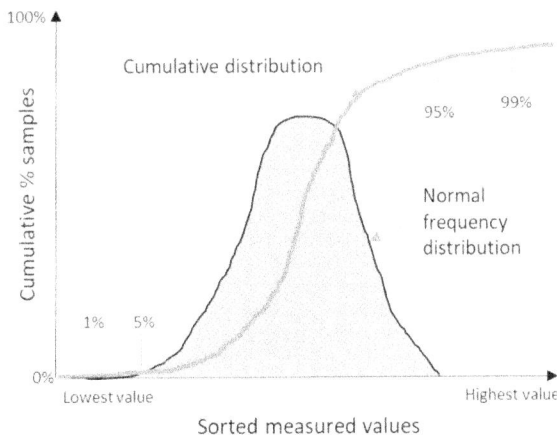

FIGURE 8-4   Double-tailed cumulative distribution with continuous data.

It does not make sense to calculate standard deviations for some measurements in an oil analysis because they are skewed heavily to one side, or we are looking for a change in the number instead of the number itself. If we graph the distributions, the numbers produce nonparametric curves.

Nonparametric Curve for High Reference Values

The nonparametric graphic for high reference values produces a single-tailed cumulative curve (see Figure 8-5). The reason is that the measurements tend to concentrate on the higher-value side. The total base number (TBN) falls within this category of graphic. We do not need to do calculations for TBN; the lab establishes these.

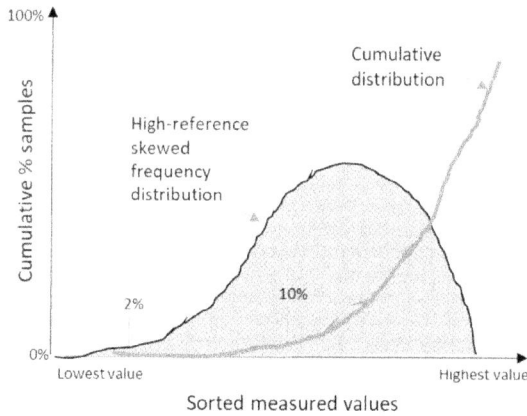

FIGURE 8-5   High reference values: single-tailed cumulative distribution with continuous data.

Nonparametric Curve for Zero Reference Values

The nonparametric graphic for zero reference values also produces a single-tailed cumulative curve (see Figure 8-6). Most measurements concentrate on the lower-value side. This is the case with the total acid number (TAN). The flagging comes from a table where a maximum TAN is set, depending on the type of lubricant.

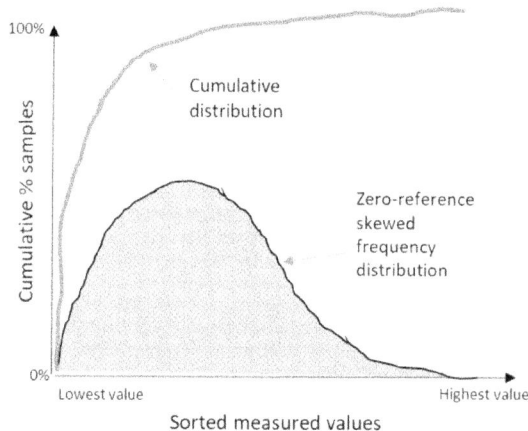

FIGURE8-6   Zero references values: single-tailed cumulative distribution with continuous data.

Measurements Handled without Standard Deviations

All measurements for high reference values and those for zero reference values that produce a single-tailed curve do not use a standard deviation calculation. The measurements in Table 8-2 fall into this category.

TABLE 8-2 Elements without Standard Deviation Calculations

Sulfation	Nitration	Oxidation
Viscosity	Viscosity shear	Glycol
TAN	TBN	Fuel
PQ index	Particle counts	Water

Mean or Median Values?

Whether to use mean or median values is the question. The *median* is the number at the exact middle of the set of values, whereas the *mean*, or *average*, is where you add up all the numbers and then divide them by the number of lines.

The mean value of a population without outliers approaches the median value. Thus, they are closely related. However, the problem is that a component will generate big wear metal numbers prior to or during a failure, and we do not want to base our maintenance decisions on a failing machine. The median is more representative of the population than the average for skewed distributions.

To clean the data, we remove any outliers exceeding 6 standard deviations from the median (see Figure 8-7).

Example of minimum to maximum values

0.45
0.47
0.49
0.52
0.55
0.55
0.6
0.61
0.62 Middle of the sorted sample set
0.7
0.74
0.76
0.8
0.91
5.3
10.4

1.53	Mean (average)
0.62	Median

FIGURE 8-7 Median and mean values.

Why Do We Need So Many Wear Tables?

A typical equipment fleet is going to have many components and many different applications, which means that many wear tables are needed for condition-based maintenance (CBM). Even the same components within different machines or performing different operations may need targeted wear tables.

Let us use the example of engines. There are no two identical engines in a product line, and although some applications may be similar, different models use different engines, no matter how much they appear to be identical. The differences between engines can be subtle, but they have an impact on oil analysis readings. Things that could be different include, but are not limited to, such things as:

- Oil sump capacity
- Power setting
- Transmission to which it is connected
- Application

Different brands of engines also bring big differences in the way they wear metals. A given Isuzu engine, for example, could have great iron numbers, but its aluminum production could be higher than that of a CAT engine, so we need to treat each engine as a different child.

Geography and Utilization

Two identical machines working in different geographic areas very possibly are going to have different wear tables. This is so because of local contaminants, altitude, humidity, and so forth. Therefore, we cannot trust that a wear table developed for a machine in Florida will be useful for a machine in Illinois.

Can We Use Wear Tables from One Country in Another Country?

Engine wear tables are extremely dependent on fuel sulfur content and engine altitude. Thus, again, we cannot blindly use tables from other countries if we want precision in the oil analysis interpretation in a specific country.

Collecting Information

We need raw data from the lab, which we can get in Excel or Tableau reports. Most labs provide this information, and some even allow you to download it to CSV files from their websites, which translate into Excel easily. Once the data are in a spreadsheet, the rest becomes very easy.

- First, check that the listing contains the intended model you want to work on and that other models or brands of equipment have not been smuggled into the list. If this is not the case, clean the list.
- The more data you have, the better results you can expect because the numbers will be representative of a larger population, and outliers will not have too much impact on the final data.

USING EXCEL TO CALCULATE
STANDARD DEVIATIONS

Excel is a great tool to easily calculate medians and standard deviations. The automatic calculation is in the Formulas menu cascade. Let Excel calculate the values in Figure 8-8.

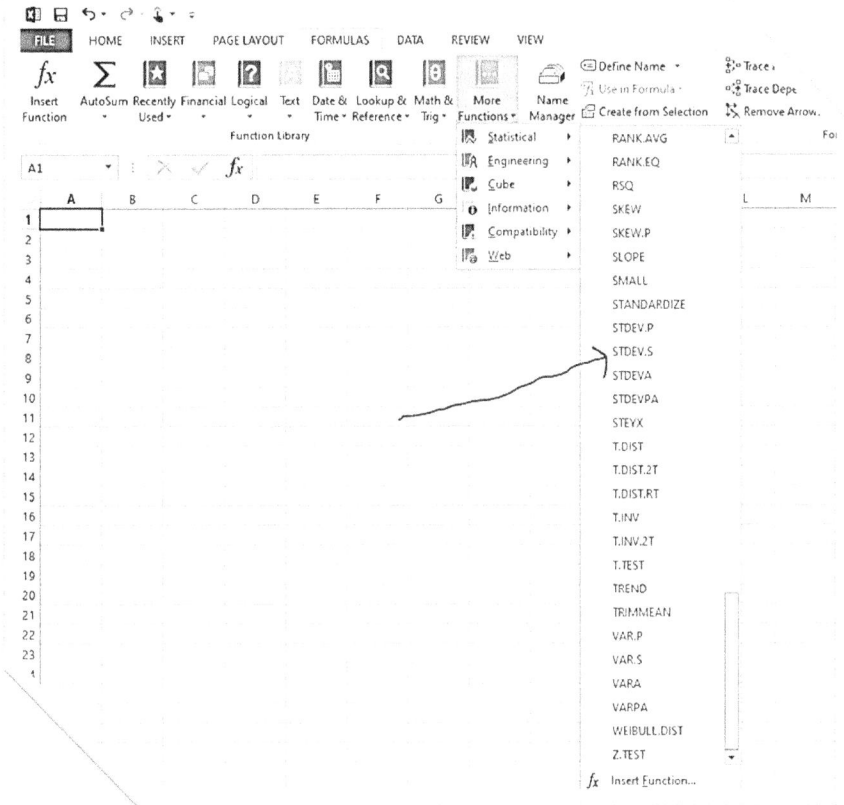

FIGURE 8-8   Standard deviation Excel formula.

The data will need some adjustment, as we will see later in this chapter. For example, the values for iron need normalization for the hours in use if you want the table to work for you. However, to start with, you need to establish the median values for wear metals, and after that, their standard deviations.

Cleaning the Data

Once you understand the basics of standard deviation calculations for the creation of wear metal tables, you need to clean the data and adjust some numbers to make the tables even more precise. First, you need to eliminate outliers resulting from abnormal conditions or poor maintenance or those that are just atypical for the model in question. If you do not clean these outliers, they will inflate the final figures and affect your maintenance decisions.

Therefore, where do we draw the line in defining outliers? The American Society for Testing and Materials (ASTM) standards recommend removing values that are equal to or more than 6 standard deviations from the median. Once you have removed the outliers from the data, you can recalculate your final median and standard deviation values. Figure 8-9 shows how outliers can affect the results if not removed. The spreadsheet in the figure provides a list of values with and without outliers. The right column is a list of numbers showing two large outliers in gray. When we remove those outliers, the mean and median values get closer.

	Without outliers	With outliers
	2368	2368
	2566	18570
	3419	3419
	3143	3143
	3621	3621
	1691	13500
	3577	3577
	2993	2993
	3248	3248
	3556	3556
	2519	2519
	3708	3708
Mean and median become closer	1709	1709
	2200	2200
	2802	2802
	2485	2485
	2488	2488
Avg.	2829	4465
Median	2802	3143
Std. Dev.	647	4482

FIGURE 8-9 | Cleaning the data.

The partial capture of a big file (11,493 lines) in Figure 8-10 shows that the standard deviation for iron (Fe) is 382 ppm in this example (arrow). If we apply the ASTM rule of removing any outlier more than 6 standard deviations above the median, then any value greater than 2292 ppm (382×6) above the initial median needs to be removed from the final calculation before establishing the median value.

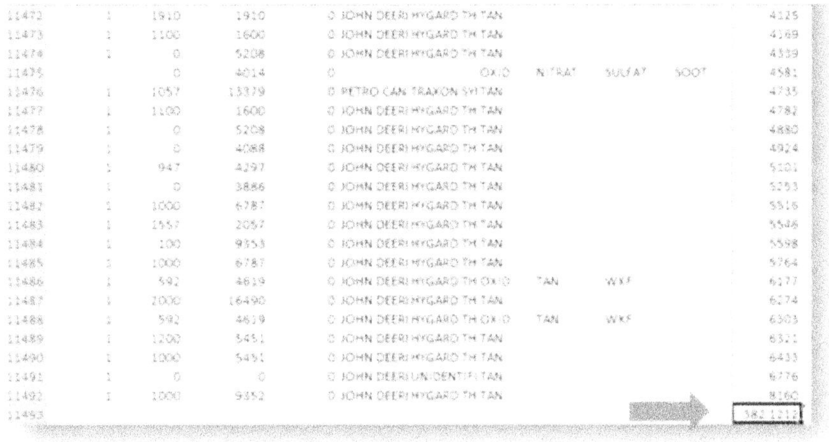

FIGURE 8-10   Example. Cleaning the outliers from the data.

Normalizing Time-Dependent Elements

Because not all oil samples from a given compartment have the same sampling interval, the data need to be normalized for iron. Iron is a time-dependent element that grows over time and is the only time-dependent element in filtered components.

For this adjustment, we use the median hours on the oil and normalize it to the desired interval. In the example in Figure 8-11, the median for oil sampling interval is 1250 hours, and we want a table that shows the wear rate per 1000 hours. We need to divide 413 ppm of iron by 1250 hours on the oil to obtain 0.33 ppm/h and then multiply this value times 1000 h, which gives a normalized value of 330 ppm for 1000 h.

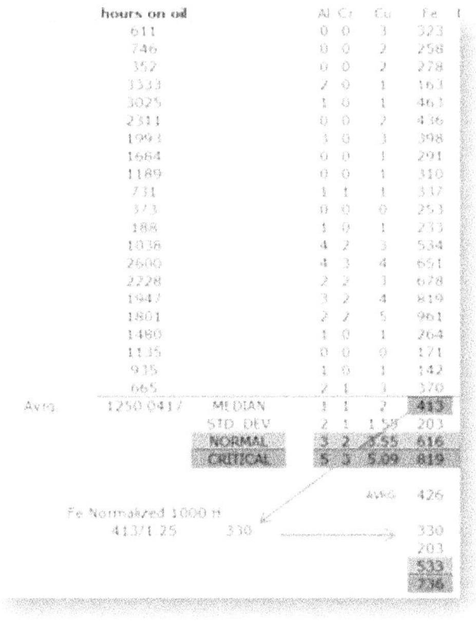

FIGURE 8-11   Normalizing iron.

Establishing Abnormal and Critical Values

After calculating the median and standard deviations, removing outliers, and getting iron readings normalized, we can proceed to establish the normal and critical values:

- Normal is the median + 1σ (1 standard deviation).
- Critical is the median + 2σ.
- All the numbers in between the preceding two are abnormal.

Wear Tables for Harsh Applications

In some applications, the standard deviation is too tight, producing abnormal and critical flags that become annoying to users. The standard deviation can be adjusted for special applications, such as where the dirt and metal production in final drives is high but does not represent a risk to the machine. This is the case with excavators, which are mostly stationary and move slowly through harsh environments without load on final drives other than propelling.

For these specific applications, it is convenient to adjust normal values to median plus 2 standard deviations and critical values to median plus 3 standard deviations.

Minor Metals and Contaminants

When handling minor metals readings, it is important to have usable tables. Some metals, such as aluminum (Al), nickel (Ni), and tin (Sn), can be present in very small amounts, and the standard deviation calculation may produce a low number. The risk with very low numbers is that readings may exceed the calculated limits too often, creating unnecessary flagging that does not mean much by itself in the operation of the machine. Adjust these values to the suggested values listed in Table 8-3 to avoid a source of continuous flagging.

TABLE 8-3   Minor Metals Adjustments

Minor metals adjustment recommendations (ppm)			
If results are	<5	5.1–7.5	7.6–10
Adjust to	5	10	15

Metals Not Typically Used by Mobile Equipment

Other metals such as titanium (Ti), vanadium (V), and silver (Ag) are not typically seen in this industry, and their presence does not indicate wear from the machine. Titanium is most likely an oil additive. Vanadium could be coming from the fuel, particularly heavy high-sulfur fuels. Silver is rare and possibly only used as a tracer. These elements are information only and do not have a threshold by themselves.

Handling of Potassium (K) and Sodium (Na)

Other contaminants, such as potassium (K) and sodium (Na), are extremely variable and do not work well with standard deviations. These readings represent the location/environment or coolant traces. They go better with preset values from the lab, which are typically set at 30 ppm for most applications. Typically, they could be from coolant, dirt, combinations with silicon (Si), or soap/fertilizer contamination.

Example of a Wear Table

The final product is a wear table that shows the normal, abnormal, and critical ranges for the component in question and could look like Table 8-4.

TABLE 8-4 Wear Table

Compound at 500 Hours	Normal (PPM)	Abnormal (PPM)	Critical (PPM)
Iron (Fe)	<31	31–70	>70
Lead (Pb)	<13	13–20	>20
Copper (Cu)	<16	16–30	>30
Chromium (Cr)	<16	16–20	>20
Aluminum (Al)	<11	11–20	>20
Nickel (Ni)	<6	6–9	>9
Silver (Ag)	<5	5–8	>8
Tin (Sn)	<5	5–10	>10
Sodium (Na)	<21	21–30	>30
Potassium (K)	<30	30–50	>50
Titanium (Ti)	<5	5–10	>10
Silicon (Si)	<16	16–25	>25
Vanadium (V)	<2	2–3	>3

Example of a Wear Table for Elements That Do Not Need Standard Deviations

Table 8-5 shows other elements that use preassigned figures and are not dependent on standard deviations.

TABLE 8-5 Elements That Do Not Need Standard Deviations

Physical properties and other contaminates	Normal (PPM)	Abnormal (PPM)	Critical (PPM)
TAN, AW hydraulic	<1	1	>2
TAN, calcium-based fluids	<5	5–6	>7
Water (%), AW hydraulic	<0.075	0.075–0.1	>0.1
Water (ppm), AW hydraulic	<750	750–1000	>1000
Water (%), calcium-based fluids	<0.10	0.10–0.15	>0.15
Water (ppm), calcium-based fluids	<1000	1000–1500	>1500
Oxidation	<20	20–25	>25
Particle count, open hydraulics	<23/19/16	23/19/16–25/21/18	>25/21/18
Viscosity at 40°C, single grade (ISO VG 32/46/68)	>±5%	>±6%	>±8%
Viscosity at 40°, multigrade (SAE 10W-40/ 15W-40)	>±30%	>±33%	>±35%
Viscosity at 40°C, tractor fluid	>±20%	>±26%	>±30%

Table for Particle Counts, PQ Index, TBN, TAN, and Viscosity Changes

Chapter 2 contains tables for other elements and physical properties that the industry uses as a guide.

CONCLUSION

Now that we have learned how to create wear tables for mobile equipment, we can move ahead and start using this knowledge in the interpretation of fluid analyses for hydraulics, engines, and powertrains in the next three chapters.

CHAPTER 9

HYDRAULIC FLUID ANALYSIS

As we cover individual systems, we will start with hydraulics. Many people tend to think of engines first when oil analysis comes to mind, but the hydraulic system on many machines costs far more to repair than the engine. It is here where condition-based maintenance (CBM) can have the biggest impact on total machine cost and profitability.

Again, we need to know the environment, the fluid, and the machine to determine the best maintenance options. Keeping fluid cleanliness in check takes a lot more than just performing the maintenance procedures in the owner's manual. The *environment* plays an important role in determining a system's cleanliness. A dusty environment may force more dirt into the reservoir than expected, and a humid environment may lift humidity beyond the desired range.

Dirt and water get into the system in small amounts through breathers, cylinder seals, and leaking valves, to name a few sources. Moreover, not all solid contaminants can be removed by onboard filtration. This is the case when small particles of dirt (silica and alumina) pass through filters repeatedly. Water will be present in reservoirs because water is present in the air. Both semi sealed and open reservoirs breathe to compensate for changes in fluid volume while in operation, drawing in air and the humidity that it contains. Of course, water also can find its way into the system via cylinders—especially if the application involves dredging or ditch cleaning.

Here is where "knowing the fluid" comes in. If the fluid uses a calcium-based additive, it will be able to hold much more water in suspension than a simple antiwear (AW) hydraulic fluid. Thus, the acceptable level of moisture and the method to get rid of it varies with the type of fluid.

Acceptable levels of contamination vary with the machine. In addition, procedures to maintain cleanliness and return the system to acceptable cleanliness levels require knowledge of the machine.

Keeping the Balance

The filtration installed in a hydraulic system needs to control ingested contaminates, resident contaminates (from manufacturing), and contaminates generated via wear. When the system is in balance it achieves a given particle count code number. If the filter is unable to maintain the balance, that is, when the cleanliness of the system goes beyond the design limits, the system is out of balance. This happens when any of the contaminants exceeds the filter's capabilities to remove them.

Our challenge is to determine the cause of the imbalance, which could be a faulty seal that allows dirt or water to enter the system or a lubricant that is no longer able to protect the system against wear.

Basic Cleanliness Levels

Ideal cleanliness levels require extensive use of filter carts or, better yet, a permanent bypass filtration system. Ideal cleanliness is not always practical, but it would be nice to have a chart that shows what normal contaminant levels are for a given machine. Unfortunately, there are many configurations of hydraulic systems using different fluids, all of which influence what is normal. You can establish your goals based on trending, but Table 3-4 gives a very general indication of normal as a starting point.

Sealed reservoirs promote better cleanliness, as indicated by lower particle counts in the table. Filtration in hydrostatic systems is superior to that in hydraulic systems, resulting in lower contamination levels. Mixing fluids can cause the 4-μm channel to spike, rendering it meaningless for cleanliness assessment purposes

Hydraulic Contamination Sources

1. Dirt
2. Water
3. Oxidation
4. High temperature
5. High total acid number (TAN)
6. Mixing
7. Bronze etching
8. Wear
   - Pump/motor
   - Cylinders
9. Microdieseling
10. Varnish formation
11. Foam formation
12. Gel formation
13. Cheese curding (see Table 9-1)

TABLE 9-1 Testing Methods for Hydraulics

Hydraulics	
What should we test?	**Test method**
Particle count	4/6/14-μm readings
Dirt	Ratio of Al/Si
Moisture	Karl Fischer titration (ASTM D6304)
Wrong fluid/mixes	Viscosity at 40°C and signature
Metals	Cu, Fe, Cr by inductively coupled plasma (ICP) spectroscopy
Acidity	TAN, KOH/g
Degradation/aging	Oxidation, Fourier transform infrared radiation (FTIR) spectroscopy
High-temperature records	Machine telematics

Testing

The suggested testing appears in order of importance to the hydraulic systems in Table 9-1. Talk with your lab personnel to ensure that these tests are included in order to obtain the maximum value from this technology. The cases discussed in this chapter will use these tests in the discussions.

Wear Thresholds

Wear tables come from adjusted standard deviation calculations of a large population of similar machines. If the lab providing oil analysis services is not offering these tables, users can develop their own tables (see Chapter 8). The machines evaluated for standard deviation calculations need to be of the same type and vintage because some metallurgy used in the machines will be characteristic of the model. Also, oil type, reservoir volume, reservoir breather type, filter beta rating, and so forth need to be similar for meaningful comparisons.

CASE DISCUSSION

Next, we will evaluate a series of cases to illustrate the various types of contaminants that are common in hydraulic systems on mobile equipment. We will use cases from real life and focus on the findings and machine health implications. Chapter 4 provides information on how to return a machine to normal conditions after contamination events.

1. Dirt and High Particle Counts in a Large Production Bulldozer with Combined Hydraulics/Hydrostatics

Oil Analysis

Figure 9-1 is an example of excessive contamination from dirt in a bulldozer hydraulic-hydrostatic combined system. Note that the constituents of dirt, silicon, and aluminum are elevated, causing the particle count to be high as well. Also notice that copper and lead levels are elevated. The report shows no evidence of cross-contamination of this fluid. TAN and viscosity are normal. The fluid just needs cleaning with a filter cart to return wear rates to normal levels.

METALS							ADDITIVES						
Iron (Fe)	Chromium Cr	Lead (Pb)	Copper (Cu)	Tin (Sn)	Aluminum (Al)	Nickel (Ni)	Magnesium (Mg)	Calcium (Ca)	Barium Ba	Phosphorus (P)	Zinc (Zn)	Molybdenum (Mo)	Boron (B)
19	<1	9	28	3	10	<1	472	1731	<1	1078	1192	181	153
27	<1	8	25	2	11	<1	462	1799	<1	1088	1210	245	180
18	<1	5	22	<1	<1	<1	454	1803	<1	1050	1180	235	175

CONTAMINANTS					WEAR		PHYSICAL PROPERTIES			INFORMATION			
Aluminum (Al)	Silicon (Si)	Sodium (Na)	Potassium (K)	Water (%)	PQ Index	Particle Counts	Viscosity (cSt @ 40C)	Oxidation	TAN	Fluid Changed	Fluid Type	Total Hours	Fluid Hours
10	22	4	5	0.04	NA	22/19/14	58.8	NA	2.35	YES	10W30	8871	3000
11	26	4	<5	0.04	NA	22/18/10	61.5	NA	1.91	YES	10W30	8345	2474
<1	27	5	<5	<0.05	NA	22/18/11	62.5	NA	1.8	YES	10W30	6345	2000

HYDRAULIC OIL ANALYSIS	LARGE PRODUCTION BULLDOZER WITH COMBINED HYDRAULICS/HYDROSTATICS. HIGH PARTICLE COUNTS AND DIRT. FLUID 10W30

FIGURE 9-1 Excessive contamination from dirt in a bulldozer hydraulic-hydrostatic combined system.

In combined systems, the hydrostatics and hydraulics share the same reservoir. While the hydrostatics operate at high pressures most of the time, the hydraulics do not require high pressure to operate the blade or the ripper. However, the hydraulics connect to the environment via hydraulic cylinders, which bring dirt into the system, especially if there is a leaking cylinder.

Wear Metal Thresholds.

See Table 9-2.

TABLE 9-2 Wear Metals Thresholds

500 hours			
Filtered System	Normal	Abnormal	Critical
Fe	<28	28	>50
Pb	<5	5	>12
Cu	<31	31	>55
Cr	<5	5	>10
Al	<9	9	>15
Si	<11	11	>16

Conclusion

Because this is a combined system in which the hydrostatics and hydraulics run from the same fluid in the same reservoir, some dirt will get into the system via blade cylinders. It is thus important to check for external leaks through cylinders and valves.

2. Silt Accumulation in a Production Hydrostatic Bulldozer with Combined Hydraulics and Hydrostatics

Oil Analysis

A large bulldozer's combined hydraulic-hydrostatic system is showing increasing copper readings with a slight jump in iron and dirt as well (see Figure 9-2). This could be a natural response to high hours on the fluid, but we do not know yet. Particle counts have been consistently high for the 6-μm channel, although the 4-μm channel is usually high on this type of fluid, which is a calcium-rich engine oil. The 14-μm channel has been inconsistent but not totally out of range.

METALS							ADDITIVES						
Iron (Fe)	Chromium (Cr)	Lead (Pb)	Copper (Cu)	Tin (Sn)	Aluminum (Al)	Nickel (Ni)	Magnesium (Mg)	Calcium (Ca)	Barium (Ba)	Phosphorus (P)	Zinc (Zn)	Molybdenum (Mo)	Boron (B)
14	3	3	23	<1	3	<1	35	2615	<1	1197	1544	85	124
15	3	3	19	<1	4	>1	43	2677	<1	1104	1602	106	116
11	2	3	14	<1	2	<1	15	3113	<1	1137	1515	87	104
4	1	1	10	<1	3	<1	15	2919	<	1150	1363	82	102
6	2	2	11	<1	3	<1	15	2941	4	1208	1346	92	113
8	1	1	10	<1	2	<1	10	2456	<1	1042	1171	63	81
CONTAMINANTS					WEAR		PHYSICAL PROPERTIES			INFORMATION			
Aluminum (Al)	Silicon (Si)	Sodium (Na)	Potassium (K)	Water (%)	PQ Index	Particle Counts	Viscosity (cSt @ 40C)	Oxidation	TAN	Fluid Changed	Fluid Type	Total Hours	Fluid Hours
3	8	4	<5	0.01	NA	21/17/12	70.8	NA	2.14	NO	10W30	9197	2800
4	6	5	<5	0.02	NA	21/17/14	72.3	NA	2.42	NO	10W30	6397	2251
2	5	6	<5	0.01	NA	21/17/14	72.6	NA	2.39	NO	10W30	4146	1140
3	4	5	<5	0.01	NA	21/19/15	73.4	NA	2.00	NO	10W30	3006	632
3	4	5	>5	0.01	NA	20/17/14	71.1	NA	2.02	NO	10W30	2374	374
2	5	5	<5	0.01	NA	20/19/15	65.4	NA	1.91	YES	10W30	2000	119
HYDRAULIC OIL ANALYSIS				LARGE PRODUCTION BULLDOZER WITH COMBINED HYDRAULICS/HYDROSTATICS. HIGH PARTICLE COUNTS AND DIRT. FLUID 10W30									

FIGURE 9-2 Silt accumulation in a production hydrostatic bulldozer with combined hydraulics and hydrostatics.

These particle count numbers are difficult to understand, and they appear to be very normal. The criticality is hard to grasp without a trending graphic display that gives us better resolution in this case. Let us see a graph with six data points from this bulldozer to try to make sense of the situation (see Figure 9-3).

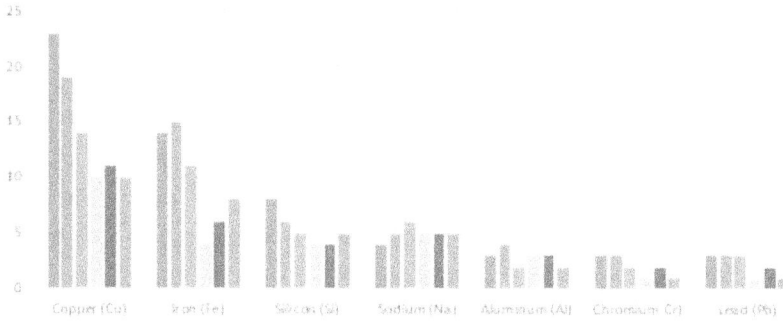

FIGURE 9-3 Trending metals and contaminants with six data points.

When the data are displayed graphically, as in Figure 9-3, it is easier to understand the situation. In this case, we can see that the silicon uptrend matches those of copper and iron. Although these numbers have not reached abnormal levels, they suggest that some silt is still there from previous fluid changes because some of the fluid remains in the system, especially if it is a combined system.

Conclusion

The lesson from this case is that combined systems, in which a high percentage of the fluid remains in cylinders and coolers, require a cleaning service with a high-efficiency filter caddy. Fluid changes alone do not resolve the challenge of keeping these systems clean.

3. Water Contamination in a Construction-Sized Excavator

A construction-sized excavator has been showing water in the hydraulic system, probably because of the type of application. The iron readings have reached an abnormal level, and the copper level is abnormal as well.

Oil Analysis

Thanks to the Karl Fischer titration test, it is feasible to get information on water levels <1,000 ppm. In high-pressure hydraulic systems, it is critical to keep humidity below 1,000 ppm. Particle counts in this case are not feasible because of the water contamination (see Figure 9-4).

METALS							ADDITIVES						
Iron (Fe)	Chromium Cr	Lead (Pb)	Copper (Cu)	Tin (Sn)	Aluminum (Al)	Nickel (Ni)	Magnesium (Mg)	Calcium (Ca)	Barium Ba	Phosphorus (P)	Zinc (Zn)	Molybdenum (Mo)	Boron (B)
35	3	2	18	2	1	<1	9	56	<1	261	316	1	<5
25	2	1	10	1	1	<1	10	60	<1	270	320	1	<5
20	<1	**<1**	4	<1	<1	<1	11	65	<1	275	328	1	<5

CONTAMINANTS					WEAR		PHYSICAL PROPERTIES			INFORMATION			
Aluminum (Al)	Silicon (Si)	Sodium (Na)	Potassium (K)	Water (%)	PQ Index	Particle Counts	Viscosity (cSt @ 40C)	Oxidation	TAN	Fluid Changed	Fluid Type	Total Hours	Fluid Hours
1	1	12	5	0.31	NA	NA	62.3	NA	0.3	YES	ISO68	3003	3003
1	1	5	<5	0.09	NA	19/17/14	63	NA	0.2	YES	ISO68	2020	2020
<1	1	<5	<5	0.05	NA	18/16/13	63.2	NA	0.2	YES	ISO68	1000	1000

HYDRAULIC OIL ANALYSIS	CONSTRUCTION SIZE EXCAVATOR . HIGH WATER CONTENT. FLUID TELLUS ISO68

FIGURE 9-4 Construction-sized excavator showing water in the hydraulic system.

Wear Metal Thresholds

See Table 9-3.

TABLE 9-3 Wear Metals Thresholds

500 hours			
Filtered System	Normal	Abnormal	Critical
Fe	<32	32	>60
Pb	<5	5	>10
Cu	<15	15	>27
Cr	<5	5	>10
Al	<7	7	>11
Si	<16	16	>25

Conclusion

The fluid signature matches the reported fluid. The contamination appears to be recent because no damage has occurred, although water was high in the previous test. Refer to Chapter 4 for procedures to clean systems.

4. High Particle Counts: Gelling and Mixing in a Construction Excavator

A large construction-sized excavator has been experiencing high particle counts. At 4,228 hours of use, the excavator is not showing any distress, and the metal readings are quite good.

Oil Analysis

There is no dirt, and the fluid has no water (see Figure 9-5). What is causing high particle counts?

METALS							ADDITIVES						
Iron (Fe)	Chromium (Cr)	Lead (Pb)	Copper (Cu)	Tin (Sn)	Aluminum (Al)	Nickel (Ni)	Magnesium (Mg)	Calcium (Ca)	Barium (Ba)	Phosphorus (P)	Zinc (Zn)	Molybdenum (Mo)	Boron (B)
6	<1	<1	2	<1	1	<1	8	39	<1	536	71	181	153
2	<1	<1	3	<1	<1	<1	2	4	<1	285	13	245	180
2	<1	<1	4	<1	<1	<1	11	5	<1	383	28	235	175
6	<1	<1	4	<1	1	<1	27	34	<1	392	90	2	<5

CONTAMINANTS				WEAR		PHYSICAL PROPERTIES			INFORMATION			
Aluminum (Al)	Silicon (Si)	Sodium (Na)	Potassium (K)	Water (%)	PQ Index	Particle Counts	Viscosity (cSt @ 40C)	Oxidation	TAN	Fluid Changed	Fluid Type	Total Hours	Fluid Hours
1	4	<1	<5	0.01	NA	23/22/18	46.4	NA	0.08	YES	ISO46	4228	2540
<1	1	<1	<5	<0.01	NA	21/20/15	43.7	NA	0.09	NO	ISO46	1688	1118
<1	1	<1	<5	<0.01	NA	19/16/12	44.0	NA	0.06	NO	ISO46	570	170
1	7	<1	<5	0.02	NA	21/19/15	43.7	NA	0.07	YES	ISO46	400	400

HYDRAULIC OIL ANALYSIS	LARGE CONSTRUCTION EXCAVATOR. HIGH PARTICLE COUNTS . MIXED FLUID. FLUID ZINC FREE HN46

FIGURE 9-5 High particle counts in a construction-sized excavator.

With the exception of one measurement at 570 hours, the particle readings appear to be outside specifications. The highlighted additives do not match those of the original factory fill. The John Deere and Hitachi factory fill should read something like Table 9-4.

TABLE 9-4 John Deere and Hitachi Factory Fill

Mg	Ca	Ba	P	Zn	Mo	B
17	25	0	410	25	0	0

As you can see, the machine has a product in its reservoir that shows molybdenum and boron. In terms of viscosity, the fluid is on target. A micropatch analysis of the fluid in Figure 9-6 shows the gel formation that results in increased particle count readings.

Normal particle counts for sealed systems (with a pressurized reservoir) using this type of fluid should be ≤23/21/16, so this machine would be compliant in three of the four samples measured (see Table 9-5).

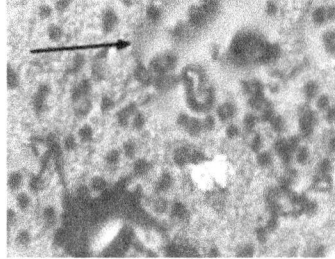

FIGURE 9-6 Micropatch analysis showing gel formation.

TABLE 9-5 Compliance in Three of the Four Samples Measured

	Normal	Abnormal	Critical
Sealed hydraulics with zinc-free fluid	≤X/21/16	≥23/21/16	≥X/22/17
Sealed hydraulics with mixed zinc-free fluid	≤X/22/17	≥X/22/17	NA

Conclusion

The last report on the fluid at 2,540 hours shows calcium, boron, and molybdenum, indicating that this fluid is a mix. For machines with mixed products, the specifications change as indicated in Table 9-5. Although mixing is not good, we can conclude that the machine is within specifications.

5. High Copper Production (Etching) in a Hydrostatic Crawler

The bulldozer in this example has a good regular maintenance plan. However, critical metals in its hydrostatics have always been high on oil analysis reports. The owner stayed with tractor hydraulic fluid (THF) instead of the manufacturer's recommendation. He does this because he buys tractor fluid in bulk and gets a good price per gallon. Fluid changes occur every 2,000 hours, but the copper readings appear to be rising over time.

Oil Analysis

The results in Figure 9-7 show copper and iron levels out of specification in every sample. The fluid is free of dirt and water, but particle counts have been high in the last two samples.

The specifications for this type of hydrostatics are copper at <21 ppm and iron at <21 ppm. The hydrostatics are not complying with specifications. Although the limits in Figure 9-8 are based on 500 hours, copper is not time dependent, and we cannot simply multiply by four to obtain the limit at 2,000 hours. Copper is the most critical metal in hydrostatic crawlers and needs to remain within specifications regardless of drain interval.

METALS							ADDITIVES						
Iron (Fe)	Chromium Cr	Lead (Pb)	Copper (Cu)	Tin (Sn)	Aluminum (Al)	Nickel (Ni)	Magnesium (Mg)	Calcium (Ca)	Barium Ba	Phosphorus (P)	Zinc (Zn)	Molybdenum (Mo)	Boron (B)
28	5	7	110	<1	13	<1	3	3307	<1	1100	1270	5	3
23	<1	5	80	1	11	<1	5	3350	<1	1155	1287	6	3
15	<1	4	57	<1	<1	<1	5	3384	<1	1200	1290	5	2

CONTAMINANTS				WEAR			PHYSICAL PROPERTIES			INFORMATION			
Aluminum (Al)	Silicon (Si)	Sodium (Na)	Potassium (K)	Water (%)	PQ Index	Particle Counts	Viscosity (cSt @ 40C)	Oxidation	TAN	Fluid Changed	Fluid Type	Total Hours	Fluid Hours
1	3	3	<5	0.03	NA	22/18/15	51.3	NA	2.35	NO	THF	7563	1563
1	2	4	<5	0.04	NA	22/17/15	50.9	NA	2.6	NO	THF	5500	1500
<1	2	5	<5	<0.05	NA	21/17/14	51.5	NA	2.5	NO	THF	3567	1567
HYDRAULIC OIL ANALYSIS				BULLDOZER WITH INDEPENDENT SEALED HYDROSTATICS. HIGH COPPER GENERATION. TRACTOR FLUID ISO68									

FIGURE 9-7 Copper and iron levels are out of specification in every sample.

Etching

Figure 9-8 shows a real case of pump rotary group etching.

FIGURE 9-8 Pump rotary group etching.

Wear Metal Thresholds

See Table 9-6.

TABLE 9-6 Wear Metals Thresholds

Filtered System	500 hours		
	Normal	Abnormal	Critical
Fe	<21	21	>30
Pb	<7	7	>12
Cu	<21	21	>32
Cr	<5	5	>10
Al	<5	5	>10
Si	<17	17	>28
Particle Counts	≤22/17/14	22/17/14	≥X/20/17

Conclusion

The bulldozer's hydrostatics are using a fluid that is not recommended for the application. The bronze alloys in the components are etching. Switching to the recommended fluid may help save this system from failure.

6. High Copper Generation (Progressive Component Failure) in a Large Production Excavator

A large production excavator has been showing signs of accelerated component wear in fluid samples. The copper and lead readings are growing together with particle counts (see Figure 9-9).

METALS							ADDITIVES						
Iron (Fe)	Chromium (Cr)	Lead (Pb)	Copper (Cu)	Tin (Sn)	Aluminum (Al)	Nickel (Ni)	Magnesium (Mg)	Calcium (Ca)	Barium (Ba)	Phosphorus (P)	Zinc (Zn)	Molybdenum (Mo)	Boron (B)
12	4	14	67	<1	4	<1	8	25	<1	225	35	5	6
10	2	12	56	<1	3	<1	2	18	<1	285	30	7	5
5	3	<1	4	<1	2	<1	11	20	<1	383	28	5	6
6	<1	<1	4	<1	3	<1	27	33	<1	390	85	2	5

CONTAMINANTS					WEAR		PHYSICAL PROPERTIES			INFORMATION			
Aluminum (Al)	Silicon (Si)	Sodium (Na)	Potassium (K)	Water (%) KF	PQ Index	Particle Counts	Viscosity (cSt @ 40C)	Oxidation	TAN	Fluid Changed	Fluid Type	Total Hours	Fluid Hours
4	4	2	<5	0.03	NA	22/18/15	46.4	NA	1.1	YES	ISO46	5500	2500
3	3	3	<5	0.04	NA	20/17/14	43.7	NA	1.2	NO	ISO46	3000	3000
2	2	<1	<5	0.02	NA	19/16/12	44.0	NA	0.06	NO	ISO46	2000	2000
3	3	<1	<5	0.03	NA	19/16/13	43.7	NA	0.07	NO	ISO46	1000	1000

HYDRAULIC OIL ANALYSIS	LARGE PRODUCTION EXCAVATOR. HIGH COPPER AND LEAD. ZINC FREE HN46 FLUID

FIGURE 9-9 Large production excavator showing signs of accelerated component wear.

Oil Analysis

Results in Figure 9-9 show no dirt, and the fluid is free of water. Particle counts are only high in the last sample. TAN is within specification. The fluid resembles the recommended zinc-free fluid. What is causing copper and lead production?

Wear Metal Thresholds

See Table 9-7.

TABLE 9-7 Wear Metals Thresholds

Large Excavator 500 hours			
Filtered System	Normal	Abnormal	Critical
Fe	<32	32	>60
Pb	<5	5	>10
Cu	<15	15	>27
Cr	<5	5	>10
Al	<7	7	>11
Si	<16	16	>25
Particle Counts	≤22/17/14	22/17/14	≥X/20/17

Conclusion

Figure 9-10 gives you an idea about how the parts fail in high-pressure hydraulic systems. Given the symptoms and lack of contaminants, water, mixes, or a high TAN, it is logical to conclude that a main component rotary group is wearing out. Most typically in an excavator, the valve plates in the main pumps are the ones that provide advance warning, and inspecting the pumps before a catastrophic failure can save users from a costly repair.

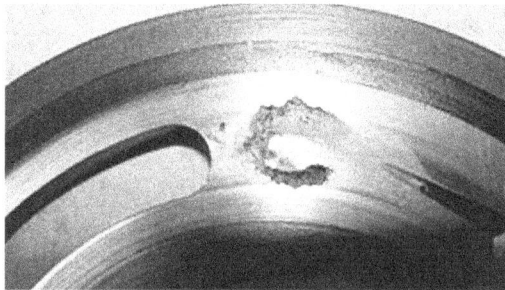

FIGURE 9-10 Part failure in a high-pressure hydraulic system.

7. High Iron Particle Production in a Medium-Sized Excavator

An excavator's hydraulic system is showing high iron levels, but the machine does not have any other symptoms to indicate that there is an issue somewhere else. The hydraulics continue to operate normally.

Oil Analysis

The results in Figure 9-11 show high particle counts, a high iron level, and an increased TAN. The machine does not have dirt, water, or a mixed fluid. The main components that produce iron are the hydraulic cylinders. If there were water in the system, the iron could be from rust, but this is not the case.

METALS							ADDITIVES						
Iron (Fe)	Chromium (Cr)	Lead (Pb)	Copper (Cu)	Tin (Sn)	Aluminum (Al)	Nickel (Ni)	Magnesium (Mg)	Calcium (Ca)	Barium (Ba)	Phosphorus (P)	Zinc (Zn)	Molybdenum (Mo)	Boron (B)
258	12	8	14	<1	7	<1	8	41	<1	215	461	2	<1
150	10	7	12	<1	6	<1	2	42	<1	219	465	1	<1
105	9	5	9	<1	5	<1	11	43	<1	225	478	2	<0
90	7	4	7	<1	5	<1	27	48	<1	233	480	2	<1

CONTAMINANTS					WEAR		PHYSICAL PROPERTIES			INFORMATION			
Aluminum (Al)	Silicon (Si)	Sodium (Na)	Potassium (K)	Water (%) KF	PQ Index	Particle Counts	Viscosity (cSt @ 40C)	Oxidation	TAN	Fluid Changed	Fluid Type	Total Hours	Fluid Hours
7	8	2	<5	0.09	NA	22/19/16	39.1	NA	1.4	YES	ISO46	2000	2000
6	7	3	<5	0.07	NA	21/18/15	40.7	NA	1.3	NO	ISO46	1500	1500
5	5	<1	<5	0.08	NA	20/17/14	42.3	NA	0.08	NO	ISO46	1000	1000
5	5	<1	<5	0.08	NA	19/16/13	44.1	NA	0.07	NO	ISO46	500	500

HYDRAULIC OIL ANALYSIS	CONSTRUCTION EXCAVATOR. HIGH IRON. AW46 FLUID

FIGURE 9-11 Report on a construction excavator showing high iron levels.

Conclusion

Typically, the cylinder most likely to produce excess iron in an excavator is the crowd cylinder, followed by the bucket cylinder. Taking individual samples from those cylinders could indicate which one shows the higher iron readings, followed by inspection and repair.

8. Fluid Oxidation and Thermal Records in a Medium-Sized Excavator

The owner of an excavator reports that the fluid has turned darker over time. The machine seems to operate normally. The fluid analysis shows fluid oxidation and a high TAN.

Oil Analysis

The report in Figure 9-12 confirms abnormal oxidation, a high TAN, and a decay of viscosity. Because there is no dirt or water, the TAN is causing the increase in metal readings. The fluid in use is an extended-life product designed to operate for 4,000 hours, but the fluid analysis shows that the fluid is no longer serviceable.

METALS							ADDITIVES						
Iron (Fe)	Chromium Cr	Lead (Pb)	Copper (Cu)	Tin (Sn)	Aluminum (Al)	Nickel (Ni)	Magnesium (Mg)	Calcium (Ca)	Barium Ba	Phosphorus (P)	Zinc (Zn)	Molybdenum (Mo)	Boron (B)
75	12	8	17	<1	12	<1	<5	14	<1	541	18	<1	3
50	10	7	15	<1	10	<1	<5	12	<1	545	12	1	1
16	9	5	10	<1	9	<1	<5	5	<1	554	10	<1	<2
15	7	4	7	<1	3	<1	<5	4	<1	560	5	<1	<1

CONTAMINANTS					WEAR		PHYSICAL PROPERTIES			INFORMATION			
Aluminum (Al)	Silicon (Si)	Sodium (Na)	Potassium (K)	Water (%) KF	PQ Index	Particle Counts	Viscosity (cSt @ 40C)	Oxidation	TAN	Fluid Changed	Fluid Type	Total Hours	Fluid Hours
12	15	2	<5	0.07	NA	NA	42.5	24	1.7	YES	ISO46	4300	300
6	9	3	<5	0.05	NA	23/19/16	43.2	21	1.5	YES	ISO46	4000	4000
9	7	<1	<5	0.06	NA	22/18/14	44.1	12	1.3	NO	ISO46	2000	2000
3	8	<1	<5	0.06	NA	22/17/14	46.1	6	0.07	NO	ISO46	100	100

HYDRAULIC OIL ANALYSIS	CONSTRUCTION EXCAVATOR. HIGH OXIDATION AND HIGH TAN. ZINC FREE ISO46 FLUID

FIGURE 9-12 Abnormal oxidation, high TAN, and a decay of viscosity in a construction excavator.

Given the information at hand, it becomes imperative to check hydraulic fluid cooling on this machine. Thanks to the machine's onboard information, there are records of hydraulic fluid temperatures to help diagnose this situation.

Excavator Hydraulic Temperature Records

Table 9-14 shows hydraulic temperature records for this machine. Starting at 3902 hours of operation, this machine's hydraulics started overheating more regularly. The readings show that for 36% of the 1,189 hours, fluid temperatures were >99°C.

TABLE 9-8 Hydraulic Temperature Records for This Machine

Hour Meter	Under 48°C	50–58°C	60–69°C	70–79°C	80–89°C	90–99°C	100–109°C	Over 110°C	Total Over 90°C
4,302	4	1.2	3.5	7.5	20.2	51.3	4.1	0.7	56.1
4,202	3.1	1.8	3.8	12.4	20.9	51.9	5.7	0.5	58.1
4,102	2.2	1.1	1.9	12.7	12.6	55.7	13.2	0.4	69.3
4,002	0.7	0.4	1.5	26.1	17.2	50	3.7	0.3	54
3,902	0.8	0.3	2.6	26	30.2	39.5	0.2	0	39.7
3,802	1.3	0.9	2.1	34	31.8	29.6	0	0	29.6
3,702	7.8	4	8.7	48.5	21.9	8.7	0	0	8.7
3,602	5.1	5.2	18.8	35.6	21.2	13.7	0	0	13.7
3,502	5.1	5.6	14.7	46.2	18.1	10	0	0	10
3,402	5.6	3.8	6.3	20.2	25.9	37.8	0	0	37.8
3,302	2.4	0.7	1.1	53	29.3	13	0	0	13
3,201	3.1	1.5	1.8	39.9	24.7	28.5	0	0	28.5
	41.2	26.5	66.8	362.1	274	389.7	26.9	1.9	418.5

Conclusion

With the three pieces of information—oil analysis, thermograph, and actual fluid color—it was feasible to diagnose the problem. Inspection confirmed that the oil cooler bypass valve spring had failed, forcing return fluid to bypass the cooler and flow directly into the reservoir.

9. Operation with Bypass Filtration in a Large Production Excavator

The owner of a large production excavator uses 1-μm bypass filtration to extend the life of fluids and components. The excavator is equipped with bypass filtration for the hydraulics. The owner is not sure that this technology actually extends the life of the fluid without compromising the hydraulic system. Let us analyze the results and reach a conclusion in this case.

Fluid Analysis

Figure 9-13 shows the fluid analysis results for this excavator's hydraulics. With 4,485 hours on the hour meter and on the fluid, the fluid provides no indication of decay or machine wear. The results show extremely good numbers that would not be feasible without bypass filtration. Iron, which is a time-dependent element, is in check, as is copper. The fluid is extremely dry at 100 ppm of water, and the TAN is normal. Viscosity is very stable.

METALS							ADDITIVES						
Iron (Fe)	Chromium (Cr)	Lead (Pb)	Copper (Cu)	Tin (Sn)	Aluminum (Al)	Nickel (Ni)	Magnesium (Mg)	Calcium (Ca)	Barium Ba)	Phosphorus (P)	Zinc (Zn)	Molybdenum (Mo)	Boron (B)
33	1	0	10	0	0	0	5	40	0	308	70	0	2
36	2	0	9	0	0	0	3	42	0	307	75	0	2
32	2	2	10	1	1	0	3	38	11	348	69	0	2
25	2	0	11	1	1	0	3	42	0	345	76	0	2

CONTAMINANTS					WEAR		PHYSICAL PROPERTIES			INFORMATION			
Aluminum (Al)	Silicon (Si)	Sodium (Na)	Potassium (K)	Water (%) KF	PQ Index	Particle Counts	Viscosity (cSt @ 40C)	Oxidation	TAN	Fluid Changed	Fluid Type	Total Hours	Fluid Hours
0	3	5	0	0.01	NA	21/15/11	44.1	7	0.45	NO	ISO46	4485	4485
0	3	4	0	0.01	NA	22/15/10	44.3	6	0.4	NO	ISO46	3980	3980
1	3	4	0	0.01	NA	21/16/11	43.9	6	0.39	NO	ISO46	3590	3590
1	3	0	0	0.02	NA	20/15/10	43.6	5	0.37	NO	ISO46	3140	3140
HYDRAULIC OIL ANALYSIS				PRODUCTION EXCAVATOR. BY PASS FILTRATION INSTALLED. ZINC FREE ISO46 FLUID									

FIGURE 9-13 Fluid analysis results for this excavator's hydraulics.

As for particle counts, the 6-μm channel is four times cleaner than the specifications require, and the 14-μm channel is six times cleaner (see Table 9-9).

TABLE 9-9   Wear Threshold Table

Large Excavator 500 hours			
Filtered System	**Normal**	**Abnormal**	**Critical**
Fe	<32	32	>60
Pb	<5	5	>10
Cu	<15	15	>27
Cr	<5	5	>10
Al	<7	7	>11
Si	<16	16	>25
Particle Counts	≤22/17/14	22/17/14	≥X/20/17

One can argue that the drawback of bypass filtration is that you cannot pinpoint when a component is about to fail because the readings are so minimal that it is difficult to realize that a critical component is producing excess metals. Reality tells a different story. When a component is about to fail for reasons not related to maintenance, the jump in metal production is registered by fluid analysis because the bypass filtration is unable to capture the jump quickly enough.

Conclusion

The physical properties of the fluid are intact, and the machine runs extremely cleanly. There is no indication that the fluid needs to be changed. Only a mishap in maintenance (mixing fluids) or problems with cooling could change the outcome. The fluid appears to be good for at least 3,000 more hours, if not more.

10. Hydraulic Fluid Foaming in a Construction-Sized Excavator

A construction- sized excavator operating with an antiwear (AW) ISO 46 fluid is showing serious issues with foaming. The operation is jerky, and response is sloppy. The problem started after a fluid change at 3,000 hours.

Oil Analysis

Results shown in Figure 9-14 indicate that this excavator is using a different fluid at 3,000 hours. There is no information on the fluid brand or type.

Use of a Fluid Mixing Calculator

To determine the amount of fluid mixing, we can build a mix calculator around an Excel spreadsheet. In an excavator with a total capacity of 66 gallons, 39 of which are in the reservoir, a suspicious fluid represents 39 gallons, or 59% of the system fluid. The original fluid that the machine had before servicing now is 27 gallons, or 41% of the total system. If we apply these percentages to the additive signature of each fluid, we come up with the additive contribution of each in the final mix (see Table 9-10).

METALS							ADDITIVES						
Iron (Fe)	Chromium Cr	Lead (Pb)	Copper (Cu)	Tin (Sn)	Aluminum (Al)	Nickel (Ni)	Magnesium (Mg)	Calcium (Ca)	Barium Ba	Phosphorus (P)	Zinc (Zn)	Molybdenum (Mo)	Boron (B)
15	3	2	12	<1	2	<1	15	1360	<1	950	1098	2	103
45	2	1	9	<1	3	<1	2	38	<1	220	459	1	<1
25	1	1	8	<1	3	<1	11	41	<1	235	465	2	<0
21	2	1	10	<1	1	<1	13	45	<1	241	476	2	<1

CONTAMINANTS					WEAR		PHYSICAL PROPERTIES			INFORMATION			
Aluminum (Al)	Silicon (Si)	Sodium (Na)	Potassium (K)	Water (%) KF	PQ Index	Particle Counts	Viscosity (cSt @ 40C)	Oxidation	TAN	Fluid Changed	Fluid Type	Total Hours	Fluid Hours
2	14	5	<5	0.08	NA	22/20/19	51.1	NA	2.5	YES	?	3050	50
3	4	4	<5	0.06	NA	22/19/14	41.3	NA	1.3	YES	AW46	3000	3000
3	3	<1	<5	0.09	NA	20/17/13	42.2	NA	0.08	NO	AW46	2000	2000
1	4	<1	<5	0.07	NA	19/16/14	45.4	NA	0.07	NO	AW46	1000	1000

HYDRAULIC OIL ANALYSIS	CONSTRUCTION EXCAVATOR. FLUID MIX. AW46 WITH SOMETHING ELSE

FIGURE 9-14 Indication that this excavator is using a different fluid at 3,000 hours.

TABLE 9-10 Contribution of Each Additive in the Final Mix

System		Ca	P	Zn	Si	Vicosity
39 gal reserv.	Suspect sign.	2270	1456	1568	19	58.4
27 gal serv. Total 66 gal	Host AW46 sign.	38	220	459	4	41.3
59% reserv.	Suspect contrib.	1339	859	925	11	34
41% serv.	Host contrib.	16	90	188	2	17
100%	Final sign. ppm	1355	949	1113	13	51

The final additive readings closely match those of the final fluid analysis report. This suggests that the machine has 39 gallons of what appears to be a tractor fluid (THF) with an initial viscosity 58.4 cSt. We also notice that this fluid formulation uses high doses of silicone as an antifoam additive.

Conclusion

The foaming situation is happening because of the incompatibility of certain fluids. In this case, an AW 46 hydraulic fluid is incompatible with a tractor fluid. When the high antifoam THF is mixed with a fluid having a small amount of antifoaming agent, its capacity to control foam is diluted, causing this situation. Additionally, when switching to a different fluid in an excavator, at least 41% of the fluid remains in lines, cylinders, coolers, motors, and so on. Refer to Chapter 4 for information on how to properly flush a hydraulic system.

11. Atypical Particle Counts in a Four-Wheel-Drive Loader

A 4WD loader reported atypical particle counts in the last two analyses. The user has not been able to resolve the issue.

Oil Analysis

The report from the user's lab detects water only by FTIR spectroscopy, and the particle counts show atypical numbers (see Figure 9-15).

METALS							ADDITIVES						
Iron (Fe)	Chromium Cr)	Lead (Pb)	Copper (Cu)	Tin (Sn)	Aluminum (Al)	Nickel (Ni)	Magnesium (Mg)	Calcium (Ca)	Barium Ba)	Phosphorus (P)	Zinc (Zn)	Molybdenum (Mo)	Boron (B)
40	3	4	18	0	12	<1	505	1820	0	1101	1200	190	160
70	4	2	25	0	10	<1	462	1799	0	1088	1210	245	180
10	2	1	3	0	8	<1	454	1803	0	1050	1180	235	175

CONTAMINANTS					WEAR		PHYSICAL PROPERTIES			INFORMATION			
Aluminum (Al)	Silicon (Si)	Sodium (Na)	Potassium (K)	Water (%) FTIR ppm	PQ Index	Particle Counts	Viscosity (cSt @ 40C)	Oxidation	TAN	Fluid Changed	Fluid Type	Total Hours	Fluid Hours
12	12	10	0	<1000	NA	22/22/22	54	NA	2.35	YES	10W30	4200	200
10	11	9	0	<1000	NA	22/22/21	54.1	NA	2.01	YES	10W30	4000	4000
8	10	10	0	<1000	NA	22/17/14	55.4	NA	1.9	NO	10W30	2000	2000

HYDRAULIC OIL ANALYSIS	CONSTRUCTION 4WD LOADER. ATYPICAL PARTICLE COUNTS. FLUID 10W30

FIGURE 9-15   A 4WD loader reporting atypical particle counts.

Testing a second sample with Karl Fischer titration produced different results. The machine's hydraulic fluid actually had 1,600 ppm of water. The standard FTIR test only detects water in engine oils at levels >1,600 ppm. The limit for water in a high-pressure system is <1000 ppm.

Wear Threshold Table

See Table 9-11.

TABLE 9-11   Wear Threshold Table

4WD Hyd. Loader Per 1,000 hours			
Filtered System	Normal	Abnormal	Critical
Fe	<51	51	>70
Pb	<13	13	>20
Cu	<21	21	>30
Cr	<16	16	>20
Al	<11	11	>20
Si	<16	16	>25

Conclusion

This case shows the importance of using the correct water detection test to pinpoint excess moisture in the fluid. Additionally, particle counts help us detect water or air by showing atypical number combinations. Ideally, particle counts should show two number differences between channels. The results on this machine show identical numbers, which is a clear indication of water or air.

12. Cheese Curding in Biodegradable Hydraulic Fluid in a Production Excavator

A large production excavator needed to do some work at a nuclear plant in Florida. Plant management does not want any oil spills in the area and requires machines to operate with a biodegradable fluid. Environmentally friendly was not enough; it needed to be a biodegradable product. The machine started using a canola oil–based biodegradable tractor fluid. The machine appeared to operate satisfactorily in the initial weeks. However, as hours mounted, the machine became jerky and uncontrollable. Inspection of the machine indicated that the fluid in the hydraulic system had coagulated, forming what appeared to be cheese curds (see Figure 9-16). These gummy formations cause the filters to plug and the valves to stick.

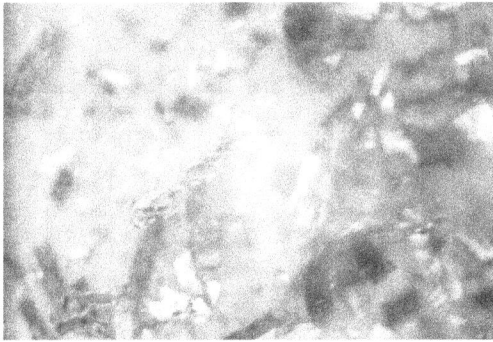

FIGURE 9-16 Cheese curd–like fluid in large production excavator.

Oil Analysis

The results in Figure 9-17 indicate that the machine was switched from the factory fill to a biodegradable product at 1,500 hours. By changing just the reservoir volume, only 62.7% of the fluid was new fluid. The machine received another change of fluid after 50 hours of work, bringing the total new fluid volume to 86%. Replacing the whole volume (99.9% new fluid) would require at least six reservoir flushes unless a different method were used. This means that with only two flushes, there is plenty of opportunity for incompatibility issues within this system.

The elemental signature with 86% biodegradable fluid in the mix (second flush) is visible in the results and matches almost perfectly the weighted average, as shown by the quick mixing calculator in Table 9-12. Also noteworthy is the fact that these results only show water by FTIR spectroscopy. The use of bioproducts requires water to be <500 ppm. The results also show that the machine is not complying with wear metal generation or particle counts. A third fluid analysis using the Karl Fischer titration test for water brought back the results. The machine's hydraulic system was operating with 1300 ppm of water.

METALS							ADDITIVES						
Iron (Fe)	Chromium (Cr)	Lead (Pb)	Copper (Cu)	Tin (Sn)	Aluminum (Al)	Nickel (Ni)	Magnesium (Mg)	Calcium (Ca)	Barium (Ba)	Phosphorus (P)	Zinc (Zn)	Molybdenum (Mo)	Boron (B)
56	4	0	45	0	8	0	0	560	0	800	412	0	0
45	3	0	55	0	10	0	0	404	0	630	280	0	0
32	2	2	10	1	1	0	0	38	11	348	69	0	0
25	2	0	11	1	1	0	0	42	0	345	76	0	0

CONTAMINANTS					WEAR		PHYSICAL PROPERTIES			INFORMATION			
Aluminum (Al)	Silicon (Si)	Sodium (Na)	Potassium (K)	Water (%) FTIR	PQ Index	Particle Counts	Viscosity (cSt @ 40C)	Oxidation	TAN	Fluid Changed	Fluid Type	Total Hours	Fluid Hours
8	8	5	0	<1500	NA	27/25/23	47	NA	2.67	YES	BIO THF	1550	50
10	14	4	0	<1500	NA	22/20/19	48.3	NA	1.56	YES	BIO THF	1500	1500
1	5	4	0	<1500	NA	21/17/13	48.5	NA	0.39	NO	ISO46	1000	1000
1	2	0	0	<1500	NA	20/16/13	49	NA	0.4	NO	ISO46	500	500

HYDRAULIC OIL ANALYSIS	LARGE PRODUCTION EXCAVATOR REFILLED WITH BIO THF INITIAL FILL ZINC FREE ISO46 FLUID

FIGURE 9-17 Results of oil analysis on a large production excavator.

TABLE 9-12 Quick Mixing Calculator

System		Ca	P	Zn	Si	Vicosity
96 gal.System	BIO Sign.	656	899	486	3	46
57 gal Serv. Total 153 gal	Host Sign.	404	630	280	3	48.3
62.7% Reserv.	BIO Contrib.	411	564	305	2	29
37.3% Serv.	Host Contrib.	151	235	104	1	18
100%	Final Sign. ppm	562	799	409	3	47

Conclusion

Switching fluids with partial flushing is not wise depending on the mixes. A total flush is always a good practice. With bioproducts, water is a serious enemy. The ability to see water at <1000 ppm with FTIR spectroscopy is not feasible; the Karl Fischer titration test is needed. To see very low water levels, fine bypass filtration becomes essential.

13. Hydraulic Hammer and Dirt in a Medium-Sized Excavator

A medium-sized excavator equipped with a hydraulic hammer experienced accelerated pump wear. The technicians believed that the hydraulic hammer pulsations caused damage to the pump.

HYDRAULIC FLUID ANALYSIS | 231

Although this will be an after-the-fact check fluid analysis, it could bring some light to the causes of this failure. The valve plate shows a clear indication that abrasive material was the cause of premature wear (see Figure 9-18).

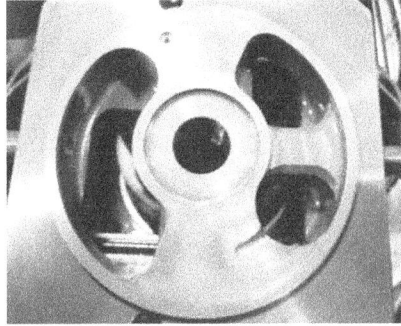

FIGURE 9-18 Abrasive material damage clearly seen in the valve plate.

Oil Analysis

The lab results show large amounts of dirt (silicon and aluminum). Dirt alone is the worst contaminant you can experience in a hydraulic system. The figures exceed the limits for dirt by fourfold (see Figure 9-19).

METALS							ADDITIVES						
Iron (Fe)	Chromium (Cr)	Lead (Pb)	Copper (Cu)	Tin (Sn)	Aluminum (Al)	Nickel (Ni)	Magnesium (Mg)	Calcium (Ca)	Barium (Ba)	Phosphorus (P)	Zinc (Zn)	Molybdenum (Mo)	Boron (B)
120	12	4	45	0	15	<1	16	2100	0	950	1112	0	<1
87	6	2	27	0	14	<1	14	1870	0	923	1023	0	0
25	2	1	8	0	3	<1	15	1950	0	935	1050	0	<2
CONTAMINANTS				WEAR			PHYSICAL PROPERTIES			INFORMATION			
Aluminum (Al)	Silicon (Si)	Sodium (Na)	Potassium (K)	Water (%) PPM	PQ Index	Particle Counts	Viscosity (cSt @ 40C)	Oxidation	TAN	Fluid Changed	Fluid Type	Total Hours	Fluid Hours
15	67	12	15	<1500	NA	22/21/19	57.5	NA	2.7	YES	TO4	3550	50
14	55	8	10	<1500	NA	22/20/19	58	NA	2.7	NO	TO4	3500	3000
3	10	4	5	<1500	NA	22/17/14	60	NA	2.3	NO	TO4	2500	500
HYDRAULIC OIL ANALYSIS			MEDIUM SIZE EXCAVATOR. HIGH DIRT AND PUMP FAILURE. FLUID TO4										

FIGURE 9-19 Lab results showing large amounts of dirt in a medium-sized excavator.

An inspection of the hydraulic hammer indicated that this hammer model has only one dust seal and one oil retention seal. A cutaway view of the hammer in Figure 9-20 shows the location of the single-seal arrangement. Additionally, the machine did not have bypass filtration, which is recommend when a machine is equipped with hydraulic hammers.

FIGURE 9-20   Location of the single-seal arrangement.

Conclusion

The pump shows clear signs of wear caused by dirt contamination. It was discovered that the hydraulic hammer was the cause of the dirt contamination.

14. Varnish in a Construction-Sized Excavator

A construction-sized excavator was showing heavy varnish all over the hydraulic system (see Figure 9-21). The varnish was affecting valve operation, making the machine unsafe to operate. Even so, it continued to be used, until an input shaft failed catastrophically. On inspection, all internal components showed heavy layers of varnish everywhere.

FIGURE 9-21   Varnish throughout the hydraulic system.

Oil Analysis

The results shown in Figure 9-22 indicate that the machine was producing large amounts of iron. The results also show that other critical metals such as copper, lead, and chromium are high. Oxidation is approaching the limit, and the TAN is over the limit.

The results suggest that heat could be the cause, but the operator did not report overheating at any time, and temperature records from the machine's onboard diagnostics do not show hydraulic overheating events. On inspection, the crowd cylinder was the culprit for the iron production because the piston nut was loose, causing the piston to produce iron (see Figure 9-23).

METALS							ADDITIVES						
Iron (Fe)	Chromium (Cr)	Lead (Pb)	Copper (Cu)	Tin (Sn)	Aluminum (Al)	Nickel (Ni)	Magnesium (Mg)	Calcium (Ca)	Barium (Ba)	Phosphorus (P)	Zinc (Zn)	Molybdenum (Mo)	Boron (B)
1560	12	8	45	<1	5	<1	<5	7	<1	525	18	<1	<1
1320	10	7	37	<1	4	<1	<5	8	<1	530	15	1	<1

CONTAMINANTS					WEAR		PHYSICAL PROPERTIES			INFORMATION			
Aluminum (Al)	Silicon (Si)	Sodium (Na)	Potassium (K)	Water (%) KF	PQ Index	Particle Counts	Viscosity (cSt @ 40C)	Oxidation	TAN	Fluid Changed	Fluid Type	Total Hours	Fluid Hours
5	4	2	<5	0.07	NA	**NA**	44.7	24	1.7	YES	ISO46	3675	100
4	3	3	<5	0.08	NA	23/22/17	45	21	1.6	NO	ISO46	3575	3575

HYDRAULIC OIL ANALYSIS	CONSTRUCTION EXCAVATOR. VARNISH FORMATION. HIGH IRON. ZINC FREE ISO46 FLUID

FIGURE 9-22 Lab results indicating large amounts of varnish in a construction excavator.

Gap between the spacer and rod due to loose nut

FIGURE 9-23 Crowd cylinder with loose piston nut causing iron production.

The sequence of events started with the loose cylinder piston nut, followed by a high-pressure, high-speed fluid internal leak, which generated static discharges, which, in turn, produced superheated microspots.

Conclusion

Static discharge is generated through friction within the fluid and can generate sparks with temperatures >10,000°C for a nanosecond, hotter than the surface of the sun. The heat generated "cooks" the fluid, speeding additive decomposition and the formation of oxidation by-products that coat all internal metal, rubber, and synthetic material surfaces with tacky layers of varnish. Refer to Chapter 4 for information on how to clean varnish from hydraulic systems.

15. Hydraulic Microdieseling in a Large Excavator

A new large Japanese excavator was exhibiting black hydraulic fluid. The machine operated normally, but the color of the fluid was creating some concern (see Figure 9-24). The machine had arrived from Japan to a West Coast port without the boom and cylinders, which came on a separate pallet. The machine had been assembled at the port.

FIGURE 9-24   Black hydraulic fluid.

Oil Analysis

A sample of the fluid collected shows the condition of the fluid. However, the test results do not show abnormal conditions aside from the atypical high particle counts (see Figure 9-25).

METALS							ADDITIVES						
Iron (Fe)	Chromium (Cr)	Lead (Pb)	Copper (Cu)	Tin (Sn)	Aluminum (Al)	Nickel (Ni)	Magnesium (Mg)	Calcium (Ca)	Barium (Ba)	Phosphorus (P)	Zinc (Zn)	Molybdenum (Mo)	Boron (B)
15	<1	0	5	<1	5	<1	16	24	<1	410	22	<1	<1
10	<1	0	4	<1	4	<1	17	25	<1	412	25	1	<1

CONTAMINANTS					WEAR		PHYSICAL PROPERTIES			INFORMATION			
Aluminum (Al)	Silicon (Si)	Sodium (Na)	Potassium (K)	Water (%) KF	PQ Index	Particle Counts	Viscosity (cSt @ 40C)	Oxidation	TAN	Fluid Changed	Fluid Type	Total Hours	Fluid Hours
2	1	<1	<5	0.04	NA	23/23/21	46.5	2	0.09	NO	ISO46	50	50
2	1	<1	<5	0.03	NA	23/23/20	46.7	1	0.09	NO	ISO46	25	25
HYDRAULIC OIL ANALYSIS			LARGE EXCAVATOR. FLUID BLACKENING. ATYPICAL PARTICLE COUNTS. ZINC FREE ISO46 FLUID										

FIGURE 9-25   Lab results from a large excavator showing black hydraulic fluid.

Conclusion

By receiving the machine with the boom disassembled, and reconnecting the components at the port, the hydraulic cylinders had air trapped inside them. While actuating the cylinders, highly compressed air caused the cylinders to burn the fluid vapor like fuel. This bears the name *dieseling* and causes the fluid to turn black. Because only a small amount of fluid was affected, the dieseling had no detrimental effect on machine performance. Particle counts will be high and atypical until the onboard filtration system is able to collect some of the burned particles (soot). Otherwise, filter caddying the machine can help.

16. Aluminum and Titanium Readings in a Backhoe

A backhoe working in an underground limestone mine showed continuous hydraulics contamination with aluminum and silicon. Its owner wanted to know what this could do to his machine.

Oil Analysis

The results in Figure 9-26 show a machine in distress. Metal production is high, with high aluminum and silicon, high iron, and even titanium, which is a metal not used in alloys for this machine. Particle counts are high even after a fluid change. The ratio of silicon to aluminum is atypical, but it still represents dirt.

METALS							ADDITIVES						
Iron (Fe)	Chromium (Cr)	Lead (Pb)	Copper (Cu)	Tin (Sn)	Aluminum (Al)	Nickel (Ni)	Magnesium (Mg)	Calcium (Ca)	Barium (Ba)	Phosphorus (P)	Zinc (Zn)	Molybdenum (Mo)	Boron (B)
67	5	4	12	<1	57	<1	10	3340	<1	1087	1260	<1	<1
120	12	7	18	<1	52	<1	17	3323	<1	987	1150	1	<1

CONTAMINANTS					WEAR	PHYSICAL PROPERTIES			INFORMATION			
Aluminum (Al)	Silicon (Si)	Sodium (Na)	Potassium (K)	Titanium (T)	Water (%) KF	Particle Counts	Viscosity (cSt @ 40C)	Oxidation	TAN	Fluid Changed	Fluid Type	Total Hours	Fluid Hours
57	21	12	<5	9	0.06	23/19/17	45.5	5	1.9	NO	THF	2000	500
52	35	11	<5	15	0.7	23/20/18	42.3	12	2.5	YES	THF	1500	1500

HYDRAULIC OIL ANALYSIS	BACKHOE LOADER. EXCESS ALUMINUM AND TITANIUM. HIGH METAL READINGS. TRACTOR FLUID ISO 68

FIGURE 9-26   Lab results showing continuous hydraulics contamination in a backhoe.

After inspecting the machine, there was no indication of leaks as possible points of dirt entry. All reservoir caps were in place. Because the machine had nine hydraulic cylinders in total plus a reservoir breather, these are the points of dirt entry.

Conclusion

Unless the machine is equipped with cylinder bellows protections (see Figure 9-27), bypass filtration, and a micron reservoir breather, it will be difficult to keep it in a healthy condition in this extreme operating application.

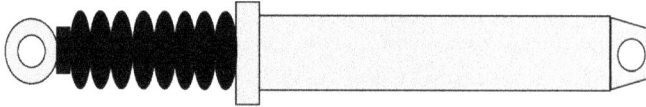

FIGURE 9-27 Cylinder bellows.

Wear Threshold Table

See Table 9-13.

TABLE 9-13 Wear Threshold Table

Backhoe Loader Per 1,000 hours			
Filtered System	**Normal**	**Abnormal**	**Critical**
Fe	<26	26	>40
Pb	<11	11	>15
Cu	<16	16	>25
Cr	<9	9	>15
Al	<9	9	>15
Si	<20	20	>45
Particle Counts	<23/19/16	23/19/16	>25/21/18

17. Sluggish Hydraulics in Winter in a Conversion Excavator

A forestry conversion excavator in northern Canada operates year-round and needs to adjust fluid viscosity for the seasons accordingly. The company uses ISO 22 fluid for the winter and ISO 68 for the summer. However, last winter the machine had trouble getting up to speed in the first hours of operation. The machine was sluggish, and its operator complained of jerky operation as well. Initially, the machine had an ISO 32 zinc-free fluid as its factory fill.

Oil Analysis

Results in Figure 9-28 show that viscosity of the machine's hydraulic fluid is at 34.1 cSt, a viscosity that is not going to work in extreme cold temperatures. The results also show the sequence of fluid changes since the machine had worked 50 hours. The machine started with a factory fill of zinc-free ISO 32 fluid. At 50 hours, the company drained and refilled the reservoir with a zinc-based AW 22 fluid for winter, giving a final viscosity of 26.3 cSt. Later, at 1,200 hours, the machine reservoir was filled with ISO 68 fluid, resulting in a viscosity of 50.2 cSt for the summer. Again, at 1,800 hours, the machine received another change back to an ISO 22 fluid, giving a final viscosity of 34.1 cSt.

The highlighted sequence in the figure shows that switching between ISO 22 and ISO 68 produced a final viscosity of 34.1 cSt, which is practically the viscosity the machine came with from the factory. This happened because the machine's reservoir represents only 61% of the total volume of the machine. After four fluid changes with three viscosities involved, the machine ended up with a viscosity that was still too high for extreme cold-weather temperatures.

METALS							ADDITIVES						
Iron (Fe)	Chromium Cr)	Lead (Pb)	Copper (Cu)	Tin (Sn)	Aluminum (Al)	Nickel (Ni)	Magnesium (Mg)	Calcium (Ca)	Barium Ba)	Phosphorus (P)	Zinc (Zn)	Molybdenum (Mo)	Boron (B)
25	2	2	35	0	10	<1	16	27	0	356	350	0	<1
22	2	1	32	0	13	<1	17	26	0	320	300	0	<1
23	3	2	27	0	4	<1	14	26	0	233	260	0	<1
10	1	1	8	0	3	<1	15	23	0	245	4	0	<2

CONTAMINANTS				WEAR		PHYSICAL PROPERTIES		INFORMATION			
Aluminum (Al)	Silicon (Si)	Sodium (Na)	Potassium (K)	Water (%) KF	PQ Index	Particle Counts	Viscosity (cSt @ 40C)	Oxidation	TAN	Fluid Changed	Fluid Type	Total Hours	Fluid Hours
10	3	<5	<5	0.07	NA	22/17/15	34.1	NA	0.06	YES	AW22	1800	600
13	4	<5	<5	0.06	NA	21/18/16	50.2	NA	0.07	YES	AW68	1200	600
4	2	<5	<5	0.07	NA	21/17/15	26.3	NA	0.08	YES	AW22	600	550
3	2	<5	<5	0.08	NA	20/17/14	32.1	NA	0.07	YES	ZF32	50	50

HYDRAULIC OIL ANALYSIS	CONVERSION EXCVATOR. SLUGGISH HYDRAULICS.AW32, AW68

FIGURE 9-28 Lab results showing the various viscosities of the hydraulic fluids of the conversion excavator.

Conclusion

By only flushing the reservoir volume each time, it was impossible to achieve the desired viscosity to handle the different seasons in this location. The user opted to switch between ISO 22 and ISO 46 fluid. Playing with viscosities between 26 and 38 cSt allowed the machine to operate between seasons without trouble.

18. Sudden Failures in a Production Excavator: Can Oil Analysis Foresee Them?

A construction excavator had a pump failure that the user was unable to predict through fluid analysis. The failure was catastrophic and spread shrapnel throughout the system. The user was disappointed that fluid analysis was unable to support the company's maintenance efforts in a more effective way. Previous fluid analyses did not show any concerns with the machine.

In this case, the pump failure shown in Figure 9-29 occurred because the sleeve of one piston became loose and caused a chain reaction. This kind of event usually occurs very quickly. A catastrophic failure of this nature is hard and expensive to fix. Chapter 4 describes the procedures to clean hydraulic systems of machines that experience this type of massive contamination.

FIGURE 9-29   Catastrophic pump failure.

Oil Analysis

The results from the oil analysis show a perfect machine in terms of maintenance (see Figure 9-30). The fluid signature is in agreement with use of a zinc-free fluid formulation, and the metal generation is well within limits. Dirt and water are also within limits, as are oxidation and TAN.

METALS							ADDITIVES						
Iron (Fe)	Chromium Cr	Lead (Pb)	Copper (Cu)	Tin (Sn)	Aluminum (Al)	Nickel (Ni)	Magnesium (Mg)	Calcium (Ca)	Barium Ba	Phosphorus (P)	Zinc (Zn)	Molybdenum (Mo)	Boron (B)
23	3	3	9	<1	4	<1	18	26	<1	389	22	<1	<1
11	2	2	5	<1	3	<1	19	28	<1	412	25	1	<1

CONTAMINANTS					WEAR		PHYSICAL PROPERTIES			INFORMATION			
Aluminum (Al)	Silicon (Si)	Sodium (Na)	Potassium (K)	Water (%) KF	PQ Index	Particle Counts	Viscosity (cSt @ 40C)	Oxidation	TAN	Fluid Changed	Fluid Type	Total Hours	Fluid Hours
4	15	<1	<5	0.06	NA	19/18/14	46.2	3	0.08	NO	ISO46	2500	2500
3	10	<1	<5	0.05	NA	19/17/14	46.3	2	0.07	NO	ISO46	500	500

HYDRAULIC OIL ANALYSIS	PRODUCTION EXCAVATOR. NORNAL RESULTS

FIGURE 9-30   Lab results after catastrophic pump failure in a production excavator.

Conclusion

ICP fluid analysis is a great tool to measure normal progressive wear, but it is blind to particles larger than 7 μm, which in this case were abundant. Often users expect oil analysis to be able to detect particles that cause catastrophic failures. This is seldom the case and usually not even feasible. One option for enhanced visibility of particles is through LaserNet Fines, which attempts to classify the type of wear based on the shape of the particles in particle counting. However, oil analysis is sometimes capable of catching small particle peaks that occur when a larger particle is causing abrasion. With luck, we can catch a catastrophic failure before it causes major failure, as in case 6 earlier in this chapter.

19. Life of Components versus Lubricant Choices

In a large and mixed equipment fleet, there are going to be lubrication requirements that are hard to comply with precisely because of difficulties in handling so many lubricants. The reality is that it is difficult to find space in lubrication trucks for all kinds of oils/fluids. Sometimes users opt for a compromise by using universal fluids to suit more than one brand, and sometimes these choices can have long-term detrimental effects in certain high-pressure applications. Case in point, a construction hydrostatic crawler was using an aggressive tractor fluid because the excavators in the fleet used the same fluid.

Oil Analysis

Although the copper production per hour was within the limits for the fleet, those limits were far from ideal, compared with applications using the right lubricant choice (see Figure 9-31).

METALS							ADDITIVES						
Iron (Fe)	Chromium Cr)	Lead (Pb)	Copper (Cu)	Tin (Sn)	Aluminum (Al)	Nickel (Ni)	Magnesium (Mg)	Calcium (Ca)	Barium Ba)	Phosphorus (P)	Zinc (Zn)	Molybdenum (Mo)	Boron (B)
8	4	12	187	4	3	0	142	3405	0	1210	1589	0	0
9	3	9	235	3	2	0	132	3361	0	1180	1550	0	0
3	2	11	190	2	2	0	140	**3400**	0	1200	1600	**0**	0
2	2	10	180	3	1	0	0	**3361**	0	1100	1270	**0**	0

CONTAMINANTS					WEAR	PHYSICAL PROPERTIES			INFORMATION			
Aluminum (Al)	Silicon (Si)	Sodium (Na)	Potassium (K)	Water (%) KF	PQ Index	Particle Counts	Viscosity (cSt @ 40C)	Oxidation	TAN	Fluid Changed	Fluid Type	Total Hours	Fluid Hours
3	8	5	0	0.06	NA	22/20/18	40.5	8	2	NO	THF	4000	1000
2	7	3	0	0.07	NA	22/19/17	38	12	2.5	YES	THF	3000	2000
2	6	4	0	0.07	NA	21/17/14	41,2	9	1.9	NO	THF	2000	1000
1	4	2	0	0.05	NA	20/16/13	41	8	1.8	YES	THF	1000	1000

HYDRAULIC OIL ANALYSIS	HYDROSTATIC CONSTRUCTION LOADER WITH THF ISO 46 FILL

FIGURE 9-31 Lab results for a construction hydrostatic crawler using an aggressive tractor fluid.

Still, the decision came from the user, and the example shows the impact of that decision on the longevity of hydrostatic components. Copper is not a time-dependent element like iron. However, in this case, copper readings were high because of an aggressive tractor fluid. Granted, a filter for a hydrostatic system is usually good to around 10 μm and 200 beta efficiency. This tells us that the filter was trapping some of the fine copper. Thus, actual copper production was >0.18 ppm per hour, which is a huge number.

The question is: How much metal is available to give away (as wear) before the hydrostatic transmission shows symptoms of lacking efficiency? Experience shows that at a copper production of 200 ppm every 1000 hours, there will be enough copper in pump and motor alloys to run for up to ~4000 hours of operation. Putting this in perspective, you can allow your machine to lose the volume of two Tylenol capsules of copper alloys before the system starts showing symptoms of poor efficiency (see Figure 9-32).

FIGURE 9-32   Tylenol capsules.

Figure 9-33 shows a rotary group from a hydrostatic crawler that suffered from this copper etching phenomenon caused by an aggressive fluid. As the figure shows, it does not take much to damage a hydrostatic component.

FIGURE 9-33   A rotary group from a hydrostatic crawler showing copper etching.

Wear Threshold Table

See Table 9-14.

TABLE 9-14 Wear Threshold Table

Construction Crawler 500 hours			
Filtered System	Normal	Abnormal	Critical
Fe	<21	21	>30
Pb	<7	7	>12
Cu	<21	21	>32
Cr	<5	5	>10
Al	<5	5	>10
Si	<17	17	>28
Particle Counts THF	≤22/17/14	22/17/14	≥X/20/17

It begs the question whether other alternatives for fluids exist for mixed fleets, and certainly they do.

20. Contamination Transfer from Scraper to Tractor in a Large Hydrostatic Crawler

A construction company uses large hydrostatic crawlers to pull scrapers for massive land filling operations. The combination crawler-scraper provides the contractor with the ideal combination in soil breaking capability and cost per cubic yard moved. The crawlers' hydraulics also operate the hydraulics on the scrapers, and the crawlers themselves use a unified hydraulic-hydrostatic system.

The fleet manager reported abnormal oil analysis readings with regard to fluid signature and the constant presence of dirt, which, in turn, elevated copper and iron readings. Let us see the oil analysis results.

Oil Analysis

The oil analysis report in Figure 9-34 shows the signature variation from the time the crawler was new and started operating the scraper. The highlighted signatures show the scraper's fluid contaminating the crawler's system, which produced an unrecognizable fluid signature. The signature issues were compounded as the tractor required top-offs and a fluid change at 1,000 hours. Fluid mixing caused particle count numbers to increase.

	METALS						ADDITIVES						
Iron (Fe)	Chromium Cr	Lead (Pb)	Copper (Cu)	Tin (Sn)	Aluminum (Al)	Nickel (Ni)	Magnesium (Mg)	Calcium (Ca)	Barium Ba	Phosphorus (P)	Zinc (Zn)	Molybdenum (Mo)	Boron (B)
65	5	7	45	0	10	<1	545	1690	0	897	867	145	167
57	4	5	38	0	13	<1	560	1700	0	905	880	150	169
23	2	4	32	0	4	<1	124	1175	0	803	875	6	14
10	>1	>1	3	0	2	<1	15	712	3	270	352	7	11

CONTAMINANTS					WEAR	PHYSICAL PROPERTIES			INFORMATION			
Aluminum (Al)	Silicon (Si)	Sodium (Na)	Potassium (K)	Water (%) KF	PQ Index	Particle Counts	Viscosity (cSt @ 40C)	Oxidation	TAN	Fluid Changed	Fluid Type	Total Hours	Fluid Hours
87	22	<5	<5	0.09	NA	22/20/18	34.1	NA	0.06	NO	10W30	2000	2000
48	15	<5	<5	0.08	NA	21/19/17	59	NA	0.07	YES	10W30	1000	1000
40	12	<5	<5	0.07	NA	21/17/15	43	NA	0.08	NO	HVI46	500	500
3	2	<5	<5	0.04	NA	18/16/13	50	NA	0.07	NO	HVI46	10	10

HYDRAULIC OIL ANALYSIS	LARGE PULLER CRAWLER WITH CAT SCRAPER. HVI AW46/10W30 AND CAT TO4.

FIGURE 9-34   Lab report showing the signature variation for a large crawler from when it was new and after it started operating a scraper.

Conclusion

The real issue here is the introduction of dirt through the scraper's cylinders. The tractor has a combined hydraulic-hydrostatic system, and the scraper adds three additional points of dirt entry that need attention. The major concern is copper generation because it represents the life of high-pressure hydrostatic components.

21. Increased Water Readings after Fluid Change in a Production Excavator

As part of company maintenance practices, the user of a large hydraulic excavator drained fluid from the bottom of the hydraulic reservoir twice a month to eliminate possible water condensation. The company does this in the morning before starting the machine. This practice has allowed the machine to show very low water readings via oil analysis. However, after changing from a calcium-free fluid to a calcium-based hydraulic fluid, water readings have increased in the oil analysis results.

Oil Analysis

The highlighted figures in the oil analysis in Figure 9-35 show the change to a calcium-based fluid at 1,500 hours and the impact on the fluid signature, viscosity, and TAN. Because only 45 of the 145 total gallons are new, the new signatures agree with these figures.

METALS							ADDITIVES						
Iron (Fe)	Chromium Cr	Lead (Pb)	Copper (Cu)	Tin (Sn)	Aluminum (Al)	Nickel (Ni)	Magnesium (Mg)	Calcium (Ca)	Barium Ba	Phosphorus (P)	Zinc (Zn)	Molybdenum (Mo)	Boron (B)
11	2	2	9	0	3	<1	147	1405	0	935	980	0	<1
16	2	1	11	0	6	<1	150	1430	0	950	975	0	<1
15	3	2	10	0	7	<1	14	26	0	233	12	0	<1
10	1	1	8	0	3	<1	15	23	0	245	4	0	<1

CONTAMINANTS					WEAR		PHYSICAL PROPERTIES			INFORMATION			
Aluminum (Al)	Silicon (Si)	Sodium (Na)	Potassium (K)	Water (%) KF	PQ Index	Particle Counts	Viscosity (cSt @ 40C)	Oxidation	TAN	Fluid Changed	Fluid Type	Total Hours	Fluid Hours
3	12	<5	<5	0.12	NA	22/19/17	53.5	NA	1.5	NO	10W30	2500	1000
6	20	6	7	0.11	NA	21/19/17	54.8	NA	1.01	YES	10W30	1500	0
7	18	<5	<5	0.06	NA	18/17/14	46.1	NA	0.08	NO	ZF46	1000	1000
3	10	<5	<5	0.07	NA	18/16/13	46.2	NA	0.07	NO	ZF46	50	500

HYDRAULIC OIL ANALYSIS	LARGE PRODUCTION EXCAVATOR. NO WATER FROM DARIN VALVE. CALCIUM FREE FLUID INITIAL FILL. ENGINE OIL 10W-30 AT FIRST CHANGE

FIGURE 9-35 Lab results showing the change to a calcium-based fluid at 1,500 hours and the impact on the fluid signature, viscosity, and TAN of a production excavator.

Water content then went to 0.12%, and the lab highlighted the last two readings. Particle counts are also in red. Aside from these changes, the machine appears to be in good health. Still, the water content needs an explanation.

When changing to a calcium-based fluid, any free water will be suspended, making its removal through the reservoir purging valve impractical. Therefore, the use of calcium-based fluids, even if it is only the reservoir volume, will cause water to be suspended and the readings in oil analysis to increase. These readings are visible to the Karl Fischer ASTM D6304 test but will not be visible to FTIR spectroscopy if the water content is <1600 PPM. With regard to particle counts, it is normal for a mixed fluid to have abnormal particle counts, as discussed in case 4 earlier in this chapter.

Conclusion

When dealing with water thresholds for hydraulic fluids, it is good to remember some guidelines for the correct interpretation of results. Table 9-15 provides some guidance. In a mixed product, it is up to the user to choose a happy medium to establish the limits based on fluid mix percentages.

TABLE 9-15 Water Content Guidelines

Type of fluid	Normal	Abnormal	Critical
AW fluid	<750	751	>1001
Universal tractor fluid	<1000	1001	>1501
Engine oil	<1000	1001	>1501
Biodegradable fluid	<500	501	>751

CONCLUSION

After discussing 21 hydraulic analysis cases and approaches to solve them, it is time to discuss engine oil analysis, which uses a different report format and tests and imposes new challenges in interpretation.

CHAPTER 10

ENGINE OIL ANALYSIS

Diesel engine manufacturers are on a never-ending quest for greater power density to provide better productivity while maintaining durability and at the same time improving fuel efficiency and lowering exhaust emissions. Modern engines are the very impressive result of this search, but with the changes come additional complications and costs (see Figure 10-1).

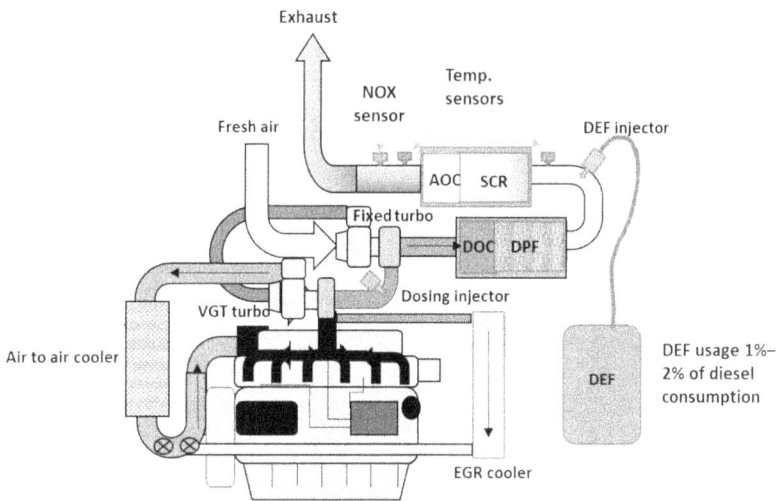

FIGURE 10-1 Tier 4 engine.

These modern marvels are less forgiving of poor maintenance practices than their predecessors were. They do not tolerate restriction from a dirty air filter or off-spec fuel as well as earlier engines. Oil analysis helps get the most out of all engines, but the cost benefit is higher in Tier 3 and 4 engines than in earlier engines. This chapter will point out the primary contaminants and metal generators in engines and help identify the root causes of failure.

Oil analysis is a great tool to monitor engine behavior and maintenance quality, but it does not work alone. It also needs coolant and fuel analysis to close the circle around engine reliability.

CASE DISCUSSIONS

1. Four-Wheel-Drive (4WD) Loader with Coolant Leak Through the Oil Cooler

The oil analysis results in Figure 10-2 are for a diesel engine with all the indications of a coolant leak. The potassium and sodium readings confirm the presence of coolant. The lead readings are also high, indicating bearing wear. The copper readings confirm that the leak could be happening through the oil cooler. A coolant analysis will reconfirm the suspicion by also showing high copper readings.

METALS							ADDITIVES							HOURS INFO	
Iron (Fe)	Chromium Cr	Lead (Pb)	Copper (Cu)	Tin (Sn)	Aluminum (Al)	Nickel (Ni)	Magnesium (Mg)	Calcium (Ca)	Barium Ba)	Phosphorus (P)	Zinc (Zn)	Molybdenum (Mo)	Boron (B)	Total Hours	Hours on Oil
82	1	12	32	2	<1	<1	18	2250	<1	1215	1123	90	31	2500	250
7	<1	1	29	<1	2	<1	20	2320	<1	1190	1685	107	132	2250	500

CONTAMINANTS							PHYSICAL PROPERTIES							OIL INFO	
Aluminum (Al)	Silicon (Si)	Sodium (Na)	Potassium (K)	Water (%)	Glycol	Fuel (%)	Soot (%) FTR	Sulphation	Nitration	Viscosity (cSt @ 100 C)	Oxidation	TBN	TAN	Type	Viscosity
<1	7	445	255	<0.05	YES	<1	0.5	8	6	15.3	10	6	NA	CJ4	10W30
2	10	172	54	<0.05	YES	<1	0.5	8	5	9.4	9	8	NA	CJ4	10W30

ENGINE OIL ANALYSIS	4WD Loader diesel engine. High sodium and potassium. Coolant present

FIGURE 10-2 Oil analysis results for a 4WD loader diesel engine.

Oil Analysis

In addition, an important point to notice is the impact of a massive coolant leak on the viscosity of the oil. The viscosity of the 10W-30 oil should not be higher than 10.5 cSt at 250 hours.

Conclusion

A coolant test on the same machine confirmed that there is a leak through the oil cooler represented by the copper readings (see Figure 10-3).

PHYSICAL/CHEMICAL										INFORMATION		
Freeze Point D3321	Glycol Content	pH	Reserve Alkalinity	Nitrites D5827	Nitrates	Molybdate	Silicates	OA UPLC	Sodium	Coolant Type	Engine Hours	Coolant Hours
°F	%	D1287	HCl	ppm	ppm	ppm	ppm	%	ppm			
-52	57	9	5.2	10	100	190	436	1.7	3000	OAT	2,500	2500
CORROSION METALS				WATER HARDNESS					VISUAL APPEARANCE			
Pb	Fe	Al	Cu	Ca	Chlorides	Sulfates	Total Hardness	TDS	Color	Clarity	Oil Layer	Sediment
ppm	ppm	ppm	ppm	ppm	ppm	ppm	ppm	ppm				
13	4	1	137	<1	45	35	50	45	Red	Opaque	Some	NO
COOLANT ANALYSIS	4WD Loader. Excess copper present.											

FIGURE 10-3 Coolant test results.

2. Backhoe Loader Coolant Leak Through Liner Pitting

Oil Analysis

When the leak occurs through the pitting of a cylinder liner, the high iron reading will confirm the coolant leak through liner cavitation, as the flowchart in Figure 7-6 suggests. Copper readings will not necessarily be abnormal. However, the coolant leak into the cylinder will wash out liner lubrication, producing copious amounts of iron, chromium from piston rings, and aluminum from pistons (see Figure 10-4).

METALS							ADDITIVES							HOURS INFO	
Iron (Fe)	Chromium (Cr)	Lead (Pb)	Copper (Cu)	Tin (Sn)	Aluminum (Al)	Nickel (Ni)	Magnesium (Mg)	Calcium (Ca)	Barium (Ba)	Phosphorus (P)	Zinc (Zn)	Molybdenum (Mo)	Boron (B)	Total Hours	Hours on Oil
172	8	5	13	<1	25	<1	28	2250	<1	1495	1250	132	52	3500	500
8	<1	1	12	<1	2	<1	35	2300	<1	1502	1287	120	45	3000	500
CONTAMINANTS								PHYSICAL PROPERTIES						OIL INFO	
Aluminum (Al)	Silicon (Si)	Sodium (Na)	Potassium (K)	Water (%)	Glycol	Fuel (%)	Soot (%) FTR	Sulphation	Nitration	Viscosity (cSt @ 100 C)	Oxidation	TBN	TAN	Type	Viscosity
25	20	123	85	<0.05	YES	<1	0.5	NA	NA	15.7	NA	4.5	NA	CJ4	15W40
2	10	10	5	<0.05	NO	<1	0.5	NA	NA	15.5	NA	7	NA	CJ4	15W40
ENGINE OIL ANALYSIS	Backhoe diesel engine. 7500 total hours. 500 hours on oil. 15W40 oil. Coolant present														

FIGURE 10-4 Oil analysis results for a backhoe diesel engine.

Conclusion

Figure 10-5 shows the failed cylinder liner. The results of the coolant test in Figure 10-6 show that the coolant is ill prepared to protect a heavy-duty diesel engine because it is lacking critical anticavitation additives like nitrites and organic acids. Chapter 12 discusses these details.

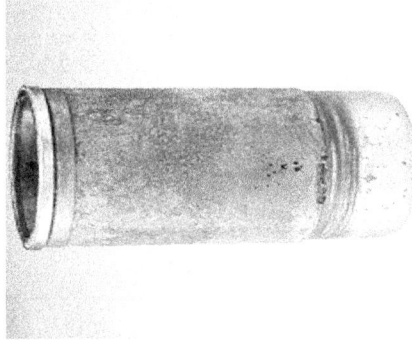

FIGURE 10-5   Failed cylinder liner.

PHYSICAL/CHEMICAL										INFORMATION		
Freeze Point D3321	Glycol Content	pH	Reserve Alkalinity	Nitrites D5827	Nitrates	Molybdate	Silicates	OA UPLC	Sodium	Coolant Type	Engine Hours	Coolant Hours
°F	%	D1287	HCl	ppm	ppm	ppm	ppm	%	ppm			
-40	52	8.1	2.5	62	477	230	266	1.2	3050	OAT	7,500	3000
CORROSION METALS				WATER HARDNESS					VISUAL APPEARANCE			
Pb	Fe	Al	Cu	Ca	Chlorides	Sulfates	Total Hardness	TDS	Color	Clarity	Oil Layer	Sediment
ppm	ppm	ppm	ppm	ppm	ppm	ppm	ppm	ppm				
10	87	9	24	<1	50	35	50	40	Gree	Opaque	NO	NO
COOLANT ANALYSIS	Backhoe Loader. Excess iron present.											

FIGURE 10-6   Coolant test results.

3. Excavator, Coolant Leak through the Exhaust Gas Recirculator (EGR)

Oil Analysis

A coolant leak through the EGR also can be detected through high iron readings without significant coolant constituent readings. This happens because the leak occurs in the combustion chamber and affects the cylinder's lubrication. Something very similar happens when the leak occurs through a cylinder head gasket. The high iron readings without significant coolant or dirt are an indication that the leak could be occurring at the top of the engine through the EGR or cylinder head gasket (see Figure 10-7).

METALS							ADDITIVES							HOURS INFO	
Iron (Fe)	Chromium Cr	Lead (Pb)	Copper (Cu)	Tin (Sn)	Aluminum (Al)	Nickel (Ni)	Magnesium (Mg)	Calcium (Ca)	Barium Ba)	Phosphorus (P)	Zinc (Zn)	Molybdenum (Mo)	Boron (B)	Total Hours	Hours on Oil
420	22	2	25	9	60	24	811	1410	<1	1019	1279	105	118	3750	250
25	2	1	15	<1	2	<1	822	1500	<1	1025	1305	98	125	3500	500

CONTAMINANTS								PHYSICAL PROPERTIES						OIL INFO	
Aluminum (Al)	Silicon (Si)	Sodium (Na)	Potassium (K)	Water (%)	Glycol	Fuel (%)	Soot (%) FTR	Sulphation	Nitration	Viscosity (cSt @ 100 C)	Oxidation	TBN	TAN	Type	Viscosity
60	35	22	40	<0.05	NO	<1	0.6	NA	NA	14.2	NA	6.3	NA	CK4	15W40
2	10	<5	<5	<0.05	NO	<1	0.5	8	5	9.4	9	8	NA	CK4	15W40

ENGINE OIL ANALYSIS | Excavator with diesel engine. High iron and nickel

FIGURE 10-7 Oil analysis results for an excavator diesel engine.

Other metals could appear because of lubrication wash. The aluminum readings come from the pistons, whereas the nickel readings very possibly come from valve guides. The ratio of silicon to aluminum at 0.58 does not suggest that it is dirt, and the lead reading from bearings is very low.

Conclusion

Because the symptoms are a little more subtle than for other coolant contamination failures, knowing the frequent need to top-up the coolant would help to confirm the diagnosis of a coolant leak.

Water

Water is always present in a diesel engine sump because water is a by-product of combustion. However, water needs to be kept within the limits the engine and oil can handle. Because coolant is typically a mixture of 50% water and 50% glycol, any leaks of coolant into the engine represent a water leak as well. Water evaporates more easily, though, and glycol stays.

Water can be damaging to an engine, causing corrosion and shortening the oil's oxidative life. It also can combine with soot to tie up detergent and dispersant additives and drag them out of the oil as sludge.

Condensation is another source of moisture, which can be significant during storage where big temperature swings occur or humidity is high. Water from condensation and the water squeezed out of the air during compression strokes can build up in the crankcase if there is insufficient heat to drive evaporation. Insufficient heat is usually the result of a malfunctioning thermostat or a short duty cycle.

A massive coolant leak will kill an engine. Fortunately, most leaks start slowly, and oil analysis often can detect a coolant leak before it causes damage.

4. 4WD Loader with Water Contamination

Oil Analysis

The case involves a 4WD loader that had gross water contamination. Therefore, the viscosity could not be measured nor other physical properties with the exception of total base number (TBN). This case shows how water can trigger the production of critical metals such as lead and tin (see Figure 10-8).

METALS							ADDITIVES							HOURS INFO	
Iron (Fe)	Chromium Cr)	Lead (Pb)	Copper (Cu)	Tin (Sn)	Aluminum (Al)	Nickel (Ni)	Magnesium (Mg)	Calcium (Ca)	Barium Ba)	Phosphorus (P)	Zinc (Zn)	Molybdenum (Mo)	Boron (B)	Total Hours	Hours on Oil
144	1	114	51	2	7	3	765	1195	<1	1121	1375	61	11	3500	500
														3000	500

CONTAMINANTS							PHYSICAL PROPERTIES					OIL INFO	
Aluminum (Al)	Silicon (Si)	Sodium (Na)	Potassium (K)	Water (%)	Glycol	Fuel (%)	Soot (%) FTR	Sulphation	Nitration	Viscosity (cSt @ 100 C)	Oxidation	TBN	TAN	Type	Viscosity
7	12	40	9	1.03	NO	NA	0.4	NA	NA	NA	NA	5.2	NA	CJ4	15W40
														CJ4	15W40

ENGINE OIL ANALYSIS	4WD Loader with diesel engine. Water present. Viscosity measurement not feasible

FIGURE 10-8   Oil analysis results for a 4WD loader with water present.

Conclusion

High water content with no trace of glycol suggests that this engine suffered a water-entry accident. Whether this comes from operating conditions or improper storage, it could represent a serious issue for piston ring corrosion and/or damage to bearings. Lead readings are high and represent serious damage to bearings. However, because tin readings are low, the bearings are potentially salvageable.

5. 4WD Loader with High Soot

This case involves an engine that has been showing increasing amounts of soot moving toward a risky level. The viscosity of the 15W-40 oil at 500 hours should be about 17.5 cSt maximum, but this engine is showing a viscosity of 36.4 cSt. At this 200% viscosity increase, the bearings and turbocharger are at risk (see Figure 10-9).

METALS							ADDITIVES							HOURS INFO	
Iron (Fe)	Chromium Cr	Lead (Pb)	Copper (Cu)	Tin (Sn)	Aluminum (Al)	Nickel (Ni)	Magnesium (Mg)	Calcium (Ca)	Barium (Ba)	Phosphorus (P)	Zinc (Zn)	Molybdenum (Mo)	Boron (B)	Total Hours	Hours on Oil
88	4	11	2	1	10	3	134	891	<1	946	805	94	88	4500	500
66	1	<1	1	<1	9	2	74	3659	<1	1323	1648	134	121	4000	500

CONTAMINANTS							PHYSICAL PROPERTIES							OIL INFO	
Aluminum (Al)	Silicon (Si)	Sodium (Na)	Potassium (K)	Water (%)	Glycol	Fuel (%)	Soot (%) FTR	Sulphation	Nitration	Viscosity (cSt @ 100 C)	Oxidation	TBN	TAN	Type	Viscosity
10	4	4	71	<0.05	NO	<1	10	10	8	36.4	12	2.4	NA	CJ4	15W40
9	7	5	33	<0.05	NO	<1	0.7	10	8	19	11	6.3	NA	CI4	15W40

ENGINE OIL ANALYSIS	4WD Loader with diesel engine. High soot and high viscosity

FIGURE 10-9 Oil analysis results for a 4WD loader with high soot and high viscosity.

Conclusion

High levels of soot interfere with oil films and antiwear (AW) additives, which could explain the increased iron level. Another point of interest in this case is the TBN and type of oil. The TBN at 2.4 is probably below the total acid number (TAN), which makes this engine more prone to internal corrosion, as indicated by the lead reading. The switch from a CI4 oil to a CJ4 oil also suggests that in this case the more recent oil is less capable of handling soot.

Soot

Soot is the result of incomplete combustion that turns diesel engine oil black and makes exhaust smoke visible. It is often the limiting factor in oil drain life.

Soot particles in engine oil have a tendency to agglomerate, or stick together like snowballs, and dispersants in oil aid in fighting this tendency. When soot builds up in an oil to the point that it overwhelms the dispersant additives, soot particles combine with each other and with resins from oil degradation to make deposits. Viscosity increases, the soot interferes with AW additives, and the useful life of the oil is over.

Again, we need to know the machine and the environment. It could be that this level of soot is normal given the hours on the machine. Starting with Tier 1–3 engines, timing retardation became the norm to reduce the level of nitrogen oxide emissions. Although effective at reducing nitrogen oxides, the retarded timing and exhaust gas recirculation created more soot, which sometimes affects the level of soot in the crankcase. Incomplete combustion also can be due to an inadequate supply of air. High altitude, air filter blockage, or inadequate turbo boost limits the amount of air entering the cylinder. Light loads or prolonged idling often prevents the engine from burning fuel efficiently. In addition, engines designed to run stoichiometrically, where fuel and air are at the precise ratio where both are consumed completely, do not always respond quickly enough to changes in throttle input.

Finally, poor injector performance could be the issue. A squirt or drip of fuel does not burn as completely as a fine mist of fuel. Modern engines inject fuel into the cylinder up to seven to ten times per combustion cycle. It does not take much of a deposit to interfere with the timing of these microbursts of fuel and affect combustion efficiency. Figure 10-10 is an example of the impact of high pressure and high temperature on injector life. The figure shows the needle with carbon buildup and erosion.

FIGURE 10-10   Injector needle showing carbon buildup and erosion.

If the soot level in the engine oil builds at a faster pace than in the past, it is worth asking the operator if there has been a loss of power or performance because it could be a mechanical problem.

If soot is the limiting factor in a healthy engine's oil change interval, that interval can be prolonged by adding a bypass filter. Bypass filtration takes a small portion of the oil stream and filters it very finely—down to 2 μm or so. At this level of filtration, the filter can effectively remove soot, along with wear metals and other particles larger than 2 μm to allow a significantly longer drain interval.

Viscosity Increase/Decrease

Viscosity measurement in an engine is the quickest way to understand whether something is not running well within that engine. Viscosity behaves in a pattern that is easy to trace in engines and helps in monitoring engine health or utilization. Every engine shows its own pattern from which we can learn, and the pattern is based on application and emissions and tier type.

Some engines tend to decrease oil viscosity during the first 125 hours as a normal sheardown of the oil additive molecules occurs, but then viscosity bounces back as soot increases. At around 500 hours, viscosity reaches a level similar to that when the oil was new. Other engines do not show any decrease in viscosity but rather a continued increase due to soot accumulation. Moreover, there will be engines that will dilute oil from the beginning of the oil change, thinning the oil down, which also needs monitoring. All these changes in viscosity are the result of engine design, application, and altitude or could be an indication of a lack of air, an internal fuel leak, or specific machine utilization mode. A large coolant leak could cause a gross viscosity increase.

Table 10-1 is an example of a guideline developed around five types of oil viscosities used in Tier 1 and 2 engines. The numbers in bold show the "magic numbers" for shear-down at 125 hours. If a given engine is off with regard to the magic numbers at 125 hours, it is telling us that something is not running well. It is also true with numbers at around 500 hours. A chart like this is useful in combination with lab results to confirm fuel dilution or high soot levels, but again, the table needs to be for the specific engine and tier type.

TABLE 10-1   Guidelines around Five Types of Viscosities in Tier 1 and 2 Engines

Oils	New	50 h	125 h	250 h	350 h	500 h	550 h
5W-30	10.8–10.4	10.3–9.4	9.3–**8.4**	9.3–10.4	10.3–11.4	11.3–12.4	12.3–13.4
10W-30	11.0–10.7	10.5–9.5	9.5–**8.5**	9.50–10.5	10.5–11.5	11.5–12.5	12.5–13.5
10W-40	14.6–14.0	14.1–13.1	13.1–**12.1**	13.1–14.1	14.1–15.1	15.1–16.1	16.1–17.1
15W-40	16.0–15.1	14.7–13.2	13.7–**12.5**	13.5–14.7	14.5–15.7	15.5–16.7	16.5–17.5
0W-40	15.8–15.2	14.2–13.5	13.2–**12.2**	13.0–14.2	13.5–14.5	14.2–15.2	15.2–16.2

6. 4WD Loader with Fuel Dilution

Oil Analysis

The oil analysis results in Figure 10-11 show fuel dilution.

METALS							ADDITIVES							HOURS INFO	
Iron (Fe)	Chromium Cr	Lead (Pb)	Copper (Cu)	Tin (Sn)	Aluminum (Al)	Nickel (Ni)	Magnesium (Mg)	Calcium (Ca)	Barium Ba)	Phosphorus (P)	Zinc (Zn)	Molybdenum (Mo)	Boron (B)	Total Hours	Hours on Oil
14	<1	<1	<1	1	2	<1	890	450	<5	1110	1398	43	<5	3500	500
3	<1	<1	<1	<1	<1	<1	912	476	<5	1128	1297	60	<5	3000	500

CONTAMINANTS								PHYSICAL PROPERTIES					OIL INFO	
Aluminum (Al)	Silicon (Si)	Sodium (Na)	Potassium (K)	Water (%)	Glycol	Fuel (%)	Soot (%) FTR	Sulphation	Nitration	Viscosity (cSt @ 100 C)	Oxidation	TBN	TAN	Type	Viscosity
2	3	2	<5	<0.05	NO	3	0.5	9	7	12.7	8	7.2	NA	CJ4	15W40
<1	2	4	<5	<0.05	NO	<1	3	8	6	14.4	7	8.7	NA	CJ4	15W40

ENGINE OIL ANALYSIS   4WD Loader with diesel engine. Fuel dilution

FIGURE 10-11   Oil analysis results showing fuel dilution in a 4WD loader.

A fuel dilution of 3% brings down the viscosity of the 15W-40 oil to that of a 10W-30 oil in about 200 hours. Although the engine is not showing any distress because of this fuel dilution, it is something to keep an eye on because in another 200 hours the viscosity could be very low.

Conclusion

Modern engines cannot completely avoid fuel dilution, and the guidelines in Table 10-2 represent this new reality. Still, fuel dilution needs monitoring to prevent the viscosity from falling to dangerous levels that could compromise lubrication.

TABLE 10-13 New Accepted Levels of Fuel Dilution for Modern Engines (Tier 3 and 4) Using Ultra-Low-Sulfur Diesel (ULSD) Fuel

Fuel Dilution Limits Tier 3 and Tier 4 Engines		
Normal	Abnormal	Critical
4%–5%	>5%–7%	>7%

Fuel Dilution Discussion

Fuel dilution worsens when the application requires continuous engine speed variations or long idling periods. Fuel dilution is even more of a challenge in engines with high-pressure common-rail injection systems. Even though the injectors have extremely tight tolerances, they are trying to hold back fuel pressures that sometimes exceed 25,000 lb/in². Some fuel forces its way through, even though the injector is closed. This postinjection leakage finds its way into the crankcase.

At normal revolutions per minute (rpms), a relatively small portion of errant fuel is consumed without consequence. At idle or low rpms, it is longer between combustion cycles, and thus there is more time for fuel to sneak past the closed injectors. Some of the raw fuel leaks into the crevice above the top ring and then makes its way into the crankcase as a liquid.

Fouled injectors, overfueling, inadequate air (e.g., high altitude, restricted air filter, issues with a turbocharger), or improper injection timing also can cause fuel dilution. However, fuel dilution from incomplete combustion is somewhat different in nature than postinjection leakage or other sources of unburned fuel.

The unburned fuel from combustion is exposed to tremendous heat that boils off the light and medium components in the fuel. New compounds are formed by thermal cracking and are added to the mix. Then the thermally degraded fuel enters the crankcase with a higher flashpoint and viscosity than regular diesel fuel.

System leakage is not exposed to combustion temperatures and enters the crankcase with properties that are similar to new fuel. Over time, the lighter components of the oil-fuel mixture vaporize as they touch the hot surfaces.

Oils resist oxidation, but diesel fuel must oxidize (burn) easily. Therefore, in a mixture, the oil and diesel fuel are at odds with each other. The instability of the fuel tends to catalyze oxidation of the oil, producing insoluble substances that promote deposits and varnish.

When using more than nominal levels of biodiesel fuel in biodiesel blends, fuel dilution will be more persistent. Biodiesel is much less volatile than diesel fuel, so it takes longer for the heat in the crankcase to drive it off. Therefore, it has a greater tendency to build up in the crankcase.

Biodiesel fuel as a crankcase contaminant reacts differently to thermal degradation than regular diesel fuel. It has a greater tendency to form insoluble compounds, which can cause it to increase its viscosity.

We need to read the oil analysis differently with high biodiesel blends compared with straight diesel fuel. A drop in flashpoint will not be as strong of an indicator as with regular diesel fuel. Biodiesel does not produce as severe of a drop in oil viscosity, and the increased insoluble compounds actually may push the viscosity up.

7. Backhoe with Oil Thickening Due to Biodiesel

Oil Analysis

This case is an example of biodiesel oil thickening (see Figure 10-12).

METALS							ADDITIVES							HOURS INFO	
Iron (Fe)	Chromium (Cr)	Lead (Pb)	Copper (Cu)	Tin (Sn)	Aluminum (Al)	Nickel (Ni)	Magnesium (Mg)	Calcium (Ca)	Barium (Ba)	Phosphorus (P)	Zinc (Zn)	Molybdenum (Mo)	Boron (B)	Total Hours	Hours on Oil
19	<1	<1	1	<1	10	<1	910	410	<1	1010	1395	55	5	1245	260
16	<1	<1	<1	<1	8	<1	925	425	<1	1050	1425	58	5	995	250
21	<1	<	<1	<1	10	<1	950	480	<1	1100	1430	65	6	745	250

CONTAMINANTS										PHYSICAL PROPERTIES				OIL INFO	
Aluminum (Al)	Silicon (Si)	Sodium (Na)	Potassium (K)	Water (%)	Glycol	Fuel (%)	Soot (%) FTR	Sulphation	Nitration	Viscosity (cSt @ 100 C)	Oxidation	TBN	TAN	Type	Viscosity
10	6	3	<5	<0.5	NO	<1	0.6	NA	NA	15.6	NA	3.9	NA	CJ4	15W40
8	4	1	<5	<0.5	NO	<1	0.8	NA	NA	15.7	NA	4.2	NA	CJ4	15W40
10	5	8	<5	<0.5	NO	<1	1	NA	NA	15.6	NA	4.5	NA	CJ4	15W40

ENGINE OIL ANALYSIS	Backhoe Loader with diesel engine. Viscosity high for the hours. Low TBN

FIGURE 10-12 Oil analysis results showing biodiesel oil thickening in a backhoe.

Although most engines can run different ratios of biodiesel fuel blends, this sometimes may come at the expense of oil thickening and corrosion. Most manufacturers have no issue with up to 20% biodiesel blends (B20), and in certain applications, up to 100% biodiesel (B100) use is permitted depending on materials, the Tier, and exhaust aftertreatment. If the engine emissions Tier is such that it tends to dilute oil with fuel, it could be a candidate for biodiesel oil thickening. Biodiesel tends to cling to oil molecules and does not evaporate easily. Biodiesel could be corrosive to bearings because its natural TAN ranges between 1 and 3.5. Viscosity also increases due to polymerization of fuel molecules.

Conclusion

Although the results show very good wear metal readings, the TBN is abnormal for only 260 hours on the oil, and the viscosity is already at the level where it should be at around 500 hours. It will be only a matter of time before the TAN exceeds the TBN, causing accelerated metal generation. The results help in tailoring the ideal oil change for this machine.

8. Bulldozer with Low TBN and High TAN

Oil Analysis

This case involves a diesel engine experiencing high iron and lead production (see Figure 10-13).

METALS							ADDITIVES							HOURS INFO	
Iron (Fe)	Chromium Cr	Lead (Pb)	Copper (Cu)	Tin (Sn)	Aluminum (Al)	Nickel (Ni)	Magnesium (Mg)	Calcium (Ca)	Barium Ba	Phosphorus (P)	Zinc (Zn)	Molybdenum (Mo)	Boron (B)	Total Hours	Hours on Oil
327	5	144	15	1	15	6	18	3923	0	1400	1550	110	160	5500	500

CONTAMINANTS										PHYSICAL PROPERTIES				OIL INFO	
Aluminum (Al)	Silicon (Si)	Sodium (Na)	Potassium (K)	Water (%)	Glycol	Fuel (%)	Soot (%) FTR	Sulphation	Nitration	Viscosity (cSt @ 100 C)	Oxidation	TBN	TAN	Type	Viscosity
15	8	14	0	<0.5	NO	<1	1.71	15	NA		NA	5.06	7.18	C/4 Syn	0W40

ENGINE OIL ANALYSIS Bulldozer with diesel engine. High TAN low TBN

FIGURE 10-13 Oil analysis results showing high iron and lead production in a bulldozer.

The engine is using an old high-calcium oil formulation. There is no water, coolant, or dirt to blame for this other than the high TAN, which is higher than the TBN. Possibly this engine was using high-sulfur fuel.

Conclusion

At present, the ideal oil analysis needs to have both TBN and TAN measurements to be able to establish the real acidity of the oil with hours of use, especially in the era of low-TBN oils and ULSD fuels.

Alkalinity Reserve and TAN

Total base number (TBN) is a measurement of basicity, which is expressed as the number of milligrams of potassium hydroxide having the equivalent amount of alkalinity as 1 g of the sample oil (mg KOH/g). The alkaline reserve indicated by the TBN is a very useful test for engines because it tells how much acid-neutralizing additive remains. With the advent of ULSD fuels, new oil formulations have a reduced detergent content to protect exhaust filters and catalyzers and to limit emissions of these metallic additives. Old formulations had TBNs of 12 and marine engine oils could have TBNs of 60 to 100 or more. Current formulations have much lower TBNs, in the range of 5. What is important these days is not to let this number fall below the TAN. When the TBN crosses below the TAN, copious amounts of metals can be produced, especially iron and lead.

The total acid number (TAN), in contrast, is a measurement of acidity that is equal to the number of milligrams of potassium hydroxide needed to neutralize the acids in 1 g of oil. It is still not customary to measure the TAN in engines, but it is highly recommended.

9. 4WD Loader with Copper Passivation

Oil Analysis

This case involves a 4WD loader engine showing an oil change and testing sequence every 500 hours where a change of lubricant occurred. Similar spikes in copper can happen when changing oil brands and formulations. In this case, the change was from an American Petroleum Institute (API) CI4 to CJ4 performance level and may have occurred because the manufacturer upgraded its product rather than from a change in brands (see Figure 10-14).

METALS							ADDITIVES							HOURS INFO	
Iron (Fe)	Chromium (Cr)	Lead (Pb)	Copper (Cu)	Tin (Sn)	Aluminum (Al)	Nickel (Ni)	Magnesium (Mg)	Calcium (Ca)	Barium (Ba)	Phosphorus (P)	Zinc (Zn)	Molybdenum (Mo)	Boron (B)	Total Hours	Hours on Oil
90	1	3	89	1	5	6	858	380	1	1350	1220	45	<2	2500	720
75	3	4	194	1	8	1	988	391	1	1410	1250	50	1	1780	500
87	2	4	28	2	7	1	21	3005	1	1110	1260	95	126	1280	500
24	3	15	15	1	5	3	24	3150	<1	1080	1255	92	124	780	500

CONTAMINANTS							PHYSICAL PROPERTIES					OIL INFO	
Aluminum (Al)	Silicon (Si)	Sodium (Na)	Potassium (K)	Water (%)	Glycol	Fuel (%)	Soot (%) FTIR	Sulphation	Nitration	Viscosity (cSt @ 100 C)	Oxidation	TBN	TAN	Type	Viscosity
5	7	6	5	<0.5	NO	<1	0.3	15	NA	13.6	NA	3.9	NA	CJ4	10W30
8	11	11	1	<0.5	NO	<1	0.7	14	NA	13.2	NA	3.8	NA	CJ4	10W30
7	12	17	7	<0.5	NO	1	0.5	10	NA	15.5	NA	6.1	NA	CI4	15W40
4	10	15	5	<0.5	NO	<1	0.4	9	NA	15.2	NA	6	NA	CI4	15W40

ENGINE OIL ANALYSIS	4WD Loader with diesel engine. Copper high

FIGURE 10-14   Oil analysis results showing copper passivation in a 4WD loader.

The change to a different type/brand of lubricant triggered oil cooler copper passivation readings, as highlighted in the figure. The last readings are getting back to normal. If the oil in service does not change in formulation, the copper readings will normalize.

Conclusion

The new oil also has a lower TBN for the service interval, which suggests that a TAN test also should be included to determine whether the oil in service is becoming acidic. Although the copper readings look alarming, there are no other indicators showing stress in this engine. As explained earlier, overproduction of copper sometimes makes it difficult to assess other impending issues that could be occurring, especially if other critical metals such as lead and tin are present.

10. Articulated Dump Truck (ADT) with Aluminum Readings from Accessory Compressor

Oil Analysis

The ADT in this example shows high aluminum in the engine oil sample (see Figure 10-15). Sometimes the metal readings are not directly from the engine itself but rather from accessories used by the engine that share lubrication with it. This case is an example of this situation. The general readings are perfectly fine, with the exception of abnormal amounts of aluminum and potassium. Both are coming from the compressor.

METALS							ADDITIVES							HOURS INFO	
Iron (Fe)	Chromium Cr	Lead (Pb)	Copper (Cu)	Tin (Sn)	Aluminum (Al)	Nickel (Ni)	Magnesium (Mg)	Calcium (Ca)	Barium Ba)	Phosphorus (P)	Zinc (Zn)	Molybdenum (Mo)	Boron (B)	Total Hours	Hours on Oil
19	<1	<1	<1	<1	28	<1	15	1690	0	1399	1657	89	2	1245	350
20	<1	2	<1	<1	15	<1	16	1650	0	1385	1705	91	<2	895	250
28	<1	1	2	<1	35	<1	18	1770	0	1279	1520	83	2	645	250

CONTAMINANTS										PHYSICAL PROPERTIES				OIL INFO	
Aluminum (Al)	Silicon (Si)	Sodium (Na)	Potassium (K)	Water (%)	Glycol	Fuel (%)	Soot (%) FTR	Sulphation	Nitration	Viscosity (cSt @ 100 C)	Oxidation	TBN	TAN	Type	Viscosity
28	2	8	75	<0.5	NO	<1	0.3	1	1	14.4	3	7.9	NA	CJ4	15W40
15	7	9	50	<0.5	NO	<1	0.2	1	1	14.7	3	8.2	NA	CJ4	15W40
35	1	9	220	<0.5	NO	<1	0.3	1	1	14.3	3	7.8	NA	CJ4	15W40

ENGINE OIL ANALYSIS	ADT with with Mercedes Benz diesel engine. High aluminum and potassium

FIGURE 10-15 Oil analysis results showing high aluminum and potassium in an ADT.

Aluminum is not in the correct ratio with silicon to indicate dirt. The aluminum is coming from the compressor's connecting rod, which does not use a Babbitt bearing but rather a direct aluminum contact.

Conclusion

Potassium in this sample does not have a corresponding amount of sodium or silicon to indicate glycol contamination. Machines equipped with air compressors often have elevated potassium readings, which could be coming from desiccant filter chemicals used to improve their drying performance.

Another case where an accessory can contribute wear metal to an engine oil sample is an inline injection pump that also shares the engine lubricant, as opposed to its own reservoir. On rare occasions, the injection pump can produce increased chromium readings.

11. 4WD Loader with High Iron and Copper and Time on Oil

Oil Analysis

This case involves a 4WD loader that has accumulated 700 hours on the oil. In addition, the sample results show high iron and copper numbers, which are the direct result of high hours and TAN becoming higher than TBN. This is a typical example showing that excessive drain intervals accelerate iron wear and low TBN/high TAN causes corrosive copper wear (see Figure 10-16).

METALS							ADDITIVES							HOURS INFO	
Iron (Fe)	Chromium (Cr)	Lead (Pb)	Copper (Cu)	Tin (Sn)	Aluminum (Al)	Nickel (Ni)	Magnesium (Mg)	Calcium (Ca)	Barium (Ba)	Phosphorus (P)	Zinc (Zn)	Molybdenum (Mo)	Boron (B)	Total Hours	Hours on Oil
124	3	5	90	1	5	6	858	1415	1	964	1075	239	271	3000	700
85	2	6	87	1	8	1	988	1435	1	972	1067	242	265	2300	600
77	2	4	25	2	7	1	976	1432	1	982	1074	235	273	1700	500
24	3	3	15	1	5	3	983	1456	<1	976	1065	256	279	1200	500

CONTAMINANTS							PHYSICAL PROPERTIES							OIL INFO	
Aluminum (Al)	Silicon (Si)	Sodium (Na)	Potassium (K)	Water (%)	Glycol	Fuel (%)	Soot (%) FTR	Sulphation	Nitration	Viscosity (cSt @ 100 C)	Oxidation	TBN	TAN	Type	Viscosity
6	8	6	<5	<0.5	NO	<1	0.3	12	NA	16.1	NA	3.7	4.1	CJ4	15W40
7	10	11	<5	<0.5	NO	<1	0.7	10	NA	15.5	NA	3.9	4.02	CJ4	15W40
6	12	16	7	<0.5	NO	1	0.5	10	NA	15.4	NA	6.1	NA	CI4	15W40
5	11	5	<5	<0.5	NO	<1	0.4	9	NA	15.2	NA	6	NA	CI4	15W40

ENGINE OIL ANALYSIS	4WD Loader with diesel engine. High iron and copper and high TAN

FIGURE 10-16 Oil analysis results showing high iron and copper numbers and high hours on the oil in a 4WD loader.

Conclusion

The iron and copper readings are the result of high hours and corrosion. Iron is a metal that accumulates with hours because iron particles are very small and heavy. They can pass through the oil filter easily, and although they do not cause visible damage, they consume corrosion-inhibitor additive.

12. 4WD Loader with Sulfation and Oxidation

Oil Analysis

This case illustrates an extreme case of sulfation and oxidation in a 4WD loader. Soot is also high, and viscosity has been high in the last two samples (see Figure 10-17).

Iron (Fe)	Chromium (Cr)	Lead (Pb)	Copper (Cu)	Tin (Sn)	Aluminum (Al)	Nickel Ni	Magnesium Mg	Calcium Ca	Barium (Ba)	Phosphorus (P)	Zinc (Zn)	Molybdenum (Mo	Boron (B)	Total Hors	Hours on Oil
154	3	11	90	1	5	6	858	1415	1	964	1075	239	271	3000	700
123	2	8	87	1	8	8	988	1435	1	972	10067	242	265	2300	600
77	2	4	25	1	7	6	976	1432	1	982	1074	235	273	1700	500
24	3	3	15	1	5	3	983	1456	<1	976	1065	256	279	1200	500

CONTAMINANTS										PHYSICAL PROPERTIES				OIL INFO	
Aluminum (Al)	Silicon Si	Sodium (Na)	Potassium (K)	Water (%)	Glycol	Fuel (%)	Soot %FTR	Sulphation	Nitration	Viscosity (cSt @ 100°C)	Oxidation	TBN	TAN	Type	Viscosity
6	8	6	<5	<0.5	NO	<1	2.5	26	20	17.2	25	3.0	3.9	CJ4	15W40
7	10	11	<5	<0.5	NO	<1	2.2	15	16	16.5	19	3.2	3.1	CJ4	15W40
6	12	16	7	<0.5	NO	1	1.5	12	17	15.5	15	3.9	3	CJ4	15W40
5	11	5	<5	<0.5	NO	<1	1.5	11	14	15.4	14	4.0	2.9	CJ4	15W40

ENGINE OIL ANALYSIS	4WD Loader with diesel engine, high iron and high TAN. High soot, oxidation and sulfation, low TBN, high TAN

FIGURE 10-17 Oil analysis results showing high sulfation and oxidation in a 4WD loader.

Although oxidation is more of a concern in gasoline engines than in diesel engines, it still can happen in diesel engines. It is generally associated with other indicators, such as the use of high-sulfur fuels, where the condemning factor is going to be sulfation before oxidation.

Conclusion

Sulfation, nitration, and oxidation cause oil viscosity to increase, and concomitant acid production will decrease the TBN faster. The extended interval at 500 hours is not working for this engine, which is possibly using a high-sulfur fuel. Reducing the drain interval will help this engine survive the corrosive attack from burning this type of fuel.

13. Bulldozer with High-Sulfur Fuel

Oil Analysis

The results of the oil analysis in Figure 10-18 show a high iron content but not dirt. The results also show sulfation. The use of high-sulfur diesel fuels is not something from the past. There are markets, usually outside the United States, where this is still a reality, and oil labs still receive samples with abnormal diesel fuels. Machines using high-sulfur fuel will experience accelerated sulfation and a faster depletion of the TBN as well as more metal generation, especially iron, lead, and copper (see Figure 10-18).

Machines in markets with high-sulfur fuels are better off with old oil formulations with higher TBNs. This case involves a bulldozer running on a high-sulfur fuel. The last sample had 139 hours, and the preceding sample had 250 hours. During the period of the first oil sample, the machine was using 4000 ppm sulfur fuel, and during the period of the second oil sample, it was using 3000 ppm sulfur fuel. The oil viscosity is already high for the hours, as is the sulfation.

METALS							ADDITIVES							HOURS INFO	
Iron (Fe)	Chromium Cr	Lead (Pb)	Copper (Cu)	Tin (Sn)	Aluminum (Al)	Nickel (Ni)	Magnesium (Mg)	Calcium (Ca)	Barium Ba)	Phosphorus (P)	Zinc (Zn)	Molybdenum (Mo)	Boron (B)	Total Hours	Hours on Oil
150	11	12	15	<1	1	<1	838	1210	<1	1093	1288	2	1	2555	139
76	4	9	2	<1	2	<1	825	1222	<1	1050	1266	2	1	2416	250

CONTAMINANTS							PHYSICAL PROPERTIES							OIL INFO	
Aluminum (Al)	Silicon (Si)	Sodium (Na)	Potassium (K)	Water (%)	Glycol	Fuel (%)	Soot (%) FTR	Sulphation	Nitration	Viscosity (cSt @ 100 c)	Oxidation	TBN	TAN	Type	Viscosity
1	12	0	16	<0.05	NO	<1	0.18	13	2	15.14	7	5.6	NA	CJ4	15W40
2	11	0	12	<0.05	NO	<1	0.3	24	3	15.4	13	5.1	NA	CJ4	15W40

ENGINE OIL ANALYSIS	Bulldozer with diesel engine. High sulphation and high viscosity for the hours

FIGURE 10-18   Oil analysis results showing a high iron content in a bulldozer.

This has increased the iron readings from corrosion but has not raised the lead readings yet because the TBN is still at a safe level. Although no TAN measurement is available for the period, it can be assumed that the TBN still protects the bearings, but not very well or for much longer. Chromium from ring wear is also elevated.

The fuel report analysis in Figure 10-19 tells the story.

PHYSICAL/CHEMICAL										
API Gravity	Calculated Cetane	Water by Karl-Fischer	Water by Distillation	Sulfur %	Biodiesel	Acid Number	Cloud Point "T	Cold Filter Plugging	Total Particulate	
D287	D4737	D6304	ASTM D95	D4294	Volume %	mgkKOH/g	D2500	D6371	D6217	
34.5	44.3	250	0.1	0.3	0	1.2	NA	NA	29	
35.1	44.1	200	0.1	0.4	0	1.3	NA	NA	25	

DISLTILLATION D86						MICROBIOLOGICAL		APPEARANCE D4176		
Initial Boiling Point	10% Recovered	50% Recovered	90% Recovered	End Point	% Recovered	Microbial Growth	Organisms per ml	Clarity	Free Water	Particulate
°F	°F	°F	°F	°F	Volume %	POS	1000			
365	419	493	607	635	98	NA	NA	Clear	NO	
370	411	489	584	633	97	NA	NA	Clear	NO	

FUEL ANALYSIS	Bulldozer with diesel engine. High sulphur content. 2239 total hours.

FIGURE 10-19   Fuel analysis report.

Conclusion

When the maintenance of a fleet is feasible using the tools available through fluid analysis, the power of the maintenance crew grows because it can easily establish with high certainty what is causing abnormal readings.

14. Backhoe Loader with Fuel Dilution and Soot Combined

Oil Analysis

There can be cases where there is a combination of soot and fuel dilution. This case involves a backhoe diesel engine displaying such a combination (see Figure 10-20).

METALS							ADDITIVES							HOURS INFO	
Iron (Fe)	Chromium Cr)	Lead (Pb)	Copper (Cu)	Tin (Sn)	Aluminum (Al)	Nickel (Ni)	Magnesium (Mg)	Calcium (Ca)	Barium Ba)	Phosphorus (P)	Zinc (Zn)	Molybdenum (Mo)	Boron (B)	Total Hours	Hours on Oil
122	5	8	87	3	23	2	830	1379	<1	990	1334	278	<5	1250	250
87	2	4	28	1	7	3	891	1492	1	960	1150	257	<5	1000	250

CONTAMINANTS								PHYSICAL PROPERTIES						OIL INFO	
Aluminum (Al)	Silicon (Si)	Sodium (Na)	Potassium (K)	Water (%)	Glycol	Fuel (%)	Soot (%) FTR	Sulphation	Nitration	Viscosity (cSt @ 100 C)	Oxidation	TBN	TAN	Type	Viscosity
23	15	10	5	<0.05	NO	2	3	NA	NA	10.5	NA	4	NA	CK4	10W30
7	12	17	7	<0.05	NO	1.8	2.8	NA	NA	10.7	NA	3.9	NA	CK4	10W30

ENGINE OIL ANALYSIS	Backhoe Loader with diesel engine. Fuel dilution and soot

FIGURE 10-20   Oil analysis results showing soot and fuel dilution in a backhoe loader.

Engines using ULSD fuel are not going to see major issues with bearing wear under these conditions in the short term, but in this example, iron and copper are high, and chromium and aluminum are somewhat elevated in the most recent sample.

Conclusion

The fuel dilutes the oil and neutralizes the viscosity increase caused by soot. For this reason, viscosity looks normal even with this level of contamination.

15. Cummins Engine: Weak Acids

This case involves a Cummins diesel engine equipped with bypass filtration that has accumulated >100,000 miles of driving.

Oil Analysis

The results in Figure 10-21 prove the concept of fine bypass filtration keeping metals and dirt at bay. However, fine filtration cannot fight the decay of the TBN unless the filter is equipped with a time-release TBN booster additive, which is also available in the market.

METALS							ADDITIVES							HOURS INFO	
Iron (Fe)	Chromium Cr	Lead (Pb)	Copper (Cu)	Tin (Sn)	Aluminum (Al)	Nickel (Ni)	Magnesium (Mg)	Calcium (Ca)	Barium Ba)	Phosphorus (P)	Zinc (Zn)	Molybdenum (Mo)	Boron (B)	Total Hours	Hours on Oil
48	3	110	10	2	5	<1	420	5248	157	1654	1889	1	<5	520,000	101,195
15	<1	16	4	<1	4	<1	427	4251	157	1654	1561	<1	<5	418,805	45,450

CONTAMINANTS							PHYSICAL PROPERTIES							OIL INFO	
Aluminum (Al)	Silicon (Si)	Sodium (Na)	Potassium (K)	Water (%)	Glycol	Fuel (%)	Soot (%) FTR	Sulphation	Nitration	Viscosity (cSt @ 100 C)	Oxidation	TBN	TAN	Type	Viscosity
5	5	9	5	<0.5	NO	<1	0.3	NA	NA	17.1	NA	4.6	NA	CK4	15W40
4	5	4	<1	<0.5	NO	<1	<0.1	NA	NA	16.7	NA	6.3	NA	CK4	15W40

ENGINE OIL ANALYSIS — Cummins Diesel Engine with by-pass filtration.

FIGURE 10-21   Oil analysis results after >100,000 miles in a Cummins diesel engine with bypass filtration.

Conclusion

The lead readings are the direct result of the low TBN/high TAN and the impact of weak acids on lead.

Weak Acids

TBN measurements are informative and easy to obtain, but they can be misleading with regard to actual oil acidity. Some detergents are not effective in neutralizing all acid species (weak acids) present in the lubricant, thereby maintaining the TBN while, in fact, the oil may no longer provide sufficient protection against bearing corrosion. Weak acid testing is involved and time-consuming and requires titration. As an alternative, the minimum standard should be the addition of a TAN test to every oil analysis of modern diesel engines.

16. Use of Oil Analysis as a Forensic Tool

Although we promote the use of oil analysis mainly as a preventative tool, it also works as a forensic tool. This case discussed is from a real-life situation. Two similar John Deere 6155 tractors with 6068 engines operating in Panama had failed twice. The turbochargers failed first, followed by the engine seizing. The turbocharger and the crankshaft bearings showed abrasive wear. Because of their similarities, we discuss only one of the two cases.

Oil Analysis

The results on this failed engine indicate that there were copious amounts of dirt in the sample, represented by silicon, aluminum, sodium, and potassium. This engine had a similar situation when the tractor had only 512 hours. With this amount of dirt, the first component to fail would be the turbocharger, followed by crankshaft bearings, exactly as occurred in this case (see Figure 10-22).

METALS							ADDITIVES							HOURS INFO	
Iron (Fe)	Chromium Cr	Lead (Pb)	Copper (Cu)	Tin (Sn)	Aluminum (Al)	Nickel (Ni)	Magnesium (Mg)	Calcium (Ca)	Barium Ba	Phosphorus (P)	Zinc (Zn)	Molybdenum (Mo)	Boron (B)	Total Hours	Hours on Oil
320	15	163	158	39	710	4	882	1494	3	907	1071	273	291	1734	120
11	<2	<2	7	<1	6	<1	745	1276	<1	833	994	228	234	1614	50
433	29	84	143	24	921	6	860	1395	3	840	1078	265	268	1564	250

CONTAMINANTS										PHYSICAL PROPERTIES				OIL INFO	
Aluminum (Al)	Silicon (Si)	Sodium (Na)	Potassium (K)	Water (%)	Glycol	Fuel (%)	Soot (%) FTR	Sulphation	Nitration	Viscosity (cSt @ 100 C)	Oxidation	TBN	IAN	Type	Viscosity
710	1067	25	18	0.05	NO	1.28	0.54	19	6	15.23	16	7.13	NA	CJ4	15W40
6	5	<5	<2	0.06	NO	0.94	0.47	19	6	10.05	15	8.59	NA	CJ4	15W30
921	1618	268	21	0.02	NO	0.91	0.12	21	6	14.96	16	6.63	NA	CJ4	15W40

ENGINE OIL ANALYSIS — John Deere 6155 with turbocharger and crankshaft bearing failures.

FIGURE 10-22   Turbocharger and crankcase bearing failures in a John Deere 6155 tractor.

Figure 10-23 shows the failed bearing with symptoms resulting from this level of contamination, and this helped us to conclude that the cause of the failure was a massive amount of dirt entry through the oil sump filling tube.

FIGURE 10-23   Failed main bearing.

Conclusion

Massive dirt contamination in an engine can occur through the oil sump filling tube. In contrast, massive dirt contamination through the inlet system is generally not feasible and does not cause turbocharger or crankshaft bearing failures. Excessive contamination through the intake causes damage to the upper part of the combustion chamber, namely piston rings and liners. Because this engine suffered two consecutive engine failures in 1,600 hours and for the same causes, it is thus reasonable to conclude that the contamination was the product of sabotage.

17. Choosing Lubricants for Extended Service Intervals (Power Stroke 6.4 and 6.7 Engines with 5W-40 Synthetic Oil)

A business operation with a large fleet of diesel trucks is interested in finding ways to extend service intervals and realize some savings in labor and materials (see Figure 10-24). The operation uses oil analysis but requires help with interpretation of the results to be able to decide on the best lubricant options.

FIGURE 10-24   Power Stroke 6.4 engine (*left*); Power Stroke 6.7 engine (*right*).

Power Stoke 6.4 with 15-Quart Sump Capacity Results

The meaningful data for this engine start at 3,000 miles on the oil. Any data below this number are not useful for the calculating wear tables (see Table 10-3).

The data suggest that this engine shows signs of stress at about 4,000 miles. The ideal oil change intervals thus would be at 4,000 miles. Oil viscosity drop and high oxidation are the limiting factors. At around 6,700 miles, the TBN falls to a critical level, and possibly the TAN (acid number) is higher than the TBN, which pushes the iron production to the calculated critical threshold. This engine has a small oil sump, and together with viscosity loss to shearing, this contributes to a short oil interval.

Power Stroke 6.7 V8 with 13-Quart Oil Sump Capacity 5W-40 Oil Results

The meaningful data for this engine start at 5,800 miles. Data below this figure do not bring value to the calculations. The engine behaves extremely well even beyond 10,000 miles on the oil. The calculation for wear tables uses an average of 8,783 miles. The performance of this engine is good, and there is little viscosity loss, producing extraordinary results despite its small oil sump. However, oil intervals should not exceed 10,000 miles because low TBN and oxidation take over, as shown in the top results in Table 10-4.

TABLE 1O-3 Data Analysis Engine Power Stoke 6.4

V8	Miles on Oil	Fuel	Soot	TBN	Viscosity	Oxidation	Nitration	Water	Aluminum	Copper	Iron
	8,059	0	1	3.7	8.8	32.1	15.7	0.14	12.18	-3.62	85.57
	8,059	0	1	3.7	8.84	32.1	15.9	0.14	11.27	-3.17	86.56
	8,059	0	1	3.7	9.19	31.8	15.7	0.14	11.72	-3.08	94.82
	6,789	0	1.2	4.92	8.97	28.7	11.8	0.188	25.51	1.25	114.13
	6,789	0	1.2	4.82	8.8	28.5	11.8	0.188	23.1	1.49	107.95
	6,789	0	1.2	4.82	9.11	28.3	12	0.188	22.16	0.62	107.26
	6601	0	1.5	7.4	10.01	15.1	4.9	0.11	33.28	16.63	139.49
	6601	0	1.5	7.4	9.09	15.2	5	0.11	34.83	8.92	145.97
	6,245	0	1.1	5.76	8.62	25.2	9.9	0.214	11.28	2.53	60.04
	6,245	0	1.1	5.76	8.43	25.5	10.1	0.214	11.86	0.81	60.81
	6,245	0	1.1	5.76	8.37	25.4	10.1	0.214	8.24	-1.53	56.76
	6,119	0	0.8	5.48	9.26	25.3	9.8	0.092	11.79	0.08	36.97
	6,119	0	0.8	5.58	8.81	25.5	9.5	0.092	13.22	-0.76	42.33
	6,119	0	0.8	5.48	8.93	25.4	9.8	0.092	13	-0.63	43.75
	5,747	0	1	6.1	8.93	23.7	8.9	0.14	7.07	1.13	50.51
	5,747	0	1	6.1	8.06	24.1	8.7	0.14	7.23	1.72	49.66
	5,747	0	1	6	8.43	24	8.9	0.14	7.81	0.41	56.06
	5,456	0	1.1	5.66	8.84	25.4	9.3	0.114	19.58	8.4	52.27
	5,456	0	1.1	5.56	8.79	25.5	9.5	0.114	15.86	5.14	46.39
	5,456	0	1.1	5.66	8.53	25.5	9.2	0.114	18.87	4.6	53.75
	5,322	0	1	6.2	9.02	22.9	8.3	0.14	9.7	2.3	54.58
	5,322	0	1	6.3	9.09	23	8	0.14	9.94	1.09	56.49
	5,322	0	1	6.3	8.85	22.9	8.3	0.14	10.09	0.83	57.72
	5,108	0	0.6	6.46	8.8	22.5	7.7	0.144	10.51	-0.55	40.29

V8	Miles on Oil	Fuel	Soot	TBN	Viscosity	Oxidation	Nitration	Water	Aluminum	Copper	Iron
	5,108	0	0.7	6.32	**8.96**	**22.8**	7.6	0.018	10.99	−0.78	33.62
	5,108	0	0.7	6.22	**8.66**	**22.9**	7.4	0.018	10.94	−1.39	28.43
	5,039	0	1.3	7.88	**9.78**	12.6	3.6	0.062	19.14	10.44	31.23
	5,039	0	1.2	7.82	**9.66**	12.5	3.3	0.088	17.78	5.32	33.6
	5,039	0	1.2	7.82	**9.7**	12.6	3.6	0.088	15.92	3.47	31.38
	4,065	0	0.7	7.52	**9.76**	**18.4**	6.5	0.118	4.17	2.89	33.39
	4,065	0	0.8	6.98	**9.35**	**19.6**	6.1	0.092	3.82	1	31.56
LIMIT	4,065	0	0.8	6.98	**9.57**	**19.9**	6.1	0.092	5.29	1	37.98
	3,991	0	0.8	8.08	12.93	8.8	3.3	0	1	13.51	14.19

TABLE 1O-4 Data Analysis Engine Power Stroke 6.7

VO	Miles on Oil	Soot	TDN	Viscosity	Oxidation	Nitration	Water	Aluminum	Copper	Iron
	13,061	-1	4.15	12.1	**15.5**	12.9	0.014	8.38	16.55	**58.51**
	13,061	-1	4.15	13.09	**16.5**	13.1	0.074	9.56	13.86	**53.02**
	13,061	-1	4.26	12.71	**16.5**	12.9	0.014	6.11	14.29	43.43
	12,246	1	4.1	12.1	**15.8**	13.3	0.14	7.85	8.43	40
	12,046	1	4.1	11.91	**15.8**	13.4	0.14	8.08	817	41.75
	12,246	1	4.1	12.29	**15.8**	13.4	0.14	7.02	7.9	41.89
	11,264	13	**3.88**	12.77	**15.5**	13.2	0.062	10.34	7.81	**65.63**
	11,264	13	**3.88**	12.67	**15.4**	13.2	0.062	6.93	9.27	**63.83**
	11,264	13	**3.88**	12.44	**15.6**	13.4	0.062	8.59	8.9	**75.91**
	10,998	C.9	4.34	12.27	**14.0**	11.3	0.066	4.18	6.78	23.93
	10,998	C.7	5.22					2.35	8.22	23.73
	10,998	C.9	4.34	11.72	**11.6**	11.2	0.066	3.78	6.51	27.22
	10,799	-1	**4.56**	12.71	**15.3**	11.4	0.014	7.24	5.38	**59.78**
	10,799	-1	**4.56**	12.1	**15.3**	11.3	0.014	6.82	5.05	**59.79**
LIMIT	10,799	-1	**4.56**	12.2	15.0	11.5	0.014	7.01	5.57	**62.12**
	9,940	-1	5.06	12.22	14.0	9.9	0.014	5.38	8.91	31.29
	9,940	-1	4.36	12.24	14.0	10	0.014	5.67	8.57	35.67
	9,940	-1	4.36	12.31	14.0	10	0.014	5.69	6.1	35.18
	9,090	1	5.1	12.99	13.0	9.3	0.04	6.56	8.34	28.91
	9,090	1	5.1	13.51	13.2	9.1	0	5.98	8.91	31.06
	9,090	1	5	12.76	13.2	9.5	0.74	5.7	8.32	29.43
	9,023	C.8	5.38	12.12	12.3	8.2	0.092	6.14	4.27	23.23
	9,023	C.8	5.38	12.25	12.3	8.1	0.092	8.25	7.01	33.16
	9,023	C.8	5.38	12.17	12.4	8	0.092	5.3	4.72	31.45
	8,108	C.9	5.34	13.03	12.2	8.2	0.066	5.92	6.31	22.82

VO	Miles on Oil	Soot	TDN	Viscosity	Oxidation	Nitration	Water	Aluminum	Copper	Iron
	8,108	0.9	5.34	12.63	12.2	8.3	0.066	6.55	7.39	23.43
	7,702	0.8	5.38	12.19	12.5	7.3	0	4.57	6.39	78.52
	7,702	0.8	5.38	12.15	12.6	7.3	0	4.58	7.15	19.37
	7,702	0.8	5.38	12.12	12.6	7.2	0	4.84	5.4	20.77
	7631	1	5.4	12.54	12.6	9.2	0.04	11.75	2.39	44.41
	7631	1	5.5	12.63	12.6	8.3	0.04	13.25	3.7	47.74
	7631	1	54	12.65	12.6	9.1	0.04	11.58	4.13	46.19
	7,389	0.7	5.52	11.78	11.2	7.7	0.018	7.06	5.07	30.55
	7,000	0.7	5.02	12.10	11.1	7.4	0.010	7.50	5.02	31.01
	7,389	0.7	5.32	12.11	11.9	7.3	0.018	7.35	8.32	33.05
	7,099	0.8	5.38	12.23	11.5	7.4	0	1.73	112	23.83
	7,099	0.8	5.38	12.64	11.6	7.4	0	2.67	2.55	23.04
	6,846	0.5	6.5	12.47	10.8	5.3	0	3.62	5.39	14.82
	6,846	0.6	6.46	12.6	10.9	6	0	3.62	4.2	15.69
	6,846	0.6	6.46	12.2	11	5.3	0	3.79	3.18	16.10
	6,555	0.9	5.34	12.46	11.6	7.3	0.066	4.08	13.27	23.33
	6,555	0.9	5.04	12.50	11.0	7.7	0.066	4.50	10.42	20.55
	6,595	0.9	5.34	12.56	11.5	7.7	0.066	3.95	10.82	28.18
	6,542	0.7	6.22	11.93	11.3	6.1	0.018	4.17	5.48	14.52
	6,542	0.7	6.32	12.1	11.3	6.2	0.018	3.83	5.26	14.35
	6,460	0.6	6.56	13.1	10.8	6.5	0.044	5.45	6.43	20.43
	6,460	0.6	6.46	12.63	11	6.4	0.044	4.04	10.9	20.45
	6,460	0.7	6.42	12.41	10.3	6.3	0.018	5.16	0.05	22.02
	5,012	0.0	0.20	12.00	10.0	0.5	0.092		1104	11.25
	5,812	0.8	6.28	12.55	10.3	6.3	0	2.14	5.09	20.75
	5,812	0.8	6.28	12.93	10.9	6.3	0	3.18	7.01	23.71

Conclusion

The case shows that the Power Stroke 6.4 engine has limitations, even on synthetic oil, whereas its sibling, the Power Stoke 6.7, behaves very differently. This case study also shows that engines have a mind of their own and that each one has different needs for the owner to satisfy.

CONCLUSION

This chapter went through the primary causes of contamination in engine oil and their likely sources. It also discussed the effective use of oil analysis to prevent failures. Then it highlighted the power of this technology to do forensic analysis and to make sensible decisions in maintenance. In Chapter 11, we will apply oil analysis expertise to analyze powertrains.

CHAPTER 11

POWERTRAIN OIL ANALYSIS

The variety of powertrain components used by today's mobile equipment is very extensive. This book tries to cover the most important components of the most relevant machine classes with regard to their maintenance with oil analysis. This chapter covers the most important tests, in order of importance, so that service people can use this guide to achieve better fleet health management.

COMPONENTS COVERED

- Powershift transmissions
  - Direct drives and torque converters
  - Loaders and backhoes
  - Loaders with locking torque converters
  - Articulated dump trucks (ADTs)
- Axles
  - Multiple disc brake systems
  - Common sumps
  - Noncommon sumps
  - Filtered and nonfiltered
  - Single disc brake systems
  - Filtered and nonfiltered
  - Large hypoid axles
  - Large nonhypoid axles
  - Articulated ADT axles

- Wet disc brakes
  – Final drives in tracked machines
- Internal final drives
  – Crawlers
- External final drives
  – Crawlers and excavators
- Gearbox drives
  – Swing gearboxes
  – Splitter boxes
  – Pump drive boxes
- Tandems and bogies
  – Motor graders
  – Forestry machines
  – Independent transfer gearboxes
- Articulated dump trucks (ADTs)

Brief Descriptions of Components

A brief description of components is a good reminder about the applications and configurations of these components.

Transmissions

Powershift transmissions, as used on some motor graders and skidders, have evolved through the years to become very reliable and sophisticated in shifting characteristics. The transmissions used by these machines are for hauling rather than shuttling back and forth, and they could come as direct drive without a torque converter or may offer a locking torque converter.

Powershift transmissions, as used in four-wheel-drive (4WD) loaders and backhoes, operate back and forth. They have fewer gears and are always equipped with torque converters. Larger 4WD loaders use converter lock-ups (see Figure 11-1).

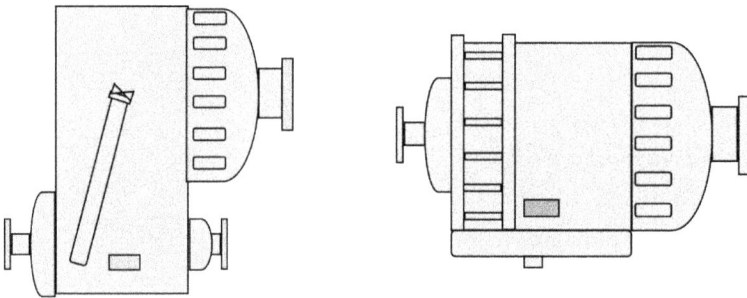

FIGURE 11-1   Transmission (*left*); powershift transmission with locking torque converter (*right*).

Axles

Axles are like the forgotten part of a machine. They hide underneath where it is not easy to see them. This sometime carries the penalty for improper servicing. Axles transfer the power to the ground and carry the weight of the machine, and because they are so close to the ground, they are closer to the elements.

Axles are typically equipped with wet disc brakes that generate heat, metals, fibers, and/or carbon. They also can carry a differential lock that can generate metals or produce internal leaks. In transport function, axles generate heat, and when decelerating, they withstand heavy loads. Depending on the application and manufacturer, they could require special lubricants and additives (see Figure 11-2).

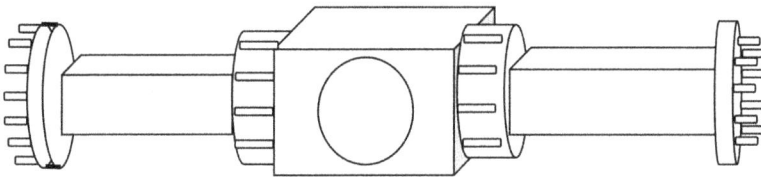

FIGURE 11-2   Axle.

Brakes in Axles

There are two approaches to brake design. In one design, multiple discs are located on the low-speed, high-torque side of the final drive, and in the other design, a single disc is located on the high-speed, low-torque side of the final drive (see Figure 11-3).

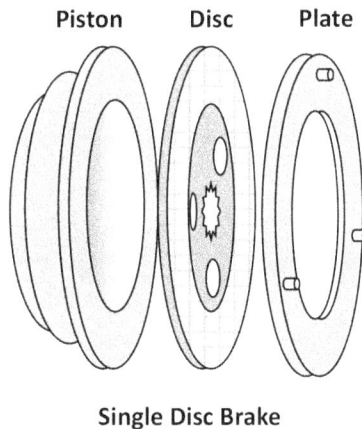

Piston   Disc   Plate

Single Disc Brake

FIGURE 11-3   Disc brake.

While the multiple-disc design has a nice braking modulation, it produces much more braking material, namely iron, and heat. When these axles are equipped with cooling and filters, they return excellent readings in oil analysis (see Figure 11-4).

FIGURE 11-4 Multidisc braking system.

Friction modifiers are essential for both types of axles. Tractor fluids already have some amount of this brake antichatter additive, but the additive can be supplemented when brake chatter persists. Some manufacturers do not use friction modifiers to reduce brake chatter but rather depend on wear materials that are less prone to chattering.

Final Drives in Tracked Machines

Final drives belong in two big groups—those that receive shock loads, as in crawler dozers/loaders, and those whose main function is to propel the machine, as in excavators. In many cases, final planetary drives work in combination with internal final drives, which could be a bull-gear type or could have double or triple planetary reductions.

Both final drive applications are close to ground, and both have similar elements to fight but are subject of different loads. Therefore, their respective oil analysis readings are interpreted differently. Because of their location and design, final drives do not typically have sight glasses to check for oil levels and lubricant color. This hinders the ability to monitor leaks, overflow, or contamination (see Figure 11-5).

FIGURE 11-5 Final drive (*left*); excavator final drive exposure (*right*).

Gearboxes, Splitters, and Pump Drives

In mobile equipment, there are plenty of small gearboxes to drive components within a machine. This is the case with pump drive, upper structure swings (as in excavators or forestry machines), splitters that drive independent hydrostatic transmissions, blade circle drives in motor graders, and so on (see Figure 11-6). The engine or a hydraulic motor drives these components. The recommended lubricant could be gear oil, engine oil, or hydraulic fluid.

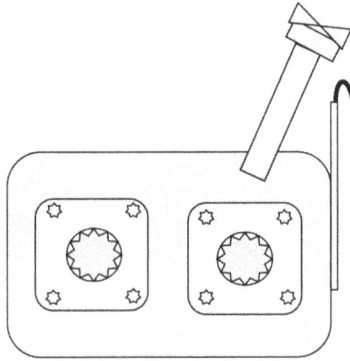

FIGURE 11-6   Pump drive.

Depending on application, these gearboxes could be prone to contamination, oil oxidation, and/or oil starvation. In some cases, these gearboxes share lubrication with the main hydraulic system. These components use low-oil-volume compartments that require maintenance from the oil analysis point of view.

Tandems and Bogies

Oscillating tandems and bogies as used in motor graders and forestry machines are a practical way to drive wheels in pairs for improved traction. In the case of motor graders, heavy-duty chains drive the wheels in an oil bath. Bogies usually have gears inside. Because of their size, they use large seals, especially at oscillating points, which are an invitation for dirt entry. They are not critical from a maintenance point of view but still require scheduled analysis (see Figure 11-7).

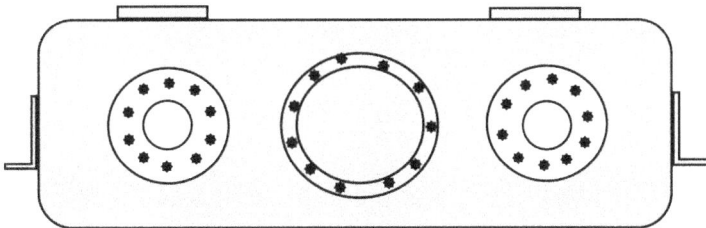

FIGURE 11-7   Motor grader tandem.

Torque Converters

Torque converters are prone to erosion from high-velocity fluid impact. If the fluid shows contamination with hard particles (i.e., dirt or metals), water, or air, some erosion of the fin edges is expected. Torque converter wear is measurable through aluminum readings. Torque converters also cause more viscosity loss from fluid shear-down. Air can be very aggressive in torque converters, causing erosion and fin failure. Dirt and water also damage torque converter fins (see Figure 11-8).

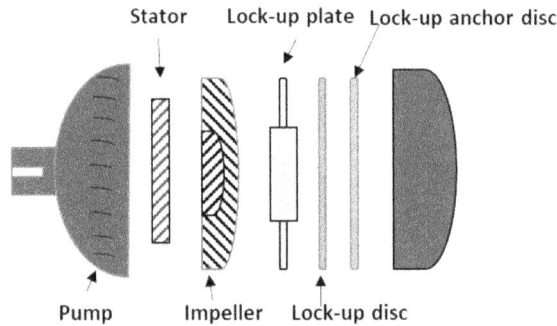

FIGURE 11-8 Torque converter.

Powertrain Contamination Sources

- Dirt
  - Through breathers
  - Through seals
  - From poor service practices
- Water
  - Through seals
  - Through breathers
  - From poor service practices
- Sediment
  - From new oil
  - Through poor filtration
  - Through lack of flushing
- From other systems
  - Through faulty seals
- Air
  - Through oil cavitation
  - Through oxidation

Metal Generators

- Break-in period
- Dirt

- Water
- High total acid number (TAN)
- Time on oil
- Slipping clutches
- Braking surfaces
- Oxidation
- Lack of limited-slip (LS) additive
- Torque converter
- Improper fluid
- Heat
- Design
- Failing component

The two preceding lists show the most typical contaminants and metal generators affecting powertrains. The lists include contaminants coming from the outside and those generated by the system's shortcomings or poor servicing. The metal generation list shows the challenges that service people face when interpreting oil analyses.

Testing
See Table 11-1.

TABLE 11-1   Testing of Powershift Transmissions and Gearboxes

What should we test?	Test method
Dirt	Ratio Al/Si
Moisture	Crackle or Karl Fischer
Acidity	TAN
Additive signature	FTIR
Degradation/aging	Oxidation
Viscosity changes	Viscosity at 40°C
Rust/corrosion/wear	Fe, Cu, Cr, Pb, Al by ICP
Particle quantification index	PQ Index
Particle counts	4/6/14 micron readings

Powershift Transmission Oil Analysis Considerations
The life of powershift transmissions depends on fluid quality and cleanliness. Additionally, because many of them depend on friction modifiers for smooth operation, TAN and oxidation become important. It is very important then to keep an eye on TAN, dirt, and moisture. Water contamination needs to stay below 1,000 ppm on machines with torque converters.

If a tractor fluid or engine oil is used in a powershift transmission, expect a quick fluid viscosity loss within the first 500 hours, especially on machines equipped with a torque converter. This is expected and considered normal. Automatic transmission fluid (ATF) generally has a lower initial viscosity but does not experience a significant drop in viscosity.

Larger 4WD loaders use torque converter lockups. This component can produce iron, aluminum, and copper readings coming from this area that otherwise could be construed as dirt. This is also true for ADTs that use lockups and retarders.

Wear Threshold Tables

The wear threshold tables used in this chapter are only for didactic purposes and do not apply to a large range of machine samples. Do not use these tables directly for interpretation of independent oil analyses for given machines. The lab or users need to calculate their own tables based on large machine population data for the same model and vintage of machine, as explained in Chapter 8.

CASE DISCUSSIONS

1. High Aluminum and Silicon in an ADT Allison HD 4500 Powershift Transmission

Oil Analysis

An ADT truck transmission is producing high metals readings that worry the user. The truck is operating normally, and there are no complaints from the operator. However, the oil lab results in Figure 11-9 show several flagged numbers that need analysis on this ADT transmission. To start with, are the iron readings high? Are the aluminum and silicon readings from dirt?

METALS							ADDITIVES						
Iron (Fe)	Chromium (Cr)	Lead (Pb)	Copper (Cu)	Tin (Sn)	Aluminum (Al)	Nickel (Ni)	Magnesium (Mg)	Calcium (Ca)	Barium (Ba)	Phosphorus (P)	Zinc (Zn)	Molybdenum (Mo)	Boron (B)
107	<1	61	29	5	66	<1	<1	9	<1	234	<1	<1	117

CONTAMINANTS					WEAR		PHYSICAL PROPERTIES			INFORMATION			
Aluminum (Al)	Silicon (Si)	Sodium (Na)	Potassium (K)	Water (%)	PQ Index	Particle Counts	Viscosity (cSt @ 40C)	Oxidation	TAN	Fluid Changed	Fluid Type	Total Hours	Fluid Hours
66	11	4	6	<0.05	20	NA	41.7	18	1.4	YES	ATF	6600	2000

POWERTRAIN OIL ANALYSIS	ADT Allison™ Powershift Transmission. TranSynd™ fluid. High iron, lead and aluminum

FIGURE 11-9   Oil analysis results from an ADT Allison powershift transmission.

Based on the wear threshold table, the iron should be <90 ppm at 1,000 hours. Because iron production is directly related to hours, at 2,000 hours on the fluid, this reading is perfectly normal.

With regard to dirt, it is important to know the signature of the fluid to establish whether or not it contains a significant increase in silicon. In this case, the fluid is TranSynd, which has only 4 ppm of silicon as foam inhibitor. Thus we can establish that from the 11 ppm shown on the oil analysis report, at least 7 ppm belong to dirt. If we establish a ratio of silicon to aluminum of 4–5:1, we can conclude that only ~1.5 ppm of aluminum is from dirt.

Conclusion

The balance of aluminum could be coming from the torque converter and/or the retarder. Because the table is for a 1,000-hour measurement, the iron at 107 PPM is not out of the ordinary, but given the high readings of lead and aluminum, a fluid flush would be ideal.

Wear Metal Thresholds

See Figure 11-10 and Table 11-2.

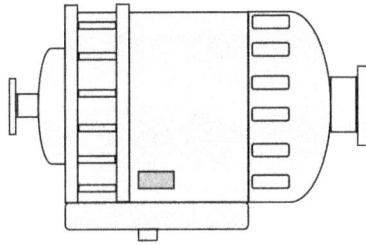

FIGURE 11-10   Allison HD 4500 powershift transmission.

TABLE 11-2   Wear Table for Allison HD 4500 Powershift Transmission

	Per 1,000 hours		
Filtered System	Normal	Abnormal	Critical
Fe	<90	90	>150
Pb	<32	32	>51
Cu	<14	14	>23
Cr	<5	5	>10
Al	<21	21	>30
Si	<8	8	>15

2. High Silicon and High TAN in a Backhoe ZF Powershift Transmission

Oil Analysis

The oil analysis Figure 11-11 is for a backhoe with high aluminum readings and viscosity sheared down to 37.5 cSt at 1,371 hours. Is the silicon reading from dirt? Is the viscosity drop abnormal? Has the TAN been abnormal?

METALS							ADDITIVES						
Iron (Fe)	Chromium Cr	Lead (Pb)	Copper (Cu)	Tin (Sn)	Aluminum (Al)	Nickel (Ni)	Magnesium (Mg)	Calcium (Ca)	Barium Ba)	Phosphorus (P)	Zinc (Zn)	Molybdenum (Mo)	Boron (B)
23	<1	<1	3	<1	<1	<1	112	3840	<1	1100	1210	4	<5
22	<1	<1	4	<1	<1	<1	109	3910	<1	1180	1430	<1	<5
18	<1	<1	3	<1	<1	<	100	3330	<1	1050	1180	2	<5

CONTAMINANTS				WEAR			PHYSICAL PROPERTIES			INFORMATION			
Aluminum (Al)	Silicon (Si)	Sodium (Na)	Potassium (K)	Water (%)	PQ Index	Particle Counts	Viscosity (cSt @ 40C)	Oxidation	TAN	Fluid Changed	Fluid Type	Total Hours	Fluid Hours
<1	33	6	<5	<0.05	<10	NA	37.5	NA	1.84	NO	THF	3371	1371
<1	33	6	<5	<0.05	<10	NA	37.8	NA	2.96	YES	THF	2000	2000
<1	27	5	<5	<0.05	<10	NA	45.7	8	3.59	NO	THF	500	500

POWERTRAIN OIL ANALYSIS	Backhoe Powershift Transmission, Shell® Tractor Fluid. High Silicon

FIGURE 11-11 Oil analysis report for a backhoe powershift transmission.

The fluid shows a high silicon reading, which is not dirt because there is not a corresponding amount of aluminum present, but it could be from silicone foam inhibitor and leaching from silicone sealant. The signature reveals that the fluid is a clone of Shell Donax TD tractor fluid. It is normal for a tractor fluid to lose viscosity in the first 500 hours of use. Table 2-12 provides information about viscosity loss per hour for different fluids in varied applications and indicates that viscosity in this case is typical. With regard to TAN, a used high-calcium fluid usually has a TAN of up to 5 without being abnormal.

Conclusion
This transmission shows excellent metal readings, a low particle quantifier (PQ) index, and no moisture, so there should not be any worries about its condition.

Wear Metal Thresholds
See Figure 11-12 and Table 11-3.

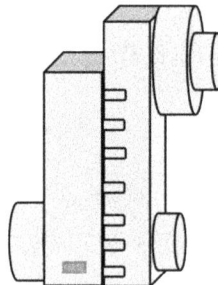

FIGURE 11-12 Backhoe powershift transmission.

TABLE 11-3 Wear Table for a Backhoe ZF Powershift Transmission

Filtered System	Per 500 hours		
	Normal	Abnormal	Critical
Fe	<50	50	>75
Pb	<7	7	>15
Cu	<35	35	>55
Cr	<5	5	>10
Al	<13	13	>20
Si	<10	10	>25

3. High Copper Generation in a Motor Grader with a Funk Transmission

Oil Analysis

The oil analysis report in Figure 11-13 shows high copper readings, and the cause needs to be determined. Are the copper readings caused by contamination, high TAN, or low viscosity? Are the copper readings above wear thresholds? Are there any other symptoms to indicate that the transmission is suffering from more than elevated copper wear?

METALS							ADDITIVES						
Iron (Fe)	Chromium Cr)	Lead (Pb)	Copper (Cu)	Tin (Sn)	Aluminum (Al)	Nickel (Ni)	Magnesium (Mg)	Calcium (Ca)	Barium Ba)	Phosphorus (P)	Zinc (Zn)	Molybdenum (Mo)	Boron (B)
29	<	2	202	<1	12	<1	86	3743	<1	1136	1327	<1	16
24	<1	5	182	<1	5	<1	98	3323	<1	1168	1632	<1	13
17	<1	2	116	<1	8	<1	90	3526	2	1117	1338	<1	16

CONTAMINANTS				WEAR		PHYSICAL PROPERTIES			INFORMATION			
Aluminum (Al)	Silicon (Si)	Sodium (Na)	Potassium (K)	Water (%)	PQ Index	Particle Counts	Viscosity (cSt @ 40C)	Oxidation	TAN	Fluid Changed	Fluid Type	Total Hours	Fluid Hours
12	8	6	<5	<0.05	<10	NA	41.3	NA	1.52	NO	THF	1500	1500
5	9	8	9	<0.05	<10	NA	41.6	NA	2.3	NO	THF	1000	1000
8	8	5	<5	<0.05	<10	NA	41.5	NA	1.59	NO	THF	500	500

POWERTRAIN OIL ANALYSIS	Motor Grader with Funk® Power Shift Transmission, Tractor Fluid.

FIGURE 11-13 Oil analysis report for a motor grader with a Funk powershift transmission.

The results show no dirt and no water and that the viscosity is within specifications for the number of hours. The copper level is normal for this machine, as shown in Table 11-4. Other metal readings are excellent, and the TAN is in good shape for a tractor fluid.

Conclusion

We can conclude that the copper likely comes from some small element made of bronze that is not having a detrimental effect on transmission life or performance. It is up to the user to decide whether a major repair justifies finding the source of these copper readings. However, consulting with the brand representative would help to clarify this point.

Wear Metal Thresholds

See Figure 11-14 and Table 11-4.

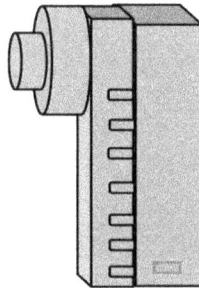

FIGURE 11-14   Funk powershift transmission.

TABLE 11-4   Wear Table for Motor Grader Funk Transmission

Per 1,OOO hours			
Filtered System	Normal	Abnormal	Critical
Fe	<53	53	>83
Pb	<8	8>	>12
Cu	<300	300	>450
Cr	<5	5	>10
Al	<7	7	>13
Si	<15	15	>30

4. High Aluminum in an ADT ZF Powershift Transmission

Oil Analysis

The oil analysis results in Figure 11-15 for an ADT transmission show high aluminum production and slight changes to the fluid signature. Where is the aluminum coming from? Is the signature change reason for worry?

The results for this transmission are not extraordinary. However, we must understand whether the aluminum production comes from dirt or from a component within the transmission. From Chapter 2, we can conclude that the fluid used appears to be a combination of ZF Ecofluid and TranSynd. We estimate that the silicon content of the fluid should be around 4 ppm, which leaves 10 ppm of silicon as coming from dirt, which exceeds the number recommended by the wear table (Table 11-5). Viscosity and TAN are within specifications, and water content is under control.

| METALS | | | | | | | ADDITIVES | | | | | | |
Iron (Fe)	Chromium Cr)	Lead (Pb)	Copper (Cu)	Tin (Sn)	Aluminum (Al)	Nickel (Ni)	Magnesium (Mg)	Calcium (Ca)	Barium Ba)	Phosphorus (P)	Zinc (Zn)	Molybdenum (Mo)	Boron (B)
78	<1	1	20	<1	141	<1	1	38	<1	248	9	<1	102
26	<1	4	15	<1	19	<1	<1	16	<1	187	9	<1	64
18	<1	2	10	<1	15	<	<1	10	<1	200	8	<1	50

| CONTAMINANTS | | | | WEAR | | | PHYSICAL PROPERTIES | | | INFORMATION | | | |
Aluminum (Al)	Silicon (Si)	Sodium (Na)	Potassium (K)	Water (%)	PQ Index	Particle Counts	Viscosity (cSt @ 40C)	Oxidation	TAN	Fluid Changed	Fluid Type	Total Hours	Fluid Hours
141	14	3	<5	<0.05	18	NA	37.9	NA	1.09	NO	ATF	3000	1000
19	8	3	<5	<0.05	15	NA	40.2	NA	0.77	YES	ATF	2000	2000
15	10	2	<5	<0.05	10	NA	37	NA	1.7	NO	ATF	1000	1000

POWERTRAIN OIL ANALYSIS — **ADT ZF® Powershift Transmission. TranSynd™ Fluid**

FIGURE 11-15 Oils analysis results for an ADT ZF powershift transmission.

Conclusion

We can conclude that 2 ppm of aluminum is coming from dirt, with the remainder coming from the torque converter and retarder. A fluid change is recommended.

Wear Metal Thresholds

See Figure 11-16 and Table 11-5.

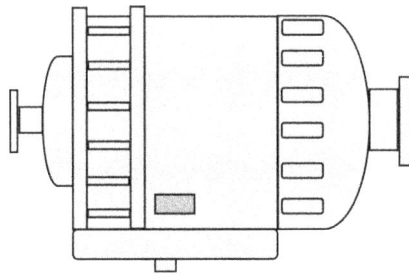

FIGURE 11-16   ADT ZF powershift transmission.

TABLE 11-5   Wear Table for ADT ZF Powershift Transmission

Filtered System	Per 1,000 hours		
	Normal	Abnormal	Critical
Fe	<90	90	>150
Pb	<32	32	>51
Cu	<14	14	>23
Cr	<5	5	>10
Al	<21	21	>30
Si	<8	8	>15

5. High Iron and Copper Readings in a 4WD Loader with ZF Powershift Transmission with Converter Locking Clutch

Oil Analysis

The oil analysis report in Figure 11-17 is for a ZF powershift transmission that shows what appears to be high iron and copper readings, dubious viscosity, and high TAN. What is causing these high iron and copper readings? Are the viscosity at 41.3 cSt and TAN at 4.43 normal?

METALS							ADDITIVES						
Iron (Fe)	Chromium Cr	Lead (Pb)	Copper (Cu)	Tin (Sn)	Aluminum (Al)	Nickel (Ni)	Magnesium (Mg)	Calcium (Ca)	Barium Ba)	Phosphorus (P)	Zinc (Zn)	Molybdenum (Mo)	Boron (B)
114	<1	2	27	<1	3	<1	149	3464	<1	1244	1473	<1	<5

CONTAMINANTS				WEAR		PHYSICAL PROPERTIES			INFORMATION			
Aluminum (Al)	Silicon (Si)	Sodium (Na)	Potassium (K)	Water (%)	PQ Index	Particle Counts	Viscosity (cSt @ 40C)	Oxidation	TAN	Fluid Changed	Fluid Type	Total Hours	Fluid Hours
3	13	41	8	<0.05	245	NA	41.3	NA	4.43	NO	THF	4500	1000

POWERTRAIN OIL ANALYSIS	4WD Loader ZF® Powershift Transmission. Tractor Fluid 57 cSt at 40° C. High iron and copper

FIGURE 11-17   Oil analysis report for a ZF powershift transmission.

The gray text elements suggest that this transmission is experiencing some wear. The first things to check for are water and dirt, which in this case do not seem to be the cause. The ratio of silicon to aluminum is at 4.33:1, which is the perfect ratio for dirt but still well below abnormal readings. Sodium is high, but this unit is from Oregon, and the reading is typical for the area, even though it is not from seawater. Next, we turn our attention to TAN and viscosity. Although both are flagged as abnormal, the viscosity drop is normal for a tractor fluid at 1,000 hours. With regard to the TAN, a calcium tractor fluid is going to have 5 as the limit for normal (see Table 2-15).

Conclusion

We realize that this transmission is equipped with a locking torque converter, which is actuated by a disc clutch. For the most part, this device works as designed, and it is automatic. Issues with a slipping clutch are evident in the oil analysis report, as we will see in the next example. A calculation of standard deviation of this vintage of machines could establish the normal readings for iron and copper and gives <50 ppm for iron as normal and <17 for copper. So this transmission is producing more iron and copper than normal, and a sampling should be done at half interval to make sure that the wear is not occurring at an accelerated speed.

Wear Metal Thresholds

See Figure 11-18 and Table 11-6.

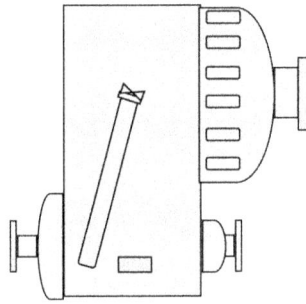

FIGURE 11-18   ZF powershift transmission.

TABLE 11-6   Wear Table for ZF Powershift Transmission

Per 1,000 hours			
Filtered System	**Normal**	**Abnormal**	**Critical**
Fe	<50	50	>80
Pb	<4	4	>8
Cu	<17	17	>30
Cr	<12	12	>20
Al	<20	20	>35
Si	<31	31	>60

6. Fluid Mixing in an ADT ZF Ecomat 2 Powershift Transmission

Oil Analysis

The oil analysis report in Figure 11-19 is for a ZF Ecomat 2 powershift transmission and shows some stress through abnormal wear metal and additive readings. What is causing the high metal readings? Is the lubricant signature within the range of the recommended fluid?

METALS							ADDITIVES						
Iron (Fe)	Chromium Cr)	Lead (Pb)	Copper (Cu)	Tin (Sn)	Aluminum (Al)	Nickel (Ni)	Magnesium (Mg)	Calcium (Ca)	Barium Ba)	Phosphorus (P)	Zinc (Zn)	Molybdenum (Mo)	Boron (B)
147	<1	61	127	14	72	<1	24	580	<1	517	310	1	106

CONTAMINANTS					WEAR		PHYSICAL PROPERTIES			INFORMATION			
Aluminum (Al)	Silicon (Si)	Sodium (Na)	Potassium (K)	Water (%)	PQ Index	Particle Counts	Viscosity (cSt @ 40C)	Oxidation	TAN	Fluid Changed	Fluid Type	Total Hours	Fluid Hours
72	14	4	<5	<0.05	NA	NA	40.1	NA	0.67	NO	ATF	4500	500

POWERTRAIN OIL ANALYSIS	ADT with Allison Ecomat™ transmission. Factory fill Ecofluid Plus A™. Service fluid TranSynd™. Mixing and metals

FIGURE 11-19   Oil analysis report for a ZF Ecomat 2 powershift transmission.

To understand abnormal metal generation, we need to focus on dirt, water, TAN, and fluid signature. Are those in line? Silicon at 14 ppm is abnormal. Although the ratio Si:Al is reversed, indicating that most of the aluminum is from torque converter and retarder wear, there is still some dirt. In contrast, there is no water, and the TAN is still very low. Viscosity at 40.1 cSt is right on the spot for Ecofluid Plus A. What about the fluid signature?

Conclusion
Ecofluid Plus A and TranSynd service fluids are zinc-free fluids. The results show the presence of calcium, zinc, and boron, which suggest that this fluid is a mix with a calcium-based fluid that possibly is less than ideal for this machine. We can conclude that there is dirt and an antagonistic mixture of incompatible fluids in this transmission that are triggering these metal readings.

Wear Metal Thresholds
See Figure 11-20 and Table 11-7.

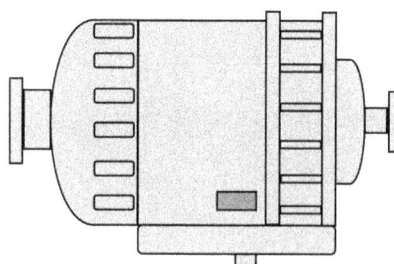

FIGURE 11-20   ZF Ecomat 2 powershift transmission.

TABLE 11-7 Wear Table for ZF Ecomat 2 Powershift Transmission

Filtered System	Per 500 hours		
	Normal	Abnormal	Critical
Fe	<70	70	>123
Pb	<13	13	>24
Cu	<28	28	>50
Cr	<5	5	>10
Al	<61	61	>105
Si	<16	16	>24

7. Water and Dirt in an ADT with an Allison HD 4500R Transmission

Oil Analysis

The oil analysis report in Figure 11-21 is for an Allison 4500R powershift transmission. The lab has flagged water and a high TAN. Also flagged are iron, lead, copper, and aluminum. Is the water causing these readings? Is the TAN causing the readings, or is the water causing the TAN to be abnormal? Is dirt an issue?

METALS							ADDITIVES							HOURS INFO	
Iron (Fe)	Chromium (Cr)	Lead (Pb)	Copper (Cu)	Tin (Sn)	Aluminum (Al)	Nickel (Ni)	Magnesium (Mg)	Calcium (Ca)	Barium Ba)	Phosphorus (P)	Zinc (Zn)	Molybdenum (Mo)	Boron (B)	Total Hous	Hours on Oil
48	3	110	10	2	5	<1	420	5248	157	1654	1889	1	<5	520,000	101,195
15	<1	16	4	<1	4	<1	427	4251	157	1654	1561	<1	<5	418,805	45,450

CONTAMINANTS								PHYSICAL PROPERTIES						OIL INFO	
Aluminum (Al)	Silicon (Si)	Sodium (Na)	Potassium (K)	Water (%)	Glycol	Fuel (%)	Soot (%) FTR	Sulphation	Nitration	Viscosity (cSt @ 100 C)	Oxidation	TBN	TAN	Type	Viscosity
5	5	9	5	<0.5	NO	<1	0.3	NA	NA	17.1	NA	4.6	NA	CK4	15W40
4	5	4	<1	<0.5	NO	<1	<0.1	NA	NA	16.7	NA	6.3	NA	CK4	15W40

ENGINE OIL ANALYSIS | Cummins Diesel Engine with by-pass filtration.

FIGURE 11-21 Oil analysis report for an Allison 4500R powershift transmission.

Conclusion

The fluid signature matches that of TranSynd, and the viscosity is a little high, as is the TAN. The threshold for silicon on this transmission is <8 ppm, and there is more than enough aluminum to make dirt a likely suspect. Because water and dirt are present, it is reasonable to conclude that this is a case of contamination that occurred during use. The presence of water has caused the TAN to increase. The metal readings are a consequence of the dirt and water contamination.

Wear Metal Thresholds

See Figure 11-22 and Table 11-8.

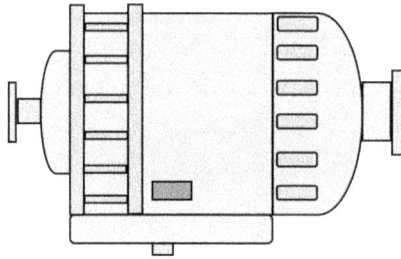

FIGURE 11-22   Allison 4500R powershift transmission.

TABLE 11-8   Wear Table for Allison 4500R Powershift Transmission

Per 500 hours			
Filtered System	**Normal**	**Abnormal**	**Critical**
Fe	<90	90	>150
Pb	<32	32	>51
Cu	<14	14	>23
Cr	<5	5	>10
Al	<21	21	>30
Si	<8	8	>15

8. Low Viscosity in an ADT ZF Transmission with Integral Transfer Case

Oil Analysis

The oil analysis report in Figure 11-23 shows excellent readings for this transmission. However, the viscosity does not appear to match that of the fluid description. This transmission uses 10W-30 engine oil, and the viscosity is in the 47-cSt range. Is the viscosity an issue?

METALS							ADDITIVES						
Iron (Fe)	Chromium (Cr)	Lead (Pb)	Copper (Cu)	Tin (Sn)	Aluminum (Al)	Nickel (Ni)	Magnesium (Mg)	Calcium (Ca)	Barium (Ba)	Phosphorus (P)	Zinc (Zn)	Molybdenum (Mo)	Boron (B)
86	<1	1	24	<1	2	<1	566	1857	<1	1191	1204	21	<5
72	<1	2	14	<1	2	<1	505	1954	<1	1235	1451	30	<5
71	<1	4	9	<1	2	<	262	2554	<1	1168	1393	4	<5

CONTAMINANTS				WEAR		PHYSICAL PROPERTIES			INFORMATION			
Aluminum (Al)	Silicon (Si)	Sodium (Na)	Potassium (K)	Water (%)	PQ Index	Particle Counts	Viscosity (cSt @ 40C)	Oxidation	TAN	Fluid Changed	Fluid Type	Total Hours	Fluid Hours
3	7	19	<5	<0.05	27	NA	46.5	NA	1.69	NO	EO	1500	1500
2	11	22	<5	<0.05	39	NA	47.3	NA	1.85	NO	EO	1000	1000
2	11	37	6	<0.05	16	NA	48.5	NA	1.62	NO	EO	500	500

POWERTRAIN OIL ANALYSIS	ADT ZF® Powershift Transmission with integral transfer case. 10W30 engine oil

FIGURE 11-23 Oil analysis report for an ADT ZF powershift transmission.

Conclusion

Table 2-12 is a table for viscosity decay of an engine oil when used in a transmission with a torque converter, and it tells us that the viscosity of SAE 10W-30 engine oil in this application at 1,000 hours should be around 47 cSt. We can conclude that this is the nature of the shear-down of engine oil in a transmission with a torque converter.

Wear Metal Thresholds

See Figure 11-24 and Table 11-9.

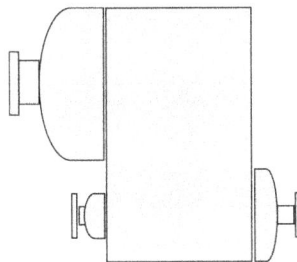

FIGURE 11-24 ADT ZF powershift transmission.

TABLE 11-9 Wear Table for ADT ZF Powershift Transmission

Per 1,000 hours			
Filtered System	Normal	Abnormal	Critical
Fe	<108	108	>135
Pb	<5	5	>10
Cu	<24	24	>30
Cr	<5	5	>10
Al	<10	10	>15
Si	<15	15	>18

9. Low Viscosity in a Backhoe ZF Transmission with High Silicon and Aluminum Readings

Oil Analysis

The oil analysis report in Figure 11-25 shows the challenges of dirt determination where aluminum from wear could mask alumina from dirt. Are the silicon readings coming from dirt? Is the viscosity too low for this fluid?

METALS							ADDITIVES						
Iron (Fe)	Chromium Cr)	Lead (Pb)	Copper (Cu)	Tin (Sn)	Aluminum (Al)	Nickel (Ni)	Magnesium (Mg)	Calcium (Ca)	Barium Ba)	Phosphorus (P)	Zinc (Zn)	Molybdenum (Mo)	Boron (B)
40	<1	<1	2	<1	16	<1	12	2095	<1	1617	1572	<1	88
35	<1	<1	1	<1	15	<1	16	2731	<1	1616	2094	<1	96
52	<1	<1	2	<1	22	<	29	2317	<1	1606	1960	<1	80

CONTAMINANTS					WEAR		PHYSICAL PROPERTIES			INFORMATION			
Aluminum (Al)	Silicon (Si)	Sodium (Na)	Potassium (K)	Water (%)	PQ Index	Particle Counts	Viscosity (cSt @ 40C)	Oxidation	TAN	Fluid Changed	Fluid Type	Total Hours	Fluid Hours
16	17	2	12	<0.05	11	NA	27.3	NA	2.48	NO	THF	3000	1000
15	13	2	8	<0.05	24	NA	29.9	NA	2.05	YES	THF	2000	2000
22	11	2	9	<0.05	17	NA	25.7	NA	1.83	NO	THF	1000	1000

POWERTRAIN OIL ANALYSIS	Backhoe ZF® Powershift Transmission. Tractor Fluid ISO 32.

FIGURE 11-25 Oil analysis report for a backhoe ZF powershift transmission.

The example is doubly challenging because the fluid in use is a clone of Shell Donax TD with known high doses of silicon foam inhibitor. This illustrates the importance of knowing the signature of the fluid in use. Instead of dirt, what we have here is aluminum from torque converter erosion combined with foam inhibitor from the fluid. This example is a real case that highlights the areas of knowledge required to perform advanced oil analysis interpretation.

Conclusion

Table 12-2 provides information on the signature of the fluid and the expected viscosity after hours of service. For 1,000 hours, the viscosity is correct for the fluid, which was originally 35 cSt. With regard to silicon readings, the fluid contains 13 ppm of antifoam additive, so we can conclude that this transmission is doing fine and that the aluminum readings are not from dirt but from torque converter erosion.

Wear Metal Thresholds

See Figure 11-26 and Table 11-10.

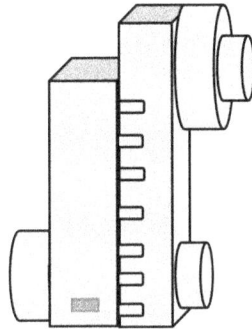

FIGURE 11-26   A backhoe ZF powershift transmission.

TABLE 11-10   Wear Table for a Backhoe ZF Powershift Transmission

Per 1,000 hours			
Filtered System	Normal	Abnormal	Critical
Fe	<50	50	>75
Pb	<7	7	>15
Cu	<35	35	>55
Cr	<5	5	>10
Al	<13	13	>20
Si	<10	10	>25

10. High Iron Content in an Unfiltered 4WD Loader Axle with Multidisc Brakes

Oil Analysis

The differential on this 4WD loader is showing copious amounts of iron at 1,000 hours (see Figure 11-27). This tendency appears on all previous tests. What is causing these high readings? Are the lead and silicon readings worrisome? The large braking area of multiple discs in machines like this produces copious amounts of iron and releases blackening frictional material into the oil, which is not visible in the oil analysis.

The readings captured by oil analyses from these unfiltered axles show high iron levels. Filtered axles produce about a quarter of the metal readings and achieve long trouble-free hours of operation. Iron in small amounts does not have a measurable negative impact, but when the amount grows, it becomes aggressive toward bearings. When iron production exceeds ~1.5 ppm per hour, wear starts to grow exponentially. This also increases the rate at which oil acidity increases.

METALS							ADDITIVES						
Iron (Fe)	Chromium (Cr)	Lead (Pb)	Copper (Cu)	Tin (Sn)	Aluminum (Al)	Nickel (Ni)	Magnesium (Mg)	Calcium (Ca)	Barium Ba)	Phosphorus (P)	Zinc (Zn)	Molybdenum (Mo)	Boron (B)
1637	6	21	10	<1	5	4	151	3497	<1	1096	1211	6	<5
3366	10	55	15	<1	10	4	140	3585	<1	1160	1220	22	<5
2999	9	54	11	<1	11	1	161	3736	<1	1285	1464	32	<5

CONTAMINANTS					WEAR		PHYSICAL PROPERTIES			INFORMATION			
Aluminum (Al)	Silicon (Si)	Sodium (Na)	Potassium (K)	Water (%)	PQ Index	Particle Counts	Viscosity (cSt @ 40C)	Oxidation	TAN	Fluid Changed	Fluid Type	Total Hours	Fluid Hours
5	26	3	5	<0.05	NA	NA	50.3	NA	0.88	YES	THF	2724	1000
10	44	8	<5	0.05	NA	NA	47.7	NA	3.22	YES	THF	1724	1000
11	44	7	<5	0.05	NA	NA	48.6	NA	3.46	YES	THF	724	724

POWERTRAIN OIL ANALYSIS	4WD Unfiltered Multi-Disc Brake Axle.

FIGURE 11-27 Oil analysis report for an unfiltered 4WD loader axle with multidisc brakes.

Conclusion

When an axle without filtration has dirt, the result is an exponentially increasing wear rate. Brake piston seals also suffer as iron sediment and brake material accumulate at the bottom of the piston annular cavity, disturbing the sealing area and scuffing the seals. A fluid flush and the installation of filters/cooling would be a sensible solution.

Wear Metal Thresholds

See Figure 11-28 and Table 11-11.

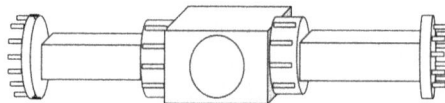

FIGURE 11-28 An unfiltered 4WD loader axle with multidisc brakes.

TABLE 11-11 Wear Table for an Unfiltered 4WD Loader Axle with Multidisc Brakes

Per 500 hours			
Nonfiltered System	Normal	Abnormal	Critical
Fe	<1100	1100	>2000
Pb	<8	8	>11
Cu	<90	90	>150
Cr	<7	7	>15
Al	<15	15	>35

11. Water in a Filtered 4WD Loader Axle with Multidisc Brakes

Oil Analysis

Machines with axle filtration are noted in their oil analyses for their clean readings. It does not take a lot of knowledge to realize that filters and cooling improve axle life severalfold.

However, they can suffer from other types of contamination. What causes water contamination in this axle? Is silicon an indication of dirt? The results in Figure 11-29 suggest this axle has been getting some water since the previous test. Together with the water, some readings of silicon are approaching the abnormal level, although there is no aluminum so as to blame it on dirt.

METALS							ADDITIVES						
Iron (Fe)	Chromium Cr)	Lead (Pb)	Copper (Cu)	Tin (Sn)	Aluminum (Al)	Nickel (Ni)	Magnesium (Mg)	Calcium (Ca)	Barium Ba)	Phosphorus (P)	Zinc (Zn)	Molybdenum (Mo)	Boron (B)
126	<1	37	2	1	<1	<1	26	2591	<1	1323	1500	83	<2
132	<1	29	1	1	<1	<1	29	2702	<1	1316	1538	80	<2
169	<1	26	2	<1	<	<	30	2764	<1	1295	1595	89	<2

CONTAMINANTS					WEAR		PHYSICAL PROPERTIES			INFORMATION			
Aluminum (Al)	Silicon (Si)	Sodium (Na)	Potassium (K)	Water (%)	PQ Index	Particle Counts	Viscosity (cSt @ 40C)	Oxidation	TAN	Fluid Changed	Fluid Type	Total Hours	Fluid Hours
<1	22	3	<5	0.15	NA	NA	53.1	NA	1.5	NO	THF	3000	1000
<1	12	2	<5	0.14	NA	NA	45.4	NA	1.8	YES	THF	2000	2000
<	11	3	<5	0.06	NA	NA	54.1	NA	1.7	NO	THF	1000	500
POWERTRAIN OIL ANALYSIS	4WD Multi-Disc Brake Axle. Tractor Fluid (57 cSt @ 40° C). Water present												

FIGURE 11-29 Oil analysis report for a water-filtered 4WD loader axle with multidisc brakes.

Conclusion

Based on Tables 2-12 and 2-13, this tractor fluid appears to be behaving as expected in terms of viscosity loss. The silicon number suggests that it is from a foam inhibitor additive in this fluid. Filtration has helped to keep dirt at bay, as well as metal production. The TAN is normal for a tractor fluid. However, the TAN will grow relatively quickly over time due to the excessive water. Consequently, this axle needs a flush.

Wear Metal Thresholds

See Figure 11-30 and Table 11-12.

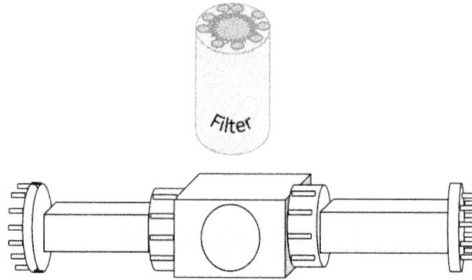

FIGURE 11-30   A filtered 4WD loader axle.

TABLE 11-12   Wear Table for a Filtered 4WD Loader Axle

Per 1,000 hours			
Filtered System	Normal	Abnormal	Critical
Fe	<250	250	>475
Pb	<35	35	>40
Cu	<25	25	>30
Cr	<5	5	>10
Al	<5	5	>10
Si	<15	15	>25

12. Lead and Copper in an Axle with Single-Disc Brakes

The single-disc-brake design produces fewer contamination by-products from disc and frictional material and runs cooler than the multidisc design. Inherent to the design is the tendency to produce chatter. However, with the correct fluid containing the proper friction modifiers, this condition improves.

Oil Analysis

The oil analysis results in Figure 11-31 indicate that this axle is producing excessive lead and copper. Where are the lead and copper readings coming from? Why is the chromium reading so small?

The results show that silicon is not from dirt because the aluminum readings are low but rather from an antifoam additive. There is no water, and the viscosity suggests that the fluid does indeed have 1,500 hours on it. So what could be causing the metal readings?

METALS							ADDITIVES						
Iron (Fe)	Chromium Cr	Lead (Pb)	Copper (Cu)	Tin (Sn)	Aluminum (Al)	Nickel (Ni)	Magnesium (Mg)	Calcium (Ca)	Barium Ba)	Phosphorus (P)	Zinc (Zn)	Molybdenum (Mo)	Boron (B)
118	1	46		<1	1	<1	127	3293	<1	1064	1327	1	<5
110	<1	42		<1	5	<1	143	3301	<1	1214	1407	1	<5
90	<1	38		<1	3	<1	135	3250	<1	1150	1400	1	<5

CONTAMINANTS				WEAR			PHYSICAL PROPERTIES			INFORMATION			
Aluminum (Al)	Silicon (Si)	Sodium (Na)	Potassium (K)	Water (%)	PQ Index	Particle Counts	Viscosity (cSt @ 40C)	Oxidation	TAN	Fluid Changed	Fluid Type	Total Hours	Fluid Hours
1	6	5	6	<0.05	21	NA	48.5	NA	3.26	YES	THF	2500	1500
5	16	5	6	<0.05	17	NA	50.6	NA	2.12	NO	THF	1000	1000
3	12	4	5	<0.05	15	NA	51	NA	3.1	NO	THF	500	500

POWERTRAIN OIL ANALYSIS	4WD Loader Filtered Single Disc Brake Axle. Tractor Fluid (57 cSt @ 40 C). Lead and Copper readings

FIGURE 11-31   Oil analysis results for a 4WD loader with a filtered single-disc brake axle.

Conclusion

Copper and lead are going to be the visible metals because spider gears within the differential use backup bronze thrust washers. This machine is borderline in terms of lead production. Any issues with contamination within this axel design will be visible through those readings. These axles may contain carburized bearings, which do not produce chromium wear readings. However, it is still important to monitor chrome readings because the axle still may contain plenty of needle bearings in final drives that could show chromium readings in the oil analysis.

Wear Metal Thresholds

See Figure 11-32 and Table 11-13.

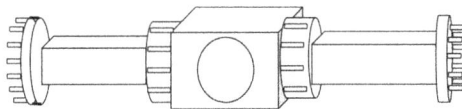

FIGURE 11-32   Filtered single-disc brake axle.

TABLE 11-13 Wear Table for a Filtered Single-Disc Brake Axle

	Per 1,000 hours		
Filtered System	Normal	Abnormal	Critical
Fe	<550	550	>850
Pb	<45	45	>80
Cu	<35	35	>60
Cr	<5	5	>10
Al	<5	5	>10
Si	<25	25	>50

13. Iron Reading/Type of Oil in a Small 4WD Loader ZF Common-Sump Multidisc-Brake Outboard Planetary Axle

ZF axles as used on small loaders have an outboard multidisc wet brake component with a unified sump. That is, final drives share oil from the differential. The axle could have a hydraulic operated differential lock, oil circulation, and filtration.

Oil Analysis

The oil analysis results in Figure 11-33 show that this axle is using a tractor fluid. The wear guidelines shown in Figure 11-34 are for this axle when using tractor fluid, even though the manufacturer of this axle recommends gear oil, not tractor fluid.

METALS							ADDITIVES						
Iron (Fe)	Chromium Cr	Lead (Pb)	Copper (Cu)	Tin (Sn)	Aluminum (Al)	Nickel (Ni)	Magnesium (Mg)	Calcium (Ca)	Barium Ba)	Phosphorus (P)	Zinc (Zn)	Molybdenum (Mo)	Boron (B)
2871	10	12	25	<1	8	<1	117	3414	<1	1042	1416	1	<5
2950	11	29	22	<1	2	3	105	3841	<1	1487	1890	6	38
724	3	10	11	<1	2	1	102	3873	<1	1286	1800	3	35
213	2	8	6	<1	2	<1	80	3569	<1	1175	1569	6	46

CONTAMINANTS					WEAR		PHYSICAL PROPERTIES			INFORMATION			
Aluminum (Al)	Silicon (Si)	Sodium (Na)	Potassium (K)	Water (%)	PQ Index	Particle Counts	Viscosity (cSt @ 40C)	Oxidation	TAN	Fluid Changed	Fluid Type	Total Hours	Fluid Hours
8	14	3	6	<0.05	NA	NA	51	NA	2.54	YES	THF	4157	1000
2	18	8	<5	<0.05	NA	NA	52.1	NA	2.2	NO	THF	2000	500
2	17	4	5	<0.05	NA	NA	55.3	NA	3.07	YES	THF	1000	1000
2	22	7	6	<0.05	NA	NA	57	NA	2.07	NO	THF	500	500
POWERTRAIN OIL ANALYSIS			Small 4WD Loader ZF® Common Sump Multi Disk Outboard Planetary Axle. Tractor Fluid (57 cSt @ 40° C)										

FIGURE 11-33 Oil analysis report for a small 4WD loader
ZF common-sump multidisc-brake outboard planetary axle.

Conclusion

The wear metal thresholds are high for these axles when they are used without filtration. Axles without filtration and using tractor fluid produce iron at incremental rates exceeding 1 ppm/hour starting at about 500 hours. Compare the results to the same type of axle that uses tractor fluid but has filtration installed (Figure 11-34). The analysis suggests that this ZF axel with filtration and cooling is using a low-viscosity tractor fluid. However, filtration and cooling have reduced wear significantly, as reflected in the metal readings.

METALS							ADDITIVES						
Iron (Fe)	Chromium Cr)	Lead (Pb)	Copper (Cu)	Tin (Sn)	Aluminum (Al)	Nickel (Ni)	Magnesium (Mg)	Calcium (Ca)	Barium Ba)	Phosphorus (P)	Zinc (Zn)	Molybdenum (Mo)	Boron (B)
88	<1	<1	19	<1	<1	2	16	2412	<1	1331	1566	8	92

CONTAMINANTS				WEAR		PHYSICAL PROPERTIES			INFORMATION			
Aluminum (Al)	Silicon (Si)	Sodium (Na)	Potassium (K)	Water (%)	PQ Index	Particle Counts	Viscosity (cSt @ 40C)	Oxidation	TAN	Fluid Changed	Fluid Type	Total Hours	Fluid Hours
<1	15	4	8	<0.05	NA	NA	31.8	NA	2.06	NO	THF	500	500

POWERTRAIN OIL ANALYSIS — Small 4WD Loader ZF® Common Sump Multi-Disc Outboard Planetary Axle. Tractor Fluid ISO 32

FIGURE 11-34 Comparison results.

Wear Metal Thresholds

See Figure 11-35 and Table 11-14.

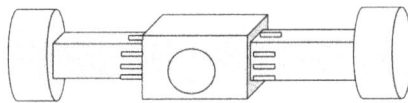

FIGURE 11-35 ZF common-sump multidisc-brake outboard planetary axle.

TABLE 11-14 Wear Table for a ZF Common-Sump Multidisc-Brake Outboard Planetary Axle

Per 500 hours			
Nonfiltered System	Normal	Abnormal	Critical
Fe	<1500	1500	>2500
Pb	<15	15	>30
Cu	<45	45	>70
Cr	<28	28	>48
Al	<15	15	>30
Si	<30	30	>40

Thresholds for machines using gear oil instead of the tractor fluid produce much better readings. The normal iron readings for axles without filtration using gear oils is <550 ppm compared with <1,500 ppm for the same axles using tractor fluids. This means that units using tractor fluids produce 2.72 times more iron than those using gear oil. Thresholds for units using gear oil are given in Table 11-15.

TABLE 11-15 Iron Readings for ZF Common-Sump
Compact Axles with Gear Oil (Multidisc Outboard Planetary)

Normal	Abnormal	Critical
<550	550	>969

14. Wrong Oil for a Liebherr-Spicer Outboard Final Drive with an Inboard Multidisc-Brake Axle

Liebherr-Spicer axles as used on compact loaders are outboard planetary units with inboard wet multidisc brakes but no common sump. No cooling or filtration is available for these axles. They use API GL5 gear oil with an LS additive. What causes the high iron readings? What lubricant is the axle in Figure 11-36 using?

METALS							ADDITIVES						
Iron (Fe)	Chromium (Cr)	Lead (Pb)	Copper (Cu)	Tin (Sn)	Aluminum (Al)	Nickel (Ni)	Magnesium (Mg)	Calcium (Ca)	Barium (Ba)	Phosphorus (P)	Zinc (Zn)	Molybdenum (Mo)	Boron (B)
1071	5	<1	17	<1	5	1	80	2502	3	1166	1136	2	<5
1227	6	<1	9	<1	5	<1	106	2500	6	1297	1296	1	<5
500	3	<1	12	<1	3	<1	110	2520	5	1301	1309	1	<5

CONTAMINANTS					WEAR		PHYSICAL PROPERTIES			INFORMATION			
Aluminum (Al)	Silicon (Si)	Sodium (Na)	Potassium (K)	Water (%)	PQ Index	Particle Counts	Viscosity (cSt @ 40C)	Oxidation	TAN	Fluid Changed	Fluid Type	Total Hours	Fluid Hours
5	36	20	<5	<0.05	489	NA	51.8	NA	2.33	NO	THF	614	614
5	29	22	6	<0.05	404	NA	52.4	NA	2.54	NO	THF	500	500
3	20	15	6	<0.05	257	NA	53.6	NA	2.42	NO	THF	100	100

POWERTRAIN OIL ANALYSIS	4WD Loader Liebherr-Spicer™ Multiple Inboard Disc Brake Axle.

FIGURE 11-36 Oil analysis report for a 4WD loader with a Liebherr-Spicer inboard multidisc-brake axle.

Oil Analysis

From the analysis, we can see that the axle is not using the proper oil. The manufacturer calls for GL5 type gear oil with a friction modifier for limited-slip differentials, and this axle is using tractor fluid. Also flagged are silicon and viscosity. However, this tractor fluid appears to be a clone of Donax TD with at least 19 ppm of silicon as a foam inhibitor. With regard to the flagged viscosity, the lab was correct in flagging it. Tractor fluids stand little chance against the inboard brakes that tear up viscosity faster than in other applications. However, even when a typical tractor fluid is new, it has less than half the viscosity of the recommended fluid, let's say a 75W-90 gear oil.

Conclusion

When using an oil of the wrong viscosity and without the recommended additives, the axle will produce copious amounts of metals.

Wear Metal Thresholds

See Figure 11-37 and Table 11-16.

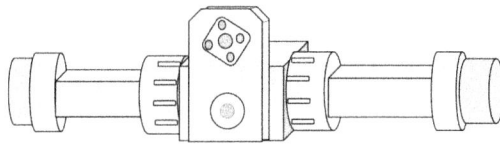

FIGURE 11-37   Liebherr-Spicer inboard multidisc-brake axle.

TABLE 11-16   Wear Table for a Liebherr-Spicer Inboard Multidisc-Brake Axle

Per 500 hours			
Nonfiltered System	**Normal**	**Abnormal**	**Critical**
Fe	<1500	1500	>2500
Pb	<15	15	>30
Cu	<45	45	>70
Cr	<28	28	>48
Al	<15	15	>40
Si	<30	30	>40

This axle uses gear oil with a Lubrizol friction modifier; the TAN numbers are as shown in Table 11-17.

TABLE 11-17   TANs for Axles with Lubrizol Friction Modifier

Normal	Abnormal	Critical
3.5–5.9	6.0–6.9	>9

15. High Iron in a ZF Backhoe Inboard Final Drive Axle

ZF axles as used on backhoes are inboard planetary axles with an inboard multidisc-brake system. The parking brake is on the input shaft and is a multidisc configuration. The axle includes a hydraulic differential lock. Gear oil is the recommended fluid for this type of axel. What caused the high iron readings in Figure 11-38?

METALS							ADDITIVES						
Iron (Fe)	Chromium Cr	Lead (Pb)	Copper (Cu)	Tin (Sn)	Aluminum (Al)	Nickel (Ni)	Magnesium (Mg)	Calcium (Ca)	Barium Ba)	Phosphorus (P)	Zinc (Zn)	Molybdenum (Mo)	Boron (B)
1007	4	1	25	<1	1	1	159	3005	<1	1110	1263	<1	<5
1993	7	2	32	4	2	<1	143	3089	<1	1054	1167	<1	<5

CONTAMINANTS				WEAR			PHYSICAL PROPERTIES			INFORMATION			
Aluminum (Al)	Silicon (Si)	Sodium (Na)	Potassium (K)	Water (%)	PQ Index	Particle Counts	Viscosity (cSt @ 40C)	Oxidation	TAN	Fluid Changed	Fluid Type	Total Hours	Fluid Hours
<1	12	3	<5	<0.05	NA	NA	54.4	NA	2.06	NO	THF	714	548
4	21	6	<5	0.05	NA	NA	55.3	NA	1.06	NO	THF	100	100

POWERTRAIN OIL ANALYSIS	Backhoe ZF® Axle. Tractor Fluid (ISO 68). High Iron Readings

FIGURE 11-38   Oil analysis report for a ZF backhoe inboard final drive axle.

Oil Analysis

This example is a result of break-in. When a machine or component is new, wear metals are increased in the beginning as rubbing surfaces lose surface asperities. Iron levels are particularly dependent on utilization during break-in.

While there are machines that do digging work all day long, others work as truck loaders and travel long distances to the job site. This will have a direct impact on iron readings. Silicon readings are also high from washing out of sealant materials.

Conclusion

This ZF axle produces more than 2 ppm/hour. Although this is the second fill, the fluid should be changed again to flush out additional break-in wear. The design of this axle, where final drives and a multidisc-brake system are installed inboard, makes it prone to running hot, contributing to the high iron production. When hauling or used as a loader, the iron production goes up. So monitoring iron production and flushing the fluid when numbers exceed certain thresholds are sound decisions.

Wear Metal Thresholds

See Figure 11-39 and Table 11-18.

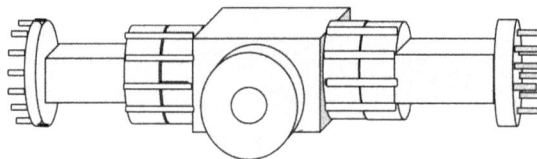

FIGURE 11-39   ZF backhoe inboard final drive axle.

TABLE 11-18 Wear Table for a ZF Backhoe Inboard Final Drive Axle

Per 500 hours			
Nonfiltered System	Normal	Abnormal	Critical
Fe	<1100	1100	>2000
Pb	<8	8	>11
Cu	<90	90	>150
Cr	<7	7	>15
Al	<15	15	>35
Si	<80	80	>140

16. Perfect Readings in a ZF Outboard Final Drive Axle

Several backhoe models on the market use ZF outboard final drive axles. This axle uses a wet multidisc-brake system but with a change in final drive location compared with the preceeding axle. This final drive location allows a lower operating temperature and less metal production for this type of axle. The ZF backhoe axle does not have a common sump; thus, every compartment needs oil level inspections and service.

Oil Analysis

The oil analysis readings in Figure 11-40 are good, except for a small viscosity loss. The axle is using a multiviscosity tractor fluid, which has sheared down as expected for this type of fluid.

METALS							ADDITIVES						
Iron (Fe)	Chromium (Cr)	Lead (Pb)	Copper (Cu)	Tin (Sn)	Aluminum (Al)	Nickel (Ni)	Magnesium (Mg)	Calcium (Ca)	Barium (Ba)	Phosphorus (P)	Zinc (Zn)	Molybdenum (Mo)	Boron (B)
36	<1	<1	22	<1	5	<1	115	3757	3	1147	1335	<1	5
20	<1	<1	11	<1	3	<1	110	3390	6	1088	1302	<1	<5
17	<1	<1	6	<1	2	<1	111	3226	5	1065	1294	<1	<5

CONTAMINANTS				WEAR		PHYSICAL PROPERTIES			INFORMATION			
Aluminum (Al)	Silicon (Si)	Sodium (Na)	Potassium (K)	Water (%)	PQ Index	Particle Counts	Viscosity (cSt @ 40C)	Oxidation	TAN	Fluid Changed	Fluid Type	Total Hours	Fluid Hours
5	17	6	<5	<0.05	34	NA	52.1	NA	1.73	NO	THF	954	954
3	16	6	<5	<0.05	<10	NA	54.9	NA	1.83	YES	THF	440	440
2	15	4	<5	<0.05	13	NA	55.7	NA	2.17	NO	THF	285	285

POWERTRAIN OIL ANALYSIS — Backhoe with ZF Outboard Multi Brake Disk Axle. Tractor Fluid (57 cSt @ 40° C)

FIGURE 11-40 Oil analysis report for ZF outboard final drive axle.

Conclusion

Not all machines are equal. They may appear similar to the eye, but each one has its own set of characteristics and needs special attention from a maintenance point of view. The rules that apply to one machine do not necessarily apply to another.

Wear Metal Thresholds

See Figure 11-41 and Table 11-19.

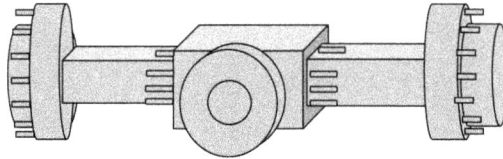

FIGURE 11-41   ZF outboard final drive axle.

TABLE 11-19   Wear Table for a ZF Outboard Final Drive Axle

Per 500 hours			
Nonfiltered System	Normal	Abnormal	Critical
Fe	<500	500	>1000
Pb	<8	8	>11
Cu	<70	70	>140
Cr	<7	7	>15
Al	<21	21	>30
Si	<30	30	>40

17. High Iron Readings and Low TAN in a ZF 240 Large Loader Axle

The ZF 240 large loader axle is a heavy-duty hypoid type of axle, but at the same time it is small. This helps with good ground clearance and installation engineering. The axel has a common-sump design, with the final drives and differential sharing the same oil. The front axle is typically equipped with an LS differential and has outboard wet multidisc brakes. The axle requires an oil pump for cooling purposes. Why did this axle produce excessive iron? How can we determine that the right oil is being used?

In a hypoid gear (see Figure 11-42), by design, the pinion gear has a very acute angle of attack. A *hypoid* is a type of spiral bevel gear whose axis does not intersect with the axis of the meshing gear. This inherent gear engagement increases both tooth sliding area and pressure between the gears. This, in turn, requires an extreme-pressure (EP) gear oil with a strong friction modifier to protect the gears from scuffing wear, and cooling is needed to prevent heat from thinning the oil excessively.

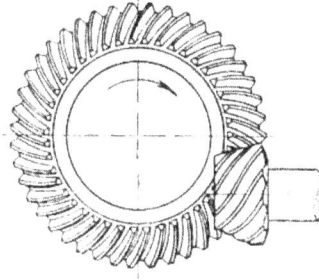

FIGURE 11-42 Hypoid gear.

Oil Analysis

The results from this axle show that at one point the lubricant had less phosphorus and a lower TAN (see Figure 11-43). This indicates that the lubricant is a normal GL5 gear oil without the LS friction modifier recommended by the manufacturer. Therefore, the axle lacks the appropriate antiscuffing protection.

METALS							ADDITIVES						
Iron (Fe)	Chromium Cr	Lead (Pb)	Copper (Cu)	Tin (Sn)	Aluminum (Al)	Nickel (Ni)	Magnesium (Mg)	Calcium (Ca)	Barium Ba	Phosphorus (P)	Zinc (Zn)	Molybdenum (Mo)	Boron (B)
101	<1	<1	2	<1	<1	<1	2	14	<1	2329	12	<1	206
554	<1	10	15	<1	3	<1	6	23	<1	1500	22	<1	238
185	1	2	2	1	1	<1	52	36	<1	2017	106	2	216

CONTAMINANTS					WEAR		PHYSICAL PROPERTIES			INFORMATION			
Aluminum (Al)	Silicon (Si)	Sodium (Na)	Potassium (K)	Water (%)	PQ Index	Particle Counts	Viscosity (cSt @ 40C)	Oxidation	TAN	Fluid Changed	Fluid Type	Total Hours	Fluid Hours
<1	6	3	<5	<0.05	<10	NA	113.6	NA	4.47	NO	GO	5454	100
3	14	1	<5	<0.05	150	NA	106.8	NA	2.2	YES	GO	4500	1000
1	4	1	<5	<0.05	31	NA	104.5	NA	3.41	YES	GO	3500	1000
POWERTRAIN OIL ANALYSIS			4WD Loader with ZF AP240™ Axle. GL5 80W90 with Specialty Friction Modifier										

FIGURE 11-43 Oil analysis results for a ZF 240 large loader hyoid axle.

Conclusion

Friction modifier additive is essential to the life of this axle, so it is critical that the correct fluid is used. These fluids typically have 1600 to 2500 ppm of phosphorus and a TAN from 3.5 to 5.9. If the correct fluid is not available, LS additives are available in the aftermarket and can be added to GL5 gear lubes, although there are different types of friction modifiers, and each application requires the correct one.

Wear Metal Thresholds

See Figure 11-44 and Table 11-20.

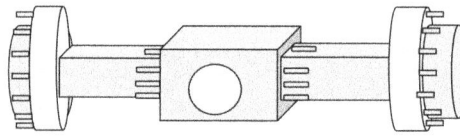

FIGURE 11-44 ZF 240 large loader hyoid axle.

TABLE 11-20 Wear Table for a ZF 240 Large Loader Hyoid Axle

Per 1,000 hours			
Nonfiltered System	Normal	Abnormal	Critical
Fe	<375	375	>680
Pb	<8	8	>15
Cu	<20	20	>35
Cr	<5	5	>15
Al	<6	6	>15
Si	<20	20	>35

18. Abnormal Additive Readings in a 4WD Loader ZF 240 Axle

Oil Analysis

The oil analysis report in Figure 11-45 is for another 4WD loader ZF axle, showing abnormal additive readings together with high iron production and dirt. Why is the lubricant showing abnormal additive readings? What is causing the high iron and copper readings?

METALS							ADDITIVES						
Iron (Fe)	Chromium (Cr)	Lead (Pb)	Copper (Cu)	Tin (Sn)	Aluminum (Al)	Nickel (Ni)	Magnesium (Mg)	Calcium (Ca)	Barium Ba)	Phosphorus (P)	Zinc (Zn)	Molybdenum (Mo)	Boron (B)
358	0	2	76	0	8	4	11	257	0	1612	136	11	146
353	0	0	89	0	11	5	7	250	0	2237	165	7	70
485	0	0	148	0	15	7	11	288	9	2294	188	11	84

CONTAMINANTS					WEAR		PHYSICAL PROPERTIES			INFORMATION			
Aluminum (Al)	Silicon (Si)	Sodium (Na)	Potassium (K)	Water (%)	PQ Index	Particle Counts	Viscosity (cSt @ 40C)	Oxidation	TAN	Fluid Changed	Fluid Type	Total Hours	Fluid Hours
8	20	1	0	<0.05	NA	NA	101	NA	2.7	YES	THF	6000	500
11	47	4	0	<0.05	NA	NA	98	NA	2.8	YES	THF	5500	1000
15	73	6	4	<0.05	NA	NA	93	NA	3	YES	THF	4500	1000
POWERTRAIN OIL ANALYSIS			4WD Loader with ZF AP420™ Axle. GL5 80W90 with abnormal additive readings and lacking friction modifier										

FIGURE 11-45 Oil analysis report for another 4WD loader ZF axle.

The results from the oil analysis show that the oil has a different signature and lower TAN than the recommended fluid. There is dirt, based on aluminum and silicon readings, and iron and copper production are beyond wear metal thresholds.

Conclusion

What is really happening here is a leak through the brake piston seals, allowing transmission fluid (tractor fluid) to fill the differential case. As seen in the preceding case, this axle type requires a gear oil with a special friction modifier. Because acid numbers do not go down so fast, the TAN shows a reduction due to contamination with another type of oil. This changes the gear oil formulation, which together with dirt produces copious amounts of iron and copper relative to the hours of use.

19. Wrong Oil and Low TAN in an ADT Dana Axle

An axle from an ADT is showing increased metal production. The user has recently changed the oil in the axle using a product supplied by his lubricant distributor. He changed the oil but also the fluid type. The new fluid lacks an LS friction-modifier additive that does not appear to be a GL5 gear oil.

Oil Analysis

The oil analysis results in Figure 11-46 show highlighted oil additive and TAN. The sudden drop in TAN suggests that the oil does not have the required friction modifier. Equally noticeable are calcium and zinc levels, which are more consistent with a tractor fluid than an EP gear lube. Although the metal production is not excessive, iron has more than doubled, and this indicates that the service people need to pay better attention to fluid selection.

METALS							ADDITIVES						
Iron (Fe)	Chromium (Cr)	Lead (Pb)	Copper (Cu)	Tin (Sn)	Aluminum (Al)	Nickel (Ni)	Magnesium (Mg)	Calcium (Ca)	Barium (Ba)	Phosphorus (P)	Zinc (Zn)	Molybdenum (Mo)	Boron (B)
120	3	3	15	<1	14	<1	4	2565	<1	839	939	2	<5
100	4	2	12	<1	3	<1	9	2561	<1	987	987	1	12
43	0	0	10	<1	3	<1	<1	101	<1	2414	56	11	6

CONTAMINANTS					WEAR		PHYSICAL PROPERTIES			INFORMATION			
Aluminum (Al)	Silicon (Si)	Sodium (Na)	Potassium (K)	Water (%)	PQ Index	Particle Counts	Viscosity (cSt @ 40C)	Oxidation	TAN	Fluid Changed	Fluid Type	Total Hours	Fluid Hours
14	7	<1	<5	<0.05	23	NA	167.5	NA	1.45	NO	GO	3133	500
3	8	2	<5	<0.05	29	NA	171.6	NA	1.67	NO	GO	2642	1000
3	11	4	<5	<0.05	33	NA	142.3	NA	4.16	YES	GO	2073	1000

POWERTRAIN OIL ANALYSIS	ADT Dana® Axle. GL5 80W90 Gear Oil with Friction Modifier. Abnormal Phosphorus Levels

FIGURE 11-46 Oil analysis report for an ADT Dana axle.

Conclusion

Trending of the results shows an axle with good maintenance until 2,073 hours. At that point, the lubricant was changed to something that does not meet the manufacturer's requirements. Although the wear results are still acceptable, making the right decision at the right time ensures longer life for components.

Wear Metal Thresholds

See Figure 11-47 and Table 11-21.

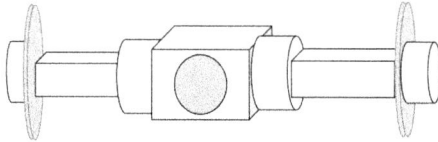

FIGURE 11-47   An ADT Dana axle.

TABLE 11-21   Wear Table for an ADT Dana Axle

Per 1,000 hours			
Nonfiltered System	Normal	Abnormal	Critical
Fe	<350	350	>600
Pb	<9	9	>17
Cu	<35	35	>65
Cr	<7	7	>15
Al	<15	15	>26
Si	<60	60	>110

20. Inconsistent Gear Oil in a Construction Crawler Final Drive

This case shows a crawler final drive in which the oil type has varied between changes. Although this is not against the rules if the manufacturer allows it, it is always better to stick to one type of lubricant for consistency in maintenance. In addition, the final drive is showing increased iron production and dirt contamination. Are these figures abnormal?

Oil Analysis

From the oil analysis results in Figure 11-48, we can see that the oil signature and viscosity have been inconsistent.

METALS							ADDITIVES						
Iron (Fe)	Chromium (Cr)	Lead (Pb)	Copper (Cu)	Tin (Sn)	Aluminum (Al)	Nickel (Ni)	Magnesium (Mg)	Calcium (Ca)	Barium (Ba)	Phosphorus (P)	Zinc (Zn)	Molybdenum (Mo)	Boron (B)
267	1	<1	<1	<1	11	1	2	53	<1	421	19	<1	42
14	<1	<1	<1	1	1	<1	146	3628	<1	1042	1260	13	23
128	5	1	<1	1	3	1	88	2762	1	894	1036	<1	<5

CONTAMINANTS				WEAR			PHYSICAL PROPERTIES			INFORMATION			
Aluminum (Al)	Silicon (Si)	Sodium (Na)	Potassium (K)	Water (%)	PQ Index	Particle Counts	Viscosity (cSt @ 40C)	Oxidation	TAN	Fluid Changed	Fluid Type	Total Hours	Fluid Hours
11	79	4	13	<0.05	207	NA	129.8	NA	1.16	YES	GO	4250	1000
1	11	3	<5	<0.05	NA	NA	63.5	NA	1.98	YES	GO	1421	222
3	17	8	<5	<0.05	NA	NA	55.2	NA	2.99	YES	GO	1199	699

POWERTRAIN OIL ANALYSIS	Construction Crawler External Final Drive. GL5 80W90 Gear Oil. Inconsistent Oil Type

FIGURE 11-48   Oil analysis report for a construction crawler final drive.

Conclusion

The iron production, based on the wear metal table for 500 hours, for this final drive suggests that the numbers are within normal limits. However, it appears that dirt has entered the system. The previous samples had practically no aluminum, but silicon and aluminum are both increased. Because dirt contains silicon and aluminum in a 4–5:1 ratio, we can presume that 40–50 ppm of silicon is coming from dirt and the remainder from foam inhibitor or silicone sealant. Because the sample has 1,000 hours, the readings are acceptable, but the final drive would benefit from dedicating a single type of fluid to it.

Wear Metal Thresholds

See Figure 11-49 and Table 11-22.

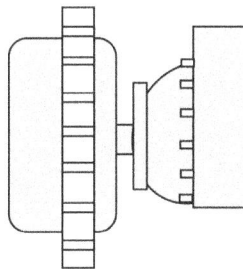

FIGURE 11-49   Construction crawler final drive.

TABLE 11-22 Wear Table for a Construction Crawler Final Drive

Per 500 hours			
Nonfiltered System	Normal	Abnormal	Critical
Fe	<480	480	>850
Pb	<5	5	>9
Cu	<15	15	>24
Cr	<12	12	>18
Al	<27	27	>50
Si	>60	60	>120

21. Internal Leak in a Liebherr Large Construction Crawler Final Drive

Crawler final drives can be equipped with parking brakes that are spring applied and hydraulically released. A hydraulically operated parking brake piston eventually can produce a fluid leak into the final drive, which shows up in the oil analysis.

Oil Analysis

The oil analysis results in Figure 11-50 confirm this final drive has an oil signature that does not match any gear oil on the market, especially because of the calcium and zinc readings.

METALS							ADDITIVES						
Iron (Fe)	Chromium (Cr)	Lead (Pb)	Copper (Cu)	Tin (Sn)	Aluminum (Al)	Nickel (Ni)	Magnesium (Mg)	Calcium (Ca)	Barium (Ba)	Phosphorus (P)	Zinc (Zn)	Molybdenum (Mo)	Boron (B)
2060	29	1	1021	<1	26	10	351	1548	<1	1394	852	97	107
185	10	2	150	3	5	<5	143	12	<1	1169	9	<1	<5

CONTAMINANTS					WEAR		PHYSICAL PROPERTIES			INFORMATION			
Aluminum (Al)	Silicon (Si)	Sodium (Na)	Potassium (K)	Water (%)	PQ Index	Particle Counts	Viscosity (cSt @ 40C)	Oxidation	TAN	Fluid Changed	Fluid Type	Total Hours	Fluid Hours
26	252	17	11	<0.05	164	NA	120.3	NA	1.08	YES	GO	7275	275
4	21	6	<5	<0.05	NA	NA	155	NA	1.06	NO	GO	6500	500
POWERTRAIN OIL ANALYSIS				Liebherr® Large Crawler Final Drive. Gear Oil 85W90. Dirt, Different Oil Signature, Viscosity Out of Range									

FIGURE 11-50 Oil analysis report for a Liebherr large construction crawler final drive.

Conclusion

This condition suggests a mixture with an oil coming from the parking brakes. The viscosity appears to be low. Additionally, the final drive has copious amounts of dirt, which increases metal readings. Hydraulic fluid leaking into the final drive from the parking brake is also bringing copper and iron.

Wear Metal Thresholds

See Figure 11-51 and Table 11-23.

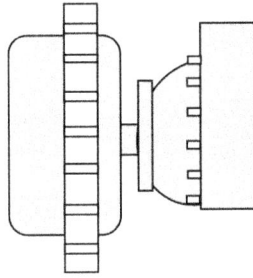

FIGURE 11-51   Liebherr large construction crawler final drive.

TABLE 11-23   Wear Table for a Liebherr Large Construction Crawler Final Drive

Per 500 hours			
Nonfiltered System	**Normal**	**Abnormal**	**Critical**
Fe	<200	200	>320
Pb	<5	5	>10
Cu	<180	180	>320
Cr	<12	12	>18
Al	<5	5	>10
Si	<17	17	>32

22. High Iron Readings in a Large Excavator Final Drive

Final drives in excavators work constantly with lateral loads as the upper structure turns in both directions. This causes the final drive seals to slide horizontally, facilitating dirt and moisture entry. To make things worse, the sprockets could spend considerable time under mud or water.

Oil Analysis

The oil analysis results in Figure 11-52 show a final drive with dirt contamination, judging by the presence of silicon, aluminum, and potassium. Not all the silicon is from dirt; some may come from silicone sealant. The PQ index is equally high.

On the positive side, these final drives do not travel under load for long distances, and large bearings can accommodate more solid contaminants than small bearings. Even so, it is a good practice to check final drives for contamination.

METALS							ADDITIVES						
Iron (Fe)	Chromium Cr)	Lead (Pb)	Copper (Cu)	Tin (Sn)	Aluminum (Al)	Nickel (Ni)	Magnesium (Mg)	Calcium (Ca)	Barium Ba)	Phosphorus (P)	Zinc (Zn)	Molybdenum (Mo)	Boron (B)
1262	14	1	2	<1	81	1	24	65	10	439	12	<1	70
417	9	<1	1	<1	47	<1	42	39	<1	422	6	<1	60

CONTAMINANTS				WEAR		PHYSICAL PROPERTIES			INFORMATION			
Aluminum (Al)	Silicon (Si)	Sodium (Na)	Potassium (K)	Water (%)	PQ Index	Particle Counts	Viscosity (cSt @ 40C)	Oxidation	TAN	Fluid Changed	Fluid Type	Total Hours	Fluid Hours
81	451	42	46	<0.05	5453	NA	134.1	NA	0.75	YES	GO	1470	450
47	241	27	24	<0.05	960	NA	186.5	NA	0.56	YES	GO	1020	1020

POWERTRAIN OIL ANALYSIS	Large Excavator Final Drive. Gear Oil 85W140. Dirt, and Metals. High PQ Index

FIGURE 11-52   Oil analysis report for a large excavator final drive.

Conclusion

The numbers meet the criteria for normal based on the wear table, but the table is based on 1,000 hours, and this sample has only 450 hours. Iron grows exponentially when it reaches 1 ppm/hour. This tells us that by 1,000 hours the iron will surpass the limit. It is an accepted rule to establish the normal threshold on excavator final drives by adding 2 standard deviations to the median value. Given the hours on the oil and the actual oil condition, flushing the final drive is the best option. However, a seal change and retorquing the planetary holding nut also would be advisable.

Wear Metal Thresholds

See Figure 11-53 and Table 11-24.

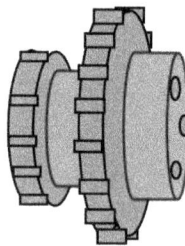

FIGURE 11-53   Large excavator final drive.

TABLE 11-24 Wear Table for a Large Excavator Final Drive

Per 1,000 hours			
Nonfiltered System	Normal	Abnormal	Critical
Fe	<1700	1700	>2500
Pb	<51	51	>80
Cu	<51	51	>80
Cr	<21	21	>36
Al	<90	90	>171
Si	<300	300	>600

23. Dirt Entry in Liebherr Hydrostatic Bulldozer Splitter Drives/Pump Drives

Splitter and pump drives attach directly to the engine to accommodate the operation of pumps or drive shafts. They also depend on their own sump for lubrication, which could consist of engine oil or gear oil. Most of them are equipped with dipsticks and filler necks that can be the source of contaminant entry. Given their small sump and proximity to the engine, their drain intervals should be short. Figure 11-54 shows the issue with filler necks when adding oil. In addition, these components have breathers that eventually could allow dirt or humidity contamination.

FIGURE 11-54 Filler neck.

Oil Analysis

The oil analysis results in Figure 11-55 provide clear indications of dirt entry and its impact on iron and chromium readings. Chromium has reached the critical limit for 500 hours, but iron is more than double the limit. Other indicators appear normal.

METALS							ADDITIVES						
Iron (Fe)	Chromium (Cr)	Lead (Pb)	Copper (Cu)	Tin (Sn)	Aluminum (Al)	Nickel (Ni)	Magnesium (Mg)	Calcium (Ca)	Barium Ba)	Phosphorus (P)	Zinc (Zn)	Molybdenum (Mo)	Boron (B)
564	12	2	33	<1	2	<1	2	264	1	585	168	10	11
216	3	1	31	<1	2	<1	3	349	1	724	227	15	16

CONTAMINANTS				WEAR			PHYSICAL PROPERTIES			INFORMATION			
Aluminum (Al)	Silicon (Si)	Sodium (Na)	Potassium (K)	Water (%)	PQ Index	Particle Counts	Viscosity (cSt @ 40C)	Oxidation	TAN	Fluid Changed	Fluid Type	Total Hours	Fluid Hours
10	45	3	<5	<0.05	NA	NA	197.5	NA	1.32	YES	GO	3021	515
8	37	2	<5	<0.05	NA	NA	185.9	NA	1.41	YES	GO	2506	506

POWERTRAIN OIL ANALYSIS	Large Liebherr® Crawler Pump Drive. Gear Oil 85W140. High Iron and Dirt

FIGURE 11-55   Oil analysis report for a Liebherr hydrostatic bulldozer.

Conclusion

Because the interval for this component is 500 hours, it is the right time to flush it and refill it with fresh oil. The oil capacity of these gearboxes is very small, making the accumulation of dirt a more serious issue.

Wear Metal Thresholds

See Figure 11-56 and Table 11-25.

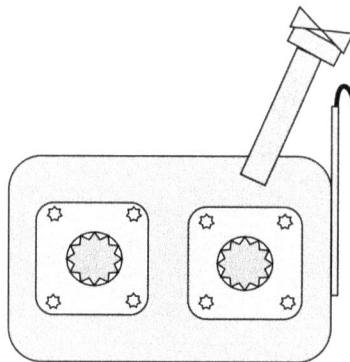

FIGURE 11-56   Pump drive or a splitter drive.

TABLE 11-25 Wear Table for a Pump Drive or a Splitter Drive

Per 500 hours			
Nonfiltered System	Normal	Abnormal	Critical
Fe	<200	200	>320
Pb	<5	5	>10
Cu	<180	180	>320
Cr	<12	12	>18
Al	<5	5	>10
Si	<17	17	>32

24. High Iron in an Excavator Swing Gearbox Drive

Used on excavators and forestry machines, among others, swing gearboxes are single or double planetary reduction gears, typically driven by a hydraulic motor. In some applications, these gearboxes get lubrication from the hydraulic system and enjoy its cooling and filtration. Nevertheless, in most applications, they depend on their own sump filled with gear oil or engine oil of some kind. Small sumps tend to degrade oil faster and can be prone to contamination through breathers, so their intervals need to be strictly kept.

Oil Analysis

The oil analysis results in Figure 11-57 show that the swing gearbox contains water and dirt. This has triggered the iron and other metal production, and water is pushing up TAN levels. Water can get into small compartments because of poor service practices and sometimes during high-pressure washing. The water content is visible, thanks to a Karl Fischer co-distillation test.

METALS							ADDITIVES						
Iron (Fe)	Chromium (Cr)	Lead (Pb)	Copper (Cu)	Tin (Sn)	Aluminum (Al)	Nickel (Ni)	Magnesium (Mg)	Calcium (Ca)	Barium (Ba)	Phosphorus (P)	Zinc (Zn)	Molybdenum (Mo)	Boron (B)
165	4	15	24	3	7	2	1	11	<1	529	13	<1	17
61	1	<1	<1	<1	2	<1	1	16	<1	349	12	<1	47

CONTAMINANTS				WEAR			PHYSICAL PROPERTIES			INFORMATION			
Aluminum (Al)	Silicon (Si)	Sodium (Na)	Potassium (K)	Water (%)	PQ Index	Particle Counts	Viscosity (cSt @ 40C)	Oxidation	TAN	Fluid Changed	Fluid Type	Total Hours	Fluid Hours
7	35	3	<5	0.13	NA	NA	141.1	NA	1.75	YES	GO	2400	1000
2	10	2	<5	<0.05	NA	NA	183.6	NA	0.64	YES	GO	1000	1000
POWERTRAIN OIL ANALYSIS			Large Excavator Swing Gearbox Drive. Gear Oil 85W140. High Water Content, High Iron and Dirt										

FIGURE 11-57 Oil analysis report for an excavator swing gearbox drive.

Conclusion

It is important to find the source of the contamination to prevent a reoccurrence of this condition. Breathers need inspection in the service practices protocol. If the oil comes from a lubrication truck, the stock oil becomes suspect and requires sampling as well.

Wear Metal Threshold

See Figure 11-58 and Table 11-26.

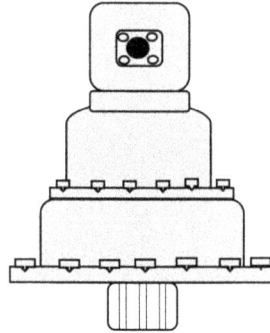

FIGURE 11-58 Excavator swing gearbox drive.

TABLE 11-26 Wear Table for an Excavator Swing Gearbox Drive

Per 500 hours			
Nonfiltered System	Normal	Abnormal	Critical
Fe	<151	151	>300
Pb	<51	51	>80
Cu	<51	51	>80
Cr	<11	11	>15
Al	<21	21	>30
Si	<31	31	>60

25. Water in a Motor Grader Tandem Drive

Motor graders use tandem drives, which are typically chain/sprocket driven in an oil bath. They tend to be easy to maintain but can be prone to contamination in dirty or humid operations, especially if there is a leak. Their typical lubricants are tractor fluids, Cat TO-4 oil, or engine oils.

Oil Analysis

The oil analysis results in Figure 11-59 show that this tandem drive has recently been contaminated with water. Although the readings do not show a tandem in distress, it will not be long before metals and TAN start rising.

METALS							ADDITIVES						
Iron (Fe)	Chromium Cr)	Lead (Pb)	Copper (Cu)	Tin (Sn)	Aluminum (Al)	Nickel (Ni)	Magnesium (Mg)	Calcium (Ca)	Barium Ba)	Phosphorus (P)	Zinc (Zn)	Molybdenum (Mo)	Boron (B)
35	<1	2	2	<1	3	<1	99	3025	<1	1086	1251	<1	12

CONTAMINANTS				WEAR			PHYSICAL PROPERTIES			INFORMATION			
Aluminum (Al)	Silicon (Si)	Sodium (Na)	Potassium (K)	Water (%)	PQ Index	Particle Counts	Viscosity (cSt @ 40C)	Oxidation	TAN	Fluid Changed	Fluid Type	Total Hours	Fluid Hours
3	16	4	8	0.89	<10	NA	61.2	NA	2.36	YES	THF	1000	1000

POWERTRAIN OIL ANALYSIS	Motor Grader Tandem. Water Present. Tractor Fluid ISO 68

FIGURE 11-59 Oil analysis results for a motor grader tandem drive.

Conclusion

Because this is a fairly new machine, it is important to establish whether the contamination occurred because of an oil leak or because of the application. The tractor fluid in use seems to be neutralizing the impact of the water contamination. However, the amount of water is way beyond the recommended level for a tractor fluid, and the component needs to be flushed.

Wear Metal Thresholds

See Figure 11-60 and Table 11-27.

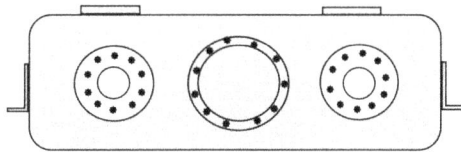

FIGURE 11-60 A motor grader tandem drive.

TABLE 11-27 Wear Table for a Motor Grader Tandem Drive

Per 500 hours			
Nonfiltered System	Normal	Abnormal	Critical
Fe	<167	167	>310
Pb	<5	5	>15
Cu	<160	160	>320
Cr	<5	5	>15
Al	<51	51	>101
Si	<15	15	>25

26. High Iron and Chromium in a Forestry Forwarder Bogie

Forwarders and tree harvesters use oscillating bogies to drive four- or eight-wheel-drive configurations. These are mainly gear-driven drives that use some type of gear oil. Wheels in pairs can use track chains to improve traction. The unit in question uses steel tracks over the tires.

Oil Analysis

The oil analysis results in Figure 11-61 shows a wear progression that needs evaluation. The jump in metals at 500 hours suggests that there is a problem that needs to be fixed before damage occurs.

METALS							ADDITIVES						
Iron (Fe)	Chromium Cr	Lead (Pb)	Copper (Cu)	Tin (Sn)	Aluminum (Al)	Nickel (Ni)	Magnesium (Mg)	Calcium (Ca)	Barium Ba	Phosphorus (P)	Zinc (Zn)	Molybdenum (Mo)	Boron (B)
328	15	15	32	3	7	<1	748	34	<1	1341	7	<1	453
150	8	7	15	2	2	<1	750	38	<1	1350	8	<1	458

CONTAMINANTS				WEAR			PHYSICAL PROPERTIES			INFORMATION			
Aluminum (Al)	Silicon (Si)	Sodium (Na)	Potassium (K)	Water (%)	PQ Index	Particle Counts	Viscosity (cSt @ 40C)	Oxidation	TAN	Fluid Changed	Fluid Type	Total Hours	Fluid Hours
7	21	2	<5	<0.05	240	NA	143.3	NA	1.66	YES	GO	1000	500
3	12	<1	<5	<0.05	200	NA	145	NA	1.6	YES	GO	500	500

POWERTRAIN OIL ANALYSIS	Forestry Forwarder Bogie. Wear Metals. Gear oil 85W90

FIGURE 11-61   Oil analysis report for a forestry forwarder bogie.

Conclusion

The initial 500 hours show a perfectly normal component without any issues. The problem does not appear to be the result of contamination or poor lubrication but rather an external factor or application. Checking the component right away can save money down the road.

Wear Metal Thresholds

See Figure 11-62 and Table 11-28.

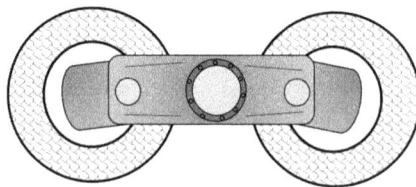

FIGURE 11-62   A forestry forwarder bogie.

TABLE 11-28 Wear Table for a Forestry Forwarder Bogie

Per 500 hours			
Nonfiltered System	Normal	Abnormal	Critical
Fe	<300	300	>600
Pb	<11	11	>20
Cu	<26	26	>50
Cr	<11	11	>15
Al	<21	21	>30
Si	<31	31	>60

27. High Metals in an ADT Nonfiltered Transfer Gearbox

ADTs may use a separate transfer case that is not contained within the transmission. Typically, these transfer gearboxes contain a differential to split the power between front and rear drive shafts, and they may contain locking mechanisms. These boxes may or may not contain filters, and the readings on oil analysis reports are going to depend on this fact. The lubricant for ADT transfer gearboxes often matches that of the transmission for ease of maintenance.

Oil Analysis

The oil analysis results in Figure 11-63 show alarming figures with regard to wear metals and dirt. The PQ index is also high. The oil capacity of these transfer gearboxes is small, which does not help in diluting contaminants for a cleaner reading.

METALS							ADDITIVES						
Iron (Fe)	Chromium (Cr)	Lead (Pb)	Copper (Cu)	Tin (Sn)	Aluminum (Al)	Nickel (Ni)	Magnesium (Mg)	Calcium (Ca)	Barium Ba)	Phosphorus (P)	Zinc (Zn)	Molybdenum (Mo)	Boron (B)
715	6	191	37	7	19	5	11	54	<1	317	57	1	38
450	3	150	31	2	14	3	12	57	<1	325	65	1	43

CONTAMINANTS				WEAR		PHYSICAL PROPERTIES			INFORMATION			
Aluminum (Al)	Silicon (Si)	Sodium (Na)	Potassium (K)	Water (%)	PQ Index	Particle Counts	Viscosity (cSt @ 40C)	Oxidation	TAN	Fluid Changed	Fluid Type	Total Hours	Fluid Hours
19	22	25	<5	<0.05	1056	NA	55.3	NA	1.5	YES	ATF	2486	1486
14	12	15	<5	<0.05	750	NA	56.2	NA	1.4	NO	ATF	1000	1000

POWERTRAIN OIL ANALYSIS | ADT Non-Filtered Transfer Gearbox. Wear Metals. ATF

FIGURE 11-63 Oil analysis report for an ADT nonfiltered transfer gearbox.

Conclusion

A small sump component without filtration is going to show high metal readings, and the iron is going to grow exponentially to produce a high PQ index. For this reason, these components need shorter drain intervals than filtered gearboxes. Let us compare these results against those of a similar machine that incorporates a filtered transfer gearbox in the next case.

Wear Metal Thresholds

See Figure 11-64 and Table 11-29.

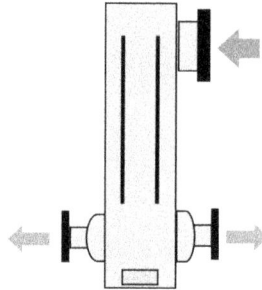

FIGURE 11-64   ADT nonfiltered transfer gearbox.

TABLE 11-29   Wear Table an ADT Nonfiltered Transfer Gearbox

Per 500 hours			
Nonfiltered System	**Normal**	**Abnormal**	**Critical**
Fe	<280	280	>500
Pb	<80	80	>130
Cu	<110	110	>200
Cr	<5	5	>10
Al	<21	21	>50
Si	<20	20	>30

28. Oil Analysis for an ADT Filtered Transfer Gearbox

There is no doubt that the results from a filtered transfer gearbox are much better than those from an unfiltered one. The results of fluid analysis results shown in Figure 11-65 at 1,000 hours are perfectly normal for this gearbox. The standard deviations for a transfer gearbox with filtration are going to show improved numbers, especially for iron, compared with nonfiltered transfer cases.

METALS							ADDITIVES						
Iron (Fe)	Chromium Cr	Lead (Pb)	Copper (Cu)	Tin (Sn)	Aluminum (Al)	Nickel (Ni)	Magnesium (Mg)	Calcium (Ca)	Barium Ba	Phosphorus (P)	Zinc (Zn)	Molybdenum (Mo)	Boron (B)
240	<1	13	62	3	4	<1	<1	36	1	269	13	<1	114
120	<1	5	31	2	2	<1	<1	38	<1	271	14	<1	115

CONTAMINANTS				WEAR			PHYSICAL PROPERTIES			INFORMATION			
Aluminum (Al)	Silicon (Si)	Sodium (Na)	Potassium (K)	Water (%)	PQ Index	Particle Counts	Viscosity (cSt @ 40C)	Oxidation	TAN	Fluid Changed	Fluid Type	Total Hours	Fluid Hours
4	20	2	<5	<0.05	54	NA	30.5	NA	0.63	YES	ATF	4200	1000
2	12	1	<5	<0.05	24	NA	31	NA	0.6	NO	ATF	3200	1000
POWERTRAIN OIL ANALYSIS			ADT Filtered Transfer Gearbox. Normal Results ATF Transynd™										

FIGURE 11-65 Oil analysis report for an ADT filtered transfer gearbox.

Conclusion

The oil analyses from both transfer gearboxes show lubricant signatures of zinc-free transmission fluid, although the viscosity of the first sample is higher than that for a typical heavy-duty ATF.

Wear Metal Thresholds

See Figure 11-66 and Table 11-30.

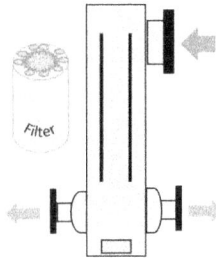

FIGURE 11-66 An ADT filtered transfer gearbox.

TABLE 11-30 Wear Table for an ADT Filtered Transfer Gearbox

Per 1,000 hours			
Filtered System	Normal	Abnormal	Critical
Fe	<241	241	>329
Pb	<55	55	>64
Cu	<77	77	>89
Cr	<5	5	>10
Al	<15	15	>17
Si	<15	15	>25

CONCLUSION

Because of the numerous types of components, lubricants, and applications, the interpretation of oil analysis results for powertrains can be challenging. To start with, there are many types of fluids and lubricants used in powertrains, and many are very specific to the component. This is why users or labs need to develop their own wear tables because the tables in this chapter cannot be widely applied. Additionally, there are several specialty fluids/friction modifiers that are essential to the life of components. The need for these additives increases complexity when interpreting powertrain results. Finally, the type of application can influence the way we interpret the results. Now let's move on to another important topic, engine coolants.

CHAPTER 12

ENGINE COOLANT ANALYSIS

Coolants contribute to roughly 30% of engine problems, yet coolant analysis is seldom in the mind of users. Coolant maintenance is as important as oil maintenance. The fact is that engine longevity depends on the health of the cooling system. Not only corrosion is at stake but also, more important, cavitation and boiling/freezing protection. Coolants are easy to control with inexpensive onsite kits combined with yearly lab testing. By the same token, water quality is equally important. Water quality is also manageable with field kits that are easy to use with little investment.

This chapter provides information about coolant history and functions and its role in an engine's life. The chapter also covers the typical tests to keep cooling systems healthy. As engines evolved and became more sophisticated, coolants also evolved to provide the added protection required by the new technologies. In particular, evolving cooling systems and coolants were designed to control temperatures and enable ever-increasing power densities. A case discussion section follows that covers an array of situations from real life. They help you to understand the behavior of different engines and applications that use a variety of heavy-duty coolants.

COOLANT DEVELOPMENT HISTORY

1859	French chemist named Charles Adolphe Wurtz discovered/obtained ethylene glycol. Wurtz treated ethylene iodide with silver acetate. Later he reacted the result from the experiment (ethylene diacetate) with potassium hydroxide.
1914–1918	During World War I, the manufacture of explosives depended on ethylene glycol.
1926	Ethylene glycol became available as a coolant and antifreeze for engines in the automotive and aviation industries.

1950   The majority of coolants were ethylene glycol with little or no additives. The life expectancy of coolants was around 1,000 hours.

1950–1980   Supplemental coolant additives (SCAs) were introduced to extend the life of a coolant. The technology included additives to control corrosion, foam, and cavitation. SCAs were added to coolants as liquids, tablets, or precharged filters to improve their life span to about 2,000–3,000 hours.

1981–2010   Coolant formulations with premixed inorganic additives came to the market. A new generation of SCAs became available to reinforce the properties of this new generation of coolants.

2011–Present   Organic and hybrid coolants become the norm. Depending on the continent from which they originate, these may have different formulations and colors. New coolants range in longevity from 4,000 to 10,000 hours depending on the application.

TYPES OF COOLANTS

The most traditional coolants use ethylene glycol, whereas some application-specific coolants use propylene glycol (safer for animals and the environment) and propylene glycol methyl ethers. Additive types for glycol coolants are inorganic low-silicate organic acids and/or hybrids.

HOW ARE COOLANTS PRODUCED TODAY?

Ethylene glycol comes from a petroleum derivate, ethylene (C_2H_4), which is a gas that converts to ethylene oxide and then to ethylene glycol $HO–CH_2CH_2–OH$. Once mixed with distilled water and additives, it becomes engine coolant. Propylene glycol comes from propylene that is converted to propylene oxide and then propylene glycol $(C_3H_8O_2)$. It is then blended with distilled water and additives to make a more environmentally friendly engine coolant.

COOLANT REQUIREMENTS

Any coolant needs to comply with the following minimum requirements for an engine:

- Transfer heat effectively (glycol/water)
- Provide freezing and boiling protection
- Provide corrosion protection through a corrosion-inhibition additive package
- Protect against liner cavitation
- Chemically stable
- Compatible with hoses, elastomers, and seals
- Relatively nonflammable and nontoxic (glycol and corrosion-inhibitor components)

HOW MUCH HEAT CAN COOLANTS REMOVE?

A coolant using a radiator and fan removes approximately 30% of the heat produced by an engine. The exhaust liberates another 30%, whereas 33% of the heat passes in the form of mechanical energy to the power take-off (PTO) or transmission. Some 7% of the heat radiates from the engine components themselves (see Figure 12-1).

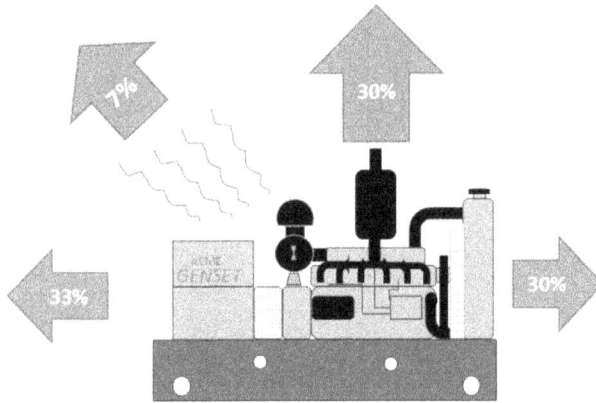

FIGURE 12-1   Coolant heat-removal process.

MARINE COOLANT OPERATION

Some marine applications use a heat-exchanger system, where the seawater cools a jacket containing the engine coolant. In this type of system, there is no fan because seawater takes the heat back to the sea (see Figure 12-2).

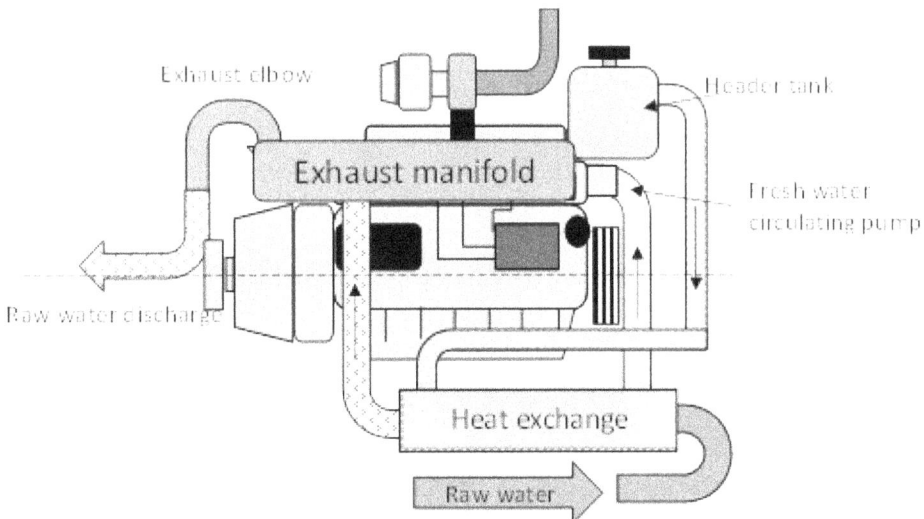

FIGURE 12-2   Marine cooling system.

SPECIFIC HEAT OF COOLANTS

Water absorbs or loses a large amount of heat for each degree of change. In contrast, the specific heat of water is higher than that of glycol, and increasing the glycol concentration in a coolant would lower the specific heat of the solution. Pure water will carry more heat than a water–glycol solution (see Table 12-1). However, using pure water is out of the question because pure water without additives will corrode the system and will not protect against freezing and/or boiling. The ideal mix of glycol and water is always a compromise in terms of cooling. However, it will provide the required protection.

TABLE 12-1   Specific Heats of Coolants

Water	1
Ethylene glycol	0.58
Water and ethylene glycol at 50% by volume	0.87

REQUIREMENTS FOR NEW ENGINE TECHNOLOGIES

New technologies are making engines more heat-intensive devices. Heat is a natural result of the adjustments to injection delay and stoichiometric combustion. Additionally, heat facilitates better combustion, which combined with higher pressure from injectors consumes higher volumes of air. This results in the need for a much more robust cooling system that not only circulates more gallons per hour but also is able to resist the added heat without compromising the protection against cavitation and corrosion (see Figure 12-3).

FIGURE 12-3   Tier 4 diesel engine.

COOLANT STANDARDS

The American Society for Testing and Materials (ASTM) D3306 specification covers glycol-based engine coolants for automotive service and light duty. The heavy-duty specifications for glycol- and propylene-based coolants are ASTM D6210 and ASTM D6211, respectively. A diesel engine should never use an automotive coolant because it very seldom contains the additives to control liner cavitation.

ENGINE COOLANT COMPOSITION

A good coolant is made of glycol, distilled water, corrosion inhibitors, foam inhibitors, scale inhibitors, and a bittering agent (to prevent children or animals from drinking it; see Figure 12-4).

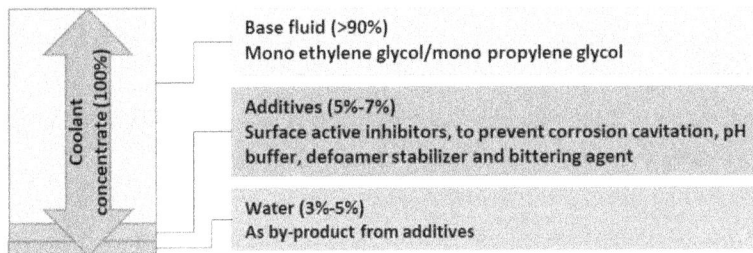

Base fluid (>90%)
Mono ethylene glycol/mono propylene glycol

Additives (5%-7%)
Surface active inhibitors, to prevent corrosion cavitation, pH buffer, defoamer stabilizer and bittering agent

Water (3%-5%)
As by-product from additives

FIGURE 12-4   Chemical composition of coolants.

CONVENTIONAL COOLANTS AND SILICATES

Silicate is effective in protecting aluminum components by forming a surface layer that prevents corrosion. However, silicate has limited solubility, and an overconcentration causes the silicate to drop out of the solution and form silica gel deposits. These deposits block radiators and plug external oil cooler tubes.

Passenger cars using large, wet aluminum parts, such as cylinder heads and radiators, use silicate-rich coolants. Engines using cast iron cylinder heads with copper or brass radiators typically use coolant formulations that are relatively low in silicates.

Traditionally, two major corrosion inhibitors were used in vehicles: silicates and phosphates. American-made vehicles traditionally use both silicates and phosphates. European vehicles used silicates and other inhibitors but no phosphates. Japanese vehicles traditionally use phosphates and other inhibitors but no silicates.

COOLANT SAMPLING

As with any other fluid, coolant sampling has its rules as well. It is important to remember that you want the coolant sample to represent what is in the cooling system during use and to contain no contaminants from poor service practices. Here is a list of sampling key points:

- Coolant needs to be warm and recently agitated. This means that the engine has been active within 10 minutes before taking the sample.
- Wait for the coolant temperature to fall below the boiling point before releasing the radiator cap. (This does not apply if you are drawing the coolant from a sampling valve on the engine.)

- Do not sample from the surge tank. Surge tank coolant is not representative of the engine coolant because it does not actively share coolant with the engine. It generally tends to be more concentrated.
- Most surge tanks are pressurized. Observe caution when removing the surge tank cap.
- Always use a clean bottle and tubing or probe. Discard the tubing and/or probe after every collection.
- When drawing a sample from a dead-end sampling valve on the engine (see Figures 12-5 and 12-6), discard the volume of one full sample bottle before taking the final sample.
- If you use a hand pump, use a color-coded pump for coolant. Never use a pump that is also used for oil sampling.
- Fill out the form as completely as possible, especially with information regarding hours on the machine and on the coolant.
- Make sure that the machine registration process is complete at your lab of preference.
- If you use barcoded bottles, use the bottle with the correct barcode for the sample you send.

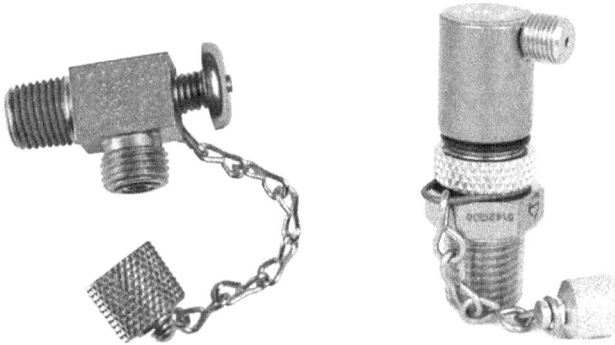

FIGURE 12-5   Sampling valves.

FIGURE 12-6   Engine coolant sampling valves (coolant on top, oil on bottom).

Organic Acid Technology (OAT) Coolants

Pure extended-life coolants (ELCs) or OAT coolants do not contain inorganic inhibitors such as nitrites, phosphates, and silicates. Their manufacturers claim that without silicates, there is no green goo formation and that without phosphates and borates, there are fewer solids that affect water pump seals over time.

Nitrite–Organic Acid Technology (NOAT) Coolants

ELCs that use nitrite as the main protection against cavitation still use some organic acids to support anticavitation and corrosion and still use silicates and molybdate but not phosphates.

Hybrid–Organic Acid Technology (HOAT) Coolants

Some manufactures use a hybrid technology in which organic acids are present, but they still use small amounts of phosphates, borates, and silicates. The decision to use one technology over another depends on individual testing and metallurgy in engines and cooling systems plus operating conditions such as component vibration and power level generation.

Heavy-Duty ELC Inorganic Signatures

Table 12-2 describes typical inorganic coolant additives for heavy-duty coolants. The boldface numbers indicate the differences with regard to the use of certain additives. The hybrid coolants, for example, do not depend on nitrite for the control of liner cavitation.

TABLE 12-2   Inorganic Coolant Additives for Heavy-Duty Coolants

Additive	Heavy-duty diesel engine coolants				
	NOAT USA	HOAT USA	OAT Europe	HOAT Europe	HOAT Japan
Chloride (mg/kg)	6	6	9	14	10
Nitrite (mg/kg)	1285	<10	50	<10	57
Nitrate (mg/kg)	829	111	760	1051	801
Phosphate (mg/kg)	<10	328	<10	<10	3093
Boron (mg/kg)	318	680	682	585	<2
Potassium (mg/kg)	16	309	168	285	1219
Sodium (mg/kg)	2644	3447	1491	5870	10,960
Molybdenum (mg/kg)	234	110	<2	<2	147
Molybdate	391	182	<2	<2	243
Silicon (mg/kg)	120	179	36	105	11
Silicate	325	485	98	285	30
pH	8.2	8.3	7.8	8.0	8.4
Freeze point, °F	−36	−36	−34	−43	−14
Reserve alkalinity	4.0	8.3	6.6	5.8	4.1
Visual assessment	Green	Yellow	Blue-green	Blue-green	Green

Inorganic Additive Functions

Table 12-3 describes the metal protection of the typical additives found in coolants.

TABLE 12-3   Traditional Coolant Inorganic Additives and Their Protection Functions

Phosphates	Iron, steel, aluminum, pH buffer
Borates	Iron, pH buffer
Nitrate, NO_3	Aluminum (pitting), solder
Nitrite, NO_2	Iron (cavitation)
Silicate	Aluminum, iron
Molybdate	Iron, aluminum, solder
Amine	Cast iron, pH buffer

Organic Additives in Heavy-Duty Coolants

See Table 12-4.

TABLE 12-4   Organic Additives in Coolants

%/Weight	Heavy Duty Diesel Engine Coolants				
	NOAT USA	HOAT USA	OAT Europe	HOAT Europe	HOAT Japan
Tolythriazole	0.054	0.09	0.01	<0.01	0.07
Benzoic acid	1.09	0.32	0.01	0.01	0.91
Sebacic acid	<0.01	0.47	0.64	0.69	12
Toluic acid	<0.01	<0.01	<0.01	<0.01	0.01
Etiexanocic acid	<0.01	<0.01	1.05	1.08	<0.01
Octanoic acid	<0.01	<0.01	<0.01	<0.01	<0.01
Tert-butyl benzoic acid	<0.01	<0.01	<0.01	<0.01	0.79
Benzotriazole	<0.01	<0.01	0.03	<0.05	<0.03
Mercaptobenzothialzole	<0.01	<0.01	<0.01	<0.01	0.04
Total	1.14	0.88	1.74	1.78	1.94

Main Organic Additive Functions

See Table 12-5.

TABLE 12-5   Organic Additive Functions in Coolants

Protection to Metals from Organic Refrigerant Additives	
Tolythriazole	Copper and Bronze
Sebacic acid	Lead and Aluminum
Etiexanocic acid	Ferrous Material
Mercaptobenzothialzole	Copper and Bronze
Benzotriazole	Copper, Bronze, Solder

HOW CORROSION INHIBITORS WORK: CONVENTIONAL COOLANTS VERSUS ELCs

Anodic Inhibitors

Anodic inhibitors (e.g., nitrite and SCAs) restrict the production of ions at the anode by combining with the metal ions that are separating from the surface. They form insoluble multilayers on the surface to shield and protect. They also stop access of the corrosion elements to the metal and prevent the oxidation of the metal as well. In other words, anodic inhibitors protect the metal by creating an insulating layer that is harder to pit with cavitation (see Figure 12-7).

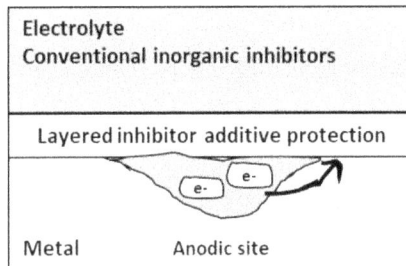

FIGURE 12-7 Anodic inhibitors.

Cathodic Inhibitors

Cathodic inhibitors complement the anodic inhibition process by capturing and removing the electrons given up by the metal atoms in becoming positively charged ions and leaving the anode. This retards corrosion by inhibiting the reduction of water to hydrogen gas and slows oxidation of the metal (see Figure 12-8).

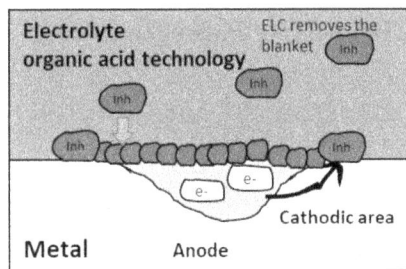

FIGURE 12-8 Cathodic inhibitors.

Adsorption Inhibitors

Adsorbed ions develop strong attractions for the adsorbing surface and share electrons to form insoluble layers. In simpler terms, the cathodic inhibitors patch the pitted surface as the metals release ions, comparable to the task of patching a road with potholes.

Coolant Freeze Protection

The proper mix of glycol and distilled water provides the best protection against freezing and the effects of corrosion and cavitation. Glycol concentrations >70% revert the freeze protection and become too dense for the water pump to operate efficiently. Ideally, a mix of 50:50 glycol and distilled water is the best compromise among freezing, boiling, and corrosion protection (see Figure 12-9).

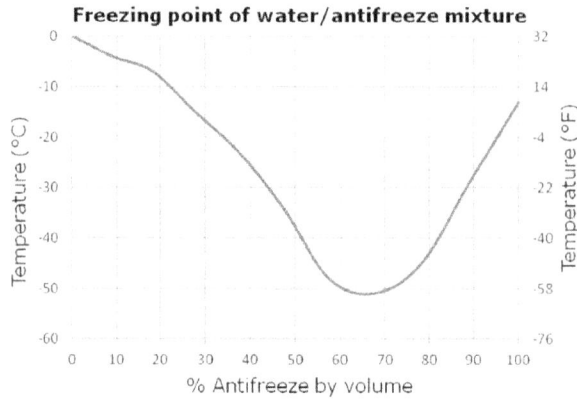

FIGURE 12-9 Freezing protection.

Boiling Protection

Compared with pure water, a coolant made of a 50:50 mixture of glycol and distilled water provides additional boiling protection in a pressurized system. Boiling protection is required to keep the coolant in a liquid state at the normal operating temperatures of modern engines. The operating temperature of modern engines far exceeds the natural boiling point of water at atmospheric pressure. Thus, a pressurized system filled with a mixture of glycol and distilled water is essential to providing engine protection. Figure 12-10 illustrates the boiling protection of coolant with different radiator cap pressures.

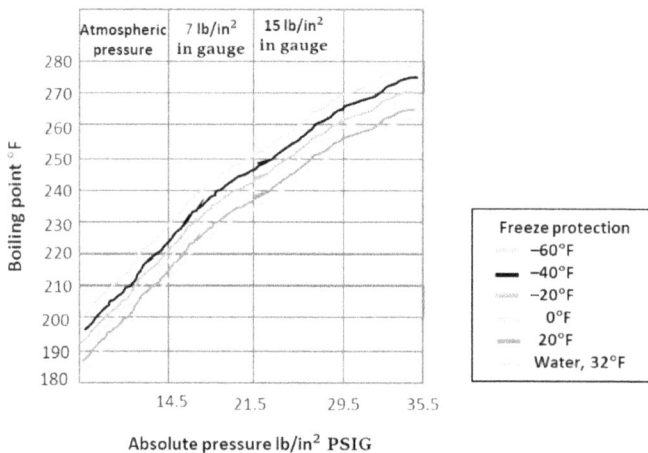

FIGURE 12-10 Boiling protection at different pressures.

What Are the Issues with Using Only Water as a Coolant?

The issues with using only water as a coolant start with the low boiling point and the little protection against freezing. There are regions in the world where freezing protection is not required, but the need for corrosion and cavitation protection is still there. In addition, formulated coolants protect against cavitation, foaming, lubricate water pump seals, and prevent deposits in hot spots.

Cavitation

Cavitation is a very misunderstood phenomenon that is present in most modern heavy-duty diesel engines and is perhaps the most serious risk to an engine. Cavitation is not only the action of corrosion but also a physical erosion of metal caused by vibration as the cylinder liner rapidly bulges from combustion and then recovers. When the cylinder recovers from bulging, negative pressure at the surface on the coolant side of the cylinder creates a vacuum that forms a bubble. When the cylinder bulges again during the next combustion stroke, the bubble is crushed and implodes violently. Bubble formations remove microportions of metal when they form and then again when they implode. The cycle repeats itself over and over, attacking the same points again and again, slowly pitting holes that eventually perforate the cylinder. To make matters worse, modern engines have floating liners for serviceability, which intensifies the vibration phenomenon. For this reason, heavy-duty engine coolants use additives to prevent cavitation (see Figure 12-11).

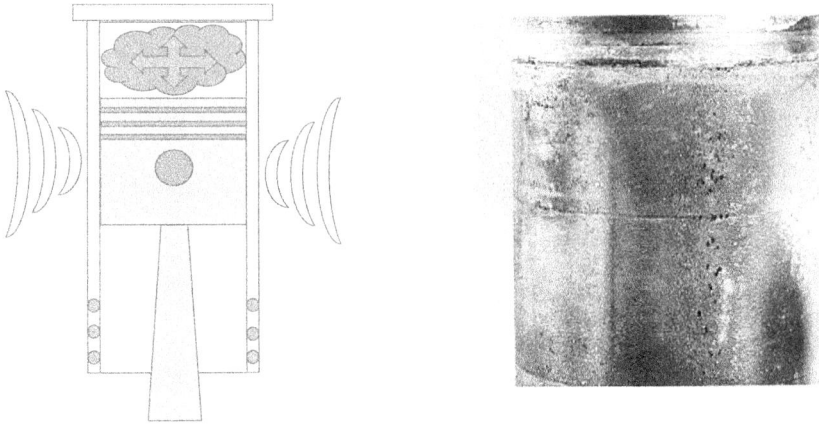

FIGURE 12-11   Diesel engine liner vibration (*left*); liner cavitation (*right*).

Coolant Specifications

In addition to original equipment manufacturer (OEM) specifications, there are basic industry specifications that are used as minimum requirements that coolants must meet. They start with ASTM D3306 for glycol-based coolants intended for automobiles and light-duty vehicles. The heavy-duty standard specifications for glycol and propylene glycol coolants are ASTM D6210 and ASTM D6211, respectively.

Traditional coolants protected against cavitation by means of nitrites, complemented later with certain organic acids. Some coolants no longer depend on nitrites for cavitation protection but rather use organic acids. Nitrites take oxygen from the solution and convert it to nitrate (NO_3) at the liner surface, creating a hard-to-pit layer. Nitrites also reduce bubble surface tension, decreasing their formation.

Conventional Coolants

Nitrite Exhaustion

One of the issues with conventional coolants is the risk of nitrite depletion. Manufacturers provide SCAs to reinforce the additive from time to time based on coolant testing. The accelerated nitrite exhaustion can be triggered by air leaking into the system, such as from a loose hose clamp or faulty radiator cap. In addition, an active corroding or pitting area in the engine will cause depletion in the nitrite levels and an increase in the levels of nitrate.

Effects of Excessive SCAs

Overconcentration of SCAs is also a problem that can produce undesired effects. One is the plugging of the system with precipitates of gelled additive, for example, silicate gelation (green goo). Solids build up in radiator tubes, solder corrodes (solder bloom), and water pump seals fail (see Figure 12-12).

FIGURE 12-12   Solder bloom.

Issues with Acidity (Low pH)

Prolonged acidity of coolants in copper core radiators can lead to copper corrosion. The copper may react with the acid and form salts of copper, such as copper sulfate ($CuSO_4$), which is generally pale green in color. As the copper sulfate deposits dry up, they turn into grayish-white powder (see Figure 12-13).

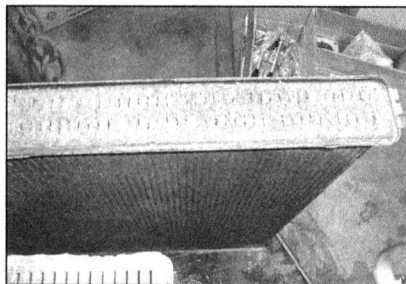

FIGURE 12-13   Copper sulfate deposits.

Basic Coolant (High Ph) in ELCs

Under severe conditions of electrolysis or stray current, the nitrites in a coolant combine with hydrogen ions to form ammonium hydroxide (NH_4OH). Ammonium hydroxide is a highly alkaline substance that increases the pH of coolant (see Figure 12-14).

FIGURE 12-14   Aluminum corrosion caused by a basic coolant.

Supplemental Coolant Additives

Over time, a coolant can show reduced additive levels or a low glycol concentration because of top-offs with water or, simply, aging. If the pH of the coolant is still within the acceptable range, an SCA can reinforce the coolant. There are different SCAs for each type of coolant. In this way, the coolant receives the correct chemistry replenishment. An organic acid coolant does not use an SCA that contains nitrite. Conversely, a nitrite-based coolant (NOAT) will use a nitrite-based SCA. Equipment manufacturers stock the respective SCAs that are correct matches for their coolants.

Degradation of Glycol

Ethylene glycol contains three elements: carbon, hydrogen, and oxygen. Under adverse operating conditions, such as extreme overheating, ethylene glycol–based coolants can degrade and oxidize to form acids such as oxalic and formic acids (both slightly acidic) called *glycolates*. These acids pull the coolant's pH down, leading to corrosion problems. The degradation of glycol also will contribute to dropout of corrosion inhibitors.

LAB AND FIELD TESTS FOR COOLANTS

Table 12-6 lists the typical tests conducted on heavy-duty coolants, whether they are nitrite based, hybrid, or organic acid based.

TABLE 12-6 Common Coolant Tests

	Lab tests	Field tests	
Test	Physical, chemical, and contaminants	HOAT/OAT	NOAT
1	pH value	X	X
2	Reserve alkalinity (mL HCl/10 mL)		
3	Glycol content (%) freeze point/boiling point	X	X
4	Nitrite (mg/L or ppm)		X
5	Total dissolved solids	X	X
6	Cumulative organic acid inhibitors	X	
7	Silicon, silicates, sodium, potassium, molybdate, nitrates, phosphates, borates, glycolates		X (molybdate)
8	Aluminum, copper, iron, lead		
9	Water hardness and chlorides	X	X

1. pH Value

We measure the level of acidity and alkalinity of a solution on the pH scale. Lower values indicate higher acidity and higher values increased alkalinity (see Figure 12-15).

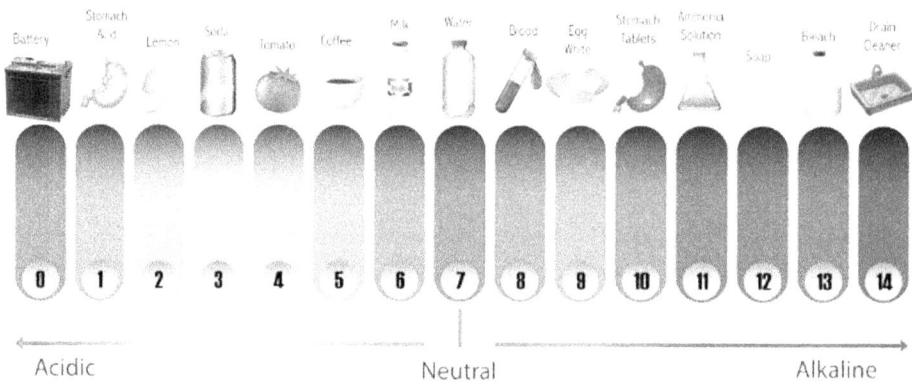

FIGURE 12-15 pH scale.

Coolants with a low pH, or increased acidity, will attack ferrous and other metallic surfaces in the cooling system. Coolants with a high pH, or increased alkalinity, will attack nonferrous surfaces such as aluminum, copper, and brass. Mineral salts cause high alkalinity in natural water. Thus:

- An acceptable pH range for conventional coolants is between 8 and 10.5.
- Average pH values for ELCs and OAT coolants are from 7 to 8.5.

2. Reserve Alkalinity

Reserve alkalinity (RA) is a measure of the acid-neutralizing ability of a coolant. At the lab, RA is determined by the number of milliliters of hydrochloric acid required to lower the pH of 10 mL of the coolant solution to 5.5. RA is primarily an indication of the amounts of alkaline additives or buffering agents present in the coolant. The coolant may become acidic due to exhaust gas leakage or oxidation of ethylene glycol, dropping the pH to the range of 2.5 to 5. Figure 12-16 shows a titrator used to determine RA.

FIGURE 12-16   Thermo Scientific titrator.

3. Glycol Concentration

There are tools that allow measuring of glycol concentration in a coolant. Refractometers (see Figure 12-17) and hygrometers are easy tools to use in the field, as are reactive test strips. Overconcentration of SCAs in coolants can cause inhibitor and silicate dropouts, which can lead to gel-type deposits in the cooling system. The silicate dropouts in the cooling system will require an alkaline-based cooling system cleaner.

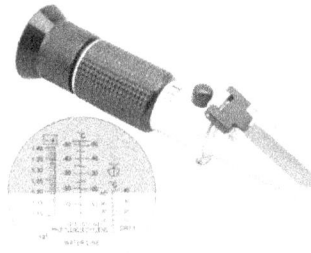

FIGURE 12-17   Refractometer.

The typical test at a laboratory includes the use of a chromatographer following the guidelines of ASTM D5827. An alternative and practical way to test for chlorides in the field is with reactive test strips.

4. Nitrites

Some ELCs have nitrites as cavitation protection. The presence of a nitrite is not an indication of contamination or mixing with a conventional coolant. Keep in mind that a good RA and an acceptable pH do not stop cavitation from occurring if nitrites are not present.

5. Total Dissolved Solids

We measure total dissolved solids (TDSs) in parts per million (ppm). They are measured in the laboratory by filtering the suspended particles, evaporating the water from the filter patch, and then weighing the solids that remain. However, a pocket tester can easily do the trick in the field.

High values of TDSs contribute to increased conductivity. High conductivity, in turn, is not a desirable condition because it becomes a conduit for stray currents that interact with the chemicals in the coolant. For instance, in a machine with poor electrical grounds, high TDSs could cause the nitrite to convert to ammonia (NH_3) and then ammonium hydroxide (NH4OH), which is caustic and increases the pH of the coolant.

6. Cumulative Organic Acids (Ultrapressure Liquid Chromatography)

We can measure the concentration of total organic acids content by a lab method called *ultrapressure liquid chromatography* (UPLC). The result allows us to determine whether the coolant needs a charge with SCAs to restore its original level of protection.

Although there are no individual organic acid limits, a total measurement of the organic acid content is desirable. Many manufactures offer an SCA to reinforce decaying numbers. A good maintenance program keeps an eye on diminishing levels of additives when they are not the result of low glycol concentrations. An alternative method to correct an organic acid concentration that is too low is by draining some coolant from the system and adding undiluted coolant. Table 12-7 lists the organic acids found in coolants.

TABLE 12-7   Coolant Organic Acids

Organic acids
Tolythriazole
Benzoic acid
Sebacic acid
Toluic acid
Ethanoic acid
Octanoic acid
Tert-butylbenzoic acid
Benzotriazole
Mercaptobenzothiazole

A practical field test for organic acid concentration is also available from OEMs. Keep in mind that these are likely to be coolant specific to the OEM's specification and may not work for other coolants.

7. Ion Chromatography

The lab can measure the following additives/contaminants using inductively coupled plasma (ICP) spectroscopy:

- Nitrites
- Nitrates
- Borates
- Molybdates
- Silicates
- Phosphates
- Glycolates

8. ICP Spectroscopy for Metals

The importance of testing for metals in coolants resides in the fact that metals show the kind of corrosion/pitting the engine is suffering. In other words, they show some failures in progress.

This is the case with iron: abnormal iron readings suggest the possibility of liner cavitation. By the same token, a large presence of copper suggests engine cooler failure. Whether the readings are from corrosion or failure is a matter of cross-checking with the oil analysis. If the same metals show up in abnormal quantities in an engine's oil analysis, you can safely assume that a component is failing. If the readings from a coolant show aluminum, it could indicate corrosion from either a lack of silicates in the case of engines that use silicate coolants or corrosion by a very alkaline coolant in the case of HOAT or OAT coolants. If the readings show lead, very possibly it is coming from radiator solder in engines that use bronze radiators and nitrite-based coolants. A coolant test combined with an oil analysis is ideal to confirm that these readings have not caused component perforation yet.

9. Water Quality (Hardness, Chlorides, and Sulfates)

Chlorides and sulfates are present in coolants as dissolved salts of calcium and magnesium, such as calcium chloride ($CaCl_2$) and magnesium chloride ($MgCl_2$). At high temperatures, the calcium and magnesium chloride salts are practically insoluble, and they precipitate, dropping out of solution as hard scale deposits. The presence of chlorides is expressed in parts per million or milligrams per liter. Maximum acceptable levels of chlorides in coolants is 100 ppm, and the same levels apply for sulfates. If the sample goes to a lab, the guidelines listed in Table 12-8 apply.

TABLE 12-8 Water Testing Guidelines for Diesel Engine Coolants (ppm)

	CAT	CUMMINS	DETROIT	JOHN DEERE	ASTM
Chlorides	50	100	40	5	40
Sulfates	50	100	100	5	100
Total dissolved solids (TDS)	250	500	340	10	340
Total hardness	100	300	170	5	40

Lab testing also can provide additional parameters regarding the water quality. Although test strips also provide pH measurements by the pass/no-pass approach, the lab standard for water pH is about 7 or neutral, neither acidic nor basic.

ON-SITE COOLANT TESTING

1. HOAT and OAT, pH, Concentration, and Organic Acid Content

The mechanism for cavitation control in organic acid and HOAT ELCs does not depend on nitrites to protect liners but mostly on organic acid content. There are field kits to test coolants for organic acid content as well as for glycol concentration and pH. Keep in mind that specific HOAT and OAT coolants may need specific test strips when it comes to testing for organic acid content. The issue of alkalinity is more important in organic acid ELCs because, by nature, these coolants have higher alkalinity.

The three-way stick is easy to use and should be included in monthly inspections. Lab services are necessary at least once a year to see if there are metals and sediments. When the test strips confirm the need for more organic acid additives and the glycol concentration appears to be correct, there are SCAs to reinforce the charge in the coolant (see Figure 12-18).

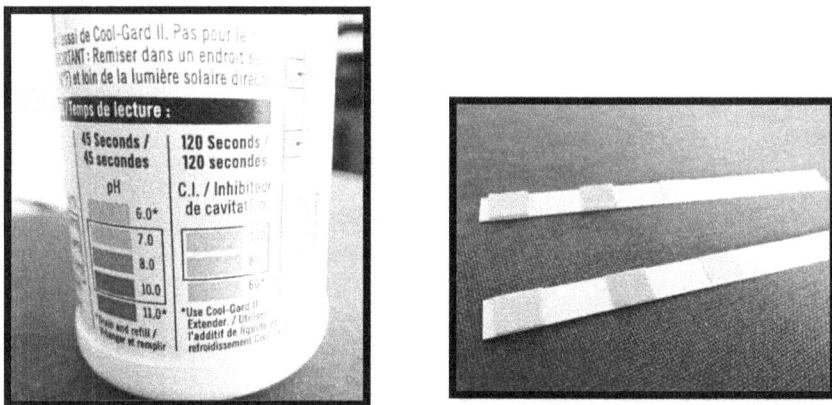

FIGURE 12-18 Kit to test organic acid level (*left*); test strips for testing pH and glycol concentration (*right*).

2. pH with Pocket Tester

We can also test coolant pH in the field with a pocket tester, which looks like the total dissolved solids tester but comes in a different color (see Figure 12-19). When the coolant pH is out of specifications, whether too acidic or too basic, the system needs flushing because there is no viable alternative for a bad pH reading.

FIGURE 12-19 pH tester.

3. NOAT, pH, Concentration, Nitrites, and Molybdates

The cavitation control test on conventional nitrite-based coolants is feasible with field kits as well. These measure molybdates, nitrites, glycol concentration, and pH. If you immerse the stick in the coolant sample and wait for about one minute, a reaction occurs that changes the color of the reactive chemicals on the strip. The resulting color should fall in the green zone of the reference chart. If it does not, then the coolant needs additional SCA. The numbers on the reference charts are ounces of SCA per gallon that the system would need to bring it back into specification (see Figure 12-20). Additionally, these strips provide a reference color for pH with a pass/no-pass scale.

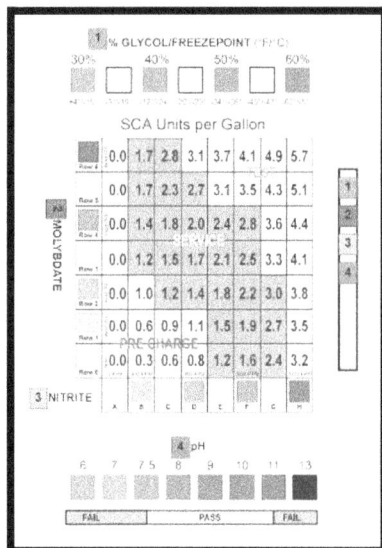

FIGURE 12-20 Nitrite-based coolant reference chart.

4. All Coolants: TDSs with a Pocket Tester

We can use a pocket TDS tester to measure total dissolved solids in the field, and the test is practical and inexpensive to perform (see Figure 12-21).

FIGURE 12-21 TDS pocket tester.

5. Water Quality Test (Hardness, Chlorides, and pH) with Three-Way Strips

Water quality is important when maintaining equipment. The idea that if water is good for human consumption, it should be good enough for engines is not correct. To operate at optimal levels, engines need mineral-free water, usually referred to as *distilled water*. The reason is that minerals cause ion activity that produces corrosion, thus accelerating corrosion inhibitor depletion. Testing for water quality is simple and inexpensive with on-site test strips or with an electronic tester.

Chlorides and sulfates are present in coolants as dissolved salts of calcium and magnesium, such as calcium chloride ($CaCl_2$) and magnesium chloride ($MgCl_2$). At high temperatures, calcium and magnesium chloride salts are practically insoluble, and they precipitate and drop out of solution as hard scale deposits. The maximum acceptable level for chlorides in coolants is 100 ppm, and the same level applies for sulfates. Test strips for chlorides, pH, and hardness look like those in Figure 12-22.

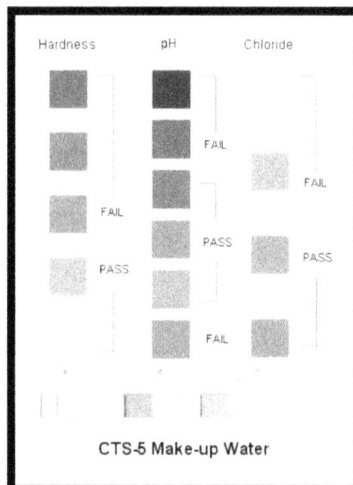

FIGURE 12-22 Water quality test strip.

ELC Coolant Guidelines

Table 12-9 can be useful when interpreting a coolant analysis. The table is a starting point but is not all-inclusive. It is always possible that certain coolants are outside the parameters of the table.

TABLE 12-9 ELC Coolant Key Guidelines

Indicator	NOAT	HOAT	Organic acid
pH value	8–10.5	8–10.5	7–8.5
Reserve alkalinity (mL HCL/10 mL)	8–20	8–20	2.5–5
Glycol content (%)	45–55	45–55	45–55
Freeze point (°F)	–40	–40	–40
Chloride ClO_2 (mg/L or ppm)	5–100	5–100	5–100
Nitrite (mg/L or ppm)	800–1400	0–50	0–100
Molybdate (ppm)	250–450	100–250	50–150
Nitrate (ppm)	500–1000	50–200	600–800
Total dissolved solids (ppm)	0–340	0–340	0–340
Cumulative organic acid (%)	0.8–2	0.8–1.5	1–2
Total hardness	<300	<300	<300
Iron (ppm)	<35	<35	<35
Copper (ppm)	<15	<15	<15
Aluminum (ppm)	<45	<45	<45
Lead (ppm)	<10	<10	<10

Coolant Analysis Reports

Coolant reports from laboratories come in many different styles and forms (see Figure12-23). Some may contain basic data and others perhaps a complete battery of tests. For the following case explanations, we will use complete sets of data. In this way, readers can become acquainted with the signatures of the coolants discussed and the key results that are important to each type of coolant.

FIGURE 12-23 Coolant analysis report.

Some of the key indicators are going to be different for each type of coolant. For example, the nitrite and molybdate in a NOAT coolant will be most important, whereas for a HOAT coolant, the organic acid content and pH level will be the focus of the evaluation.

CASE DISCUSSIONS

The following case discussions are real cases taken from the industry. Each case will have an analysis, a conclusion as to what is wrong, and how we can resolve the situation.

1. Low Freeze Protection in an Excavator with Low Glycol Content

The excavator in question, originally using a NOAT coolant, shows coolant results that do not meet the standards for ELC readings.

Coolant Analysis

The results of the coolant analysis in Figure 12-24 show that there is no protection from freezing, the glycol content is low, and there is no protection from cavitation through nitrites. Nitrates are high. Silicates and organic acids are low. The pH is also low, indicating an acidic coolant that is triggering corrosion. TDSs are also high.

PHYSICAL/CHEMICAL										INFORMATION		
Freeze Point D3321	Glycol Content	pH	Reserve Alkalinity	Nitrites D5827	Nitrates	Molybdate	Silicates	OA UPLC	Sodium	Coolant Type	Engine Hours/Miles	Coolant Hours/Miles
°F	%	D1287	HCl	ppm	ppm	ppm	ppm	%	mg/Kg			
3	31	6.9	1	962	588	<100	<10	0.73	1800	NOAT	5,487	998
-42	52	8.2	3.5	1638	259	200	220	1.2	2760	NOAT	4,875	386
-46	56	8.6	4	1547	270	234	234	1.5	2800	NOAT	4489	2,000

CORROSION METALS				WATER HARDNESS					VISUAL APPEARANCE			
Pb	Fe	Al	Cu	Ca	Chlorides	Sulfates	Total Hardness	TDS	Color	Clarity	Oil Layer	Sediment
ppm	ppm	ppm	ppm	ppm	ppm	ppm	ppm	ppm				
2	34	14	38	NA	25	145	80	350	Clear	Clear	None	None
3	21	12	35	NA	20	75	56	<200	Green	Clear	None	None
5	15	10	26	NA	23	70	64	<200	Green	Clear	None	None

COOLANT ANALYSIS	Excavator, low glycol, low pH, low nitrites, high nitrates and low OA.

FIGURE 12-24 Coolant analysis report for an excavator.

Conclusion

It is possible that this machine suffered a coolant leak and was topped off with water instead of glycol. Given the lack of glycol and low SCA, the system needs a flush and refill with fresh coolant that complies with machine specifications. The increased level of nitrates suggests that the machine could have had an air leak. The cooling system needs to get rid of air bubbles that affect nitrate production. The excess nitrate produced the low and acidic pH level. Corrosion, as suggested by the iron and copper readings, is a consequence of the lack of corrosion inhibitors.

2. Overconcentration in a CAT Off-Road Truck

The operator of this off-road truck is not reporting issues with operation, but there are concerns with the coolant test results.

Coolant Analysis

The coolant analysis results for this off-road truck in Figure 12-25 show a high SCA concentration. The gray numbers indicate excessive nitrite, molybdate, and organic acid concentrations. Consequently, nitrates are also high.

PHYSICAL/CHEMICAL										INFORMATION		
Freeze Point D3321	Glycol Content	pH	Reserve Alkalinity	Nitrites D5827	Nitrates	Molybdate	Silicates	OA UPLC	Sodium	Coolant Type	Engine Hours	Coolant Hours
°F	%	D1287	HCl	ppm	ppm	ppm	ppm	%	mg/Kg			
-40	52	8.1	8.1	900	650	220	300	2.5	3900	HOAT	30,900	2000
-40	52	8.1	8	774	466	145	280	1.2	3280	HOAT	28,900	2000
CORROSION METALS				WATER HARDNESS					VISUAL APPEARANCE			
Pb	Fe	Al	Cu	Ca	Chlorides	Sulfates	Total Hardness	TDS	Color	Clarity	Oil Layer	Sediment
ppm	ppm	ppm	ppm	ppm	ppm	ppm	ppm	ppm				
2	12	9	3	NA	25	48	36	180	Green	Clear	None	None
3	15	12	4	NA	23	45	35	176	Green	None	None	None
COOLANT ANALYSIS	Off-road CAT dump truck with concentrated nitrite											

FIGURE 12-25 Coolant analysis report for a CAT off-road truck.

Conclusion

With certain vintage machines, the Supplemental Coolant Additve (SCA) was transferred to the coolant via a pre-charged filter, which allowed the slow delivery of the additive (SCA). However, if the coolant already had plenty of SCA, using an SCA pre-charged filter could produce an over-concentration of additives, as in this case. The solution is to install a non-SCA filter until the numbers go back to normal. Luckily, in this case the pH is at the right level.

3. Low pH, High Acidity in a CAT SR4B Generator Set

A CAT standby generator set has been used sporadically, and the hours have accumulated over a long period. There have been some leaks, which over time diluted the SCA package. The coolant analysis results worry the operator.

Coolant Analysis

The coolant analysis results in Figure 12-26 show a low SCA concentration and a low pH (acidic). Hardness and TDSs are high. The results also show some level of corrosion and low freeze protection.

PHYSICAL/CHEMICAL										INFORMATION		
Freeze Point D3321	Glycol Content	pH	Reserve Alkalinity	Nitrites D5827	Nitrates	Molybdate	Silicates	OA UPLC	Sodium	Coolant Type	Engine Hours/Miles	Coolant Hours/Miles
°F	%	D1287	HCl	ppm	ppm	ppm	ppm	%	mg/Kg			
26	10	6.7	1	**220**	**200**	<100	<10	**<1**	2200	NOAT	3,322	428
-38	52	7.99	4	625	165	200	220	2.2	2760	NOAT	2,894	417
-36	51	7.84	5	875	178	234	234	2.3	2800	NOAT	2500	394

CORROSION METALS				WATER HARDNESS					VISUAL APPEARANCE			
Pb	Fe	Al	Cu	Ca	Chlorides	Sulfates	Total Hardness	TDS	Color	Clarity	Oil Layer	Sediment
ppm	ppm	ppm	ppm	ppm	ppm	ppm	ppm	ppm				
2	45	14	40	NA	75	89	150	<350	Clear	Hazy	None	Brown
3	21	12	35	NA	28	43	60	235	Green	Hazy	None	None
5	15	10	26	NA	25	35	65	220	Green	Clear	None	None

COOLANT ANALYSIS	CAT SR4B Generator, low glycol concentration

FIGURE 12-26 Coolant analysis report for a CAT SR4B generator set.

Conclusion

Because this coolant has a low glycol concentration and appears hazy, it is possible that the coolant leaked out of the engine and received water as a top-off. This has eliminated the freeze protection and has increased corrosion. Because the pH is low, this confirms an acidic condition; there are no other options than to restore this coolant to its original formulation by flushing the system and refilling it with fresh coolant.

4. High pH in an Excavator with OAT Coolant

A Japanese excavator with OAT coolant is showing increased aluminum readings, and the user has noticed some corrosion in aluminum connectors, which are leaking at some connections and hoses.

Coolant Analysis

The coolant analysis results in Figure 12-27 indicate that the coolant has become too basic with a pH of 10.6. Aluminum and TDSs are also high, and the coolant shows haziness.

PHYSICAL/CHEMICAL										INFORMATION		
Freeze Point D3321	Glycol Content	pH	Reserve Alkalinity	Nitrites D5827	Nitrates	Molybdate	Silicates	OA UPLC	Sodium	Coolant Type	Engine Hours/Miles	Coolant Hours/Miles
°F	%	D1287	HCl	ppm	ppm	ppm	ppm	%	mg/Kg			
-40	50	10.6	3	<30	857	<100	<10	1.5	9567	OAT	5,500	2,000
-38	52	8.1	3.5	40	820	200	23	1.9	9780	OAT	3,500	1,000
-36	51	8	4	55	800	240	25	1.7	9900	OAT	2500	2,500

CORROSION METALS				WATER HARDNESS					VISUAL APPEARANCE			
Pb	Fe	Al	Cu	Ca	Chlorides	Sulfates	Total Hardness	TDS	Color	Clarity	Oil Layer	Sediment
ppm	ppm	ppm	ppm	ppm	ppm	ppm	ppm	ppm				
2	25	55	40	NA	19	89	90	345	Green	Clear	None	None
3	18	34	36	NA	14	43	55	235	Green	Clear	None	None
5	14	10	32	NA	15	35	48	220	Green	Clear	None	None

COOLANT ANALYSIS	Excavator, OAT coolant with high pH, Al corrosion

FIGURE 12-27 Coolant analysis report for an excavator with OAT coolant.

Conclusion

With OAT coolants, it is important to check that the pH does not exceed 8.5 because their tendency is to become basic and not acidic. Coolants that are too basic corrode aluminum, especially connectors or tubing made of aluminum. Because of this coolant condition, the addition of an SCA is not going to solve the pH issue. Therefore, the best path to resolve this case is to flush the coolant and fill the system with new coolant.

5. Oil Contamination in an Aggregate Plant Crusher V12 Detroit Engine

A Detroit V12 stationary engine is showing hazy coolant with an oil trace, as indicated by a sheen. The operator is questioning whether the coolant is still usable.

Coolant Analysis

The coolant analysis results in Figure 12-28 show high TDSs as well as high sulfates and chlorides. The engine is showing signs of corrosion and SCAs are low for this kind of engine. The engine has had a history of high pH values as well. The coolant report also indicates the presence of an oil sheen.

PHYSICAL/CHEMICAL										INFORMATION		
Freeze Point D3321	Glycol Content	pH	Reserve Alkalinity	Nitrites D5827	Nitrates	Molybdate	Silicates	OA UPLC	Sodium	Coolant Type	Engine Hours/Miles	Coolant Hours/Miles
°F	%	D1287	HCl	ppm	ppm	ppm	ppm	%	mg/Kg			
-19	43	9.6	4	375	650	220	300	1.2	2600	?	3,322	428
-17	43	10	4	400	466	145	280	1	2760	?	2,894	417
-23	46	10.25	4.5	875	476	135	255	1.1	2800	?	2500	394

CORROSION METALS				WATER HARDNESS					VISUAL APPEARANCE			
Pb	Fe	Al	Cu	Ca	Chlorides	Sulfates	Total Hardness	TDS	Color	Clarity	Oil Layer	Sediment
ppm	ppm	ppm	ppm	ppm	ppm	ppm	ppm	ppm				
2	45	25	47	NA	65	120	300	1083	Orange	Hazy	Sheen	None
3	15	50	45	NA	50	87	210	1237	Pink	Hazy	Sheen	None
5	14	45	34	NA	45	80	220	1261	Pink	Clear	None	None

COOLANT ANALYSIS	Detroit 12 V engine, oil contamination. Automotive coolant suspected

FIGURE 12-28 Coolant analysis report for an aggregate plant crusher V12 Detroit engine.

Conclusion

The presence of an oil sheen overrides the rest of the conditions. The copper readings suggest that the oil cooler could be leaking. The current SCA level is low, and its variability throughout the life of the engine suggests some coolant mixing. It is advisable to fix the oil leak first and then flush the system. Use of the correct coolant for this engine would prevent further corrosion.

6. High Lead Readings in a Motor Grader

The coolant analysis results from an American-made motor grader using a nitrite-based NOAT coolant are showing copper and lead readings in the analysis at 2,500 hours on the coolant.

Coolant Analysis

The coolant analysis results in Figure 12-29 show increased nitrites, nitrates, molybdates, organic acids, and sodium, suggesting that this coolant system has excess SCAs. Excessive nitrates could cause acidic attack on solder, leading to the lead and copper readings. TDSs are also high, and chlorides and sulfates are high as well. The pH is marginal and borderline acidic.

PHYSICAL/CHEMICAL										INFORMATION		
Freeze Point D3321	Glycol Content	pH	Reserve Alkalinity	Nitrites D5827	Nitrates	Molybdate	Silicates	OA UPLC	Sodium	Coolant Type	Engine Hours/Miles	Coolant Hours /Miles
°F	%	D1287	HCl	ppm	ppm	ppm	ppm	%	mg/Kg			
-36	50	**7.5**	5	1530	900	450	380	2	2800	NOAT	6,500	2,500
-55	55	8	3.7	1150	835	350	23	1.9	2450	NOAT	4,000	2,000
-36	51	8.2	4	1240	840	395	25	1.7	2670	NOAT	2000	2,000

CORROSION METALS				WATER HARDNESS					VISUAL APPEARANCE			
Pb	Fe	Al	Cu	Ca	Chlorides	Sulfates	Total Hardness	TDS	Color	Clarity	Oil Layer	Sediment
ppm	ppm	ppm	ppm	ppm	ppm	ppm	ppm	ppm				
18	25	25	42	NA	12	95	80	45	Green	Clear	None	None
3	16	20	36	NA	10	90	75	42	Green	Clear	None	None
5	15	15	32	NA	9	80	80	35	Green	Clear	None	None

COOLANT ANALYSIS	Motor Grader, NOAT coolant with high nitrite readings

FIGURE 12-29   Coolant analysis report for a motor grader.

Conclusion

There are two alternative fixes for this problem. One is by flushing and refilling the system with the correct coolant. The second is to drain some of the coolant and add fresh coolant. The first approach helps to get rid of the lead and copper readings that could affect coolant analysis interpretation in the future. However, high TDSs are a concern, and a partial drain and refill are not going to solve the issue.

7. High Glycol/High SCA Concentrations in a Production Bulldozer

This production bulldozer using a HOAT coolant is not presenting any operational issues, but the user has concerns with the coolant results from the lab.

Coolant Analysis

The coolant analysis results in Figure 12-30 show subtle signs of SCA overconcentration. The glycol content is at 70%, giving an impressive –60°F freeze protection. However, the SCA constituents are high, also resulting in a high pH number. The coolant is a hybrid HOAT coolant, which is low in nitrates and silicates, but it is showing high glycol content.

PHYSICAL/CHEMICAL										INFORMATION		
Freeze Point D3321	Glycol Content	pH	Reserve Alkalinity	Nitrites D5827	Nitrates	Molybdate	Silicates	OA UPLC	Sodium	Coolant Type	Engine Hours/Miles	Coolant Hours/Miles
°F	%	D1287	HCl	ppm	ppm	ppm	ppm	%	mg/Kg			
-60	70	10	9	20	120	200	500	2.2	3700	HOAT	5,000	2,000
-36	50	8	8	<10	100	174	470	1.9	3450	HOAT	4,000	1,000
-36	51	8.2	8.3	<10	90	180	485	1.7	3455	HOAT	3000	3,000

CORROSION METALS				WATER HARDNESS					VISUAL APPEARANCE			
Pb	Fe	Al	Cu	Ca	Chlorides	Sulfates	Total Hardness	TDS	Color	Clarity	Oil Layer	Sediment
ppm	ppm	ppm	ppm	ppm	ppm	ppm	ppm	ppm				
4	17	12	9	NA	10	85	60	55	Orange	Clear	None	None
4	15	10	7	NA	9	75	45	46	Orange	Clear	None	None
3	14	11	3	NA	7	60	35	37	Orange	Clear	None	None

COOLANT ANALYSIS	Production Bulldozer, HOAT coolant with high SCA readings, High pH

FIGURE 12-30   Coolant analysis report for a production bulldozer.

Conclusion

A high pH is not desirable for engines using components with aluminum alloys. A high glycol content is not desirable either, because the flow of coolant within the engine is more restricted. Although there is no indication that this engine is suffering any corrosion or pitting at this time, it is a good time to bring the numbers to normal levels. This is easy to attain by draining some of the coolant and adding distilled water.

8. Low SCA Concentration in a European Tree Harvester

A European tree harvester with only 2,000 hours on the coolant is showing low SCA numbers.

Coolant Analysis

The coolant analysis results in Figure 12-31 show that the glycol concentration is at 50% but that most of the measurable SCA constituents are low compared with the OAT signature. The pH is approaching the upper limits. The rest of the coolant indicators are still fine.

PHYSICAL/CHEMICAL										INFORMATION		
Freeze Point D3321	Glycol Content	pH	Reserve Alkalinity	Nitrites D5827	Nitrates	Molybdate	Silicates	OA UPLC	Sodium	Coolant Type	Engine Hours/Miles	Coolant Hours /Miles
°F	%	D1287	HCl	ppm	ppm	ppm	ppm	%	mg/Kg			
-36	50	8.9	3.5	20	530	0	56	1.2	1200	OAT	6,000	2,000
-41	51	8	3.9	<10	750	0	90	1.8	1390	OAT	4,000	4,000
-40	50	8.2	4	<10	770	0	112	1.8	1423	OAT	2000	2,000

CORROSION METALS				WATER HARDNESS					VISUAL APPEARANCE			
Pb	Fe	Al	Cu	Ca	Chlorides	Sulfates	Total Hardness	TDS	Color	Clarity	Oil Layer	Sediment
ppm	ppm	ppm	ppm	ppm	ppm	ppm	ppm	ppm				
4	18	18	10	NA	11	90	62	65	Orange	Clear	None	None
4	16	17	8	NA	9	75	45	58	Orange	Clear	None	None
3	14	9	3	NA	7	60	42	45	Orange	Clear	None	None

COOLANT ANALYSIS	Tree Hasverter, OAT coolant with low SCA readings

FIGURE 12-31 Coolant analysis report for a European tree harvester.

Conclusion

An easy corrective action can solve this issue, which otherwise could bring trouble in the near future. Because the glycol content is acceptable and the pH is manageable, the addition of some OAT SCA could restore this coolant signature. The OAT SCA also can help reduce the pH level. Adjustments to the level of additional SCAs are feasible using field test strips to measure the SCAs needed. In this way, the final signature can be tuned to the right numbers.

9. High Copper Readings in a Construction 4WD Loader

A construction 4WD loader is showing increased copper numbers in the coolant analysis. The machine is using a hybrid HOAT coolant with little nitrite.

Coolant Analysis

The coolant analysis results in Figure 12-32 appear very normal with only copper as abnormal. RA is good for an OAT, glycol concentration is good, and SCA levels are also good. Because a cross-check with the oil analysis is mandatory for cases such as this, the results could show whether the copper reading is also present in oil analysis, together with some coolant constituents.

PHYSICAL/CHEMICAL										INFORMATION		
Freeze Point D3321	Glycol Content	pH	Reserve Alkalinity	Nitrites D5827	Nitrates	Molybdate	Silicates	OA UPLC	Sodium	Coolant Type	Engine Hours/Miles	Coolant Hours /Miles
°F	%	D1287	HCl	ppm	ppm	ppm	ppm	%	mg/Kg			
-41	51	8.2	8.2	11	110	179	475	1.9	3430	HOAT	4,500	1,000
-40	50	8.4	8	9	135	174	500	1.8	3450	HOAT	3,500	3,500
-41	52	8.1	8.3	12	115	180	485	1.9	3500	HOAT	2,500	2,500

CORROSION METALS				WATER HARDNESS					VISUAL APPEARANCE			
Pb	Fe	Al	Cu	Ca	Chlorides	Sulfates	Total Hardness	TDS	Color	Clarity	Oil Layer	Sediment
ppm	ppm	ppm	ppm	ppm	ppm	ppm	ppm	ppm				
1	17	12	35	NA	10	85	60	55	Orange	Clear	None	None
2	15	14	25	NA	9	75	45	46	Orange	Clear	None	None
1	14	12	9	NA	7	60	35	37	Orange	Clear	None	None

COOLANT ANALYSIS	Construction Loader, HOAT coolant with high copper readings

FIGURE 12-32   Coolant analysis report for a construction 4WD loader.

Oil Analysis Cross-Check

The presence of sodium, potassium, and increased silicon and boron confirms that there is a coolant leak into the oil. Judging by the copper reading, the coolant leak very possibly is coming from the oil cooler (see Figure 12-33).

METALS							ADDITIVES						
Iron (Fe)	Chromium Cr)	Lead (Pb)	Copper (Cu)	Tin (Sn)	Aluminum (Al)	Nickel (Ni)	Magnesium (Mg)	Calcium (Ca)	Barium Ba)	Phosphorus (P)	Zinc (Zn)	Molybdenum (Mo)	Boron (B)
110	12	11	55	<1	4	<1	32	2950	<1	1100	1197	2	12
35	3	7	2	<1	3	<1	26	3057	<1	1178	1235	2	1

CONTAMINANTS								PHYSICAL PROPERTIES				
Aluminum (Al)	Silicon (Si)	Sodium (Na)	Potassium (K)	Water (%)	Glycol	Fuel (%)	Soot (%) FTR	Sulphation	Nitration	Viscosity (cSt @ 100 C)	Oxidation	TBN	TAN
1	18	45	65	<0.05	NO	<1	1.2	13	2	16.3	7	3.1	NA
2	8	12	28	<0.05	NO	<1	0.9	24	3	15.4	13	5.6	NA

ENGINE OIL ANALYSIS	4WD loader with 4500 hours, 500 hours on oil. 15W40 CJ4 oil

FIGURE 12-33   Oil analysis cross-check.

Conclusion

By cross-checking the coolant and oil analyses, it is easy to pinpoint the troubled area. In this case, it is a failed oil cooler. The damaged oil cooler is possibly a vibration type of failure and not related to the coolant type or condition, which appears to be normal.

10. High Iron Readings and Cavitation in a Production Crawler

A production crawler using a NOAT coolant is showing increased iron readings.

Coolant Analysis

The coolant has been in the bulldozer for 5,000 hours, and the nitrite level is low (see Figure 12-34). Additionally, the pH is low, suggesting an acidic coolant. At this point, the high iron reading needs assessment because it could be the result of cavitation in an acidic low-pH, low-nitrite environment. Additionally, the haziness of the coolant and its high sulfates require installation of fresh coolant. However, a most important cross-check needs to be performed, that is, checking the oil analysis to find whether traces of coolant are present in the oil.

PHYSICAL/CHEMICAL										INFORMATION		
Freeze Point D3321	Glycol Content	pH	Reserve Alkalinity	Nitrites D5827	Nitrates	Molybdate	Silicates	OA UPLC	Sodium	Coolant Type	Engine Hours/Miles	Coolant Hours/Miles
°F	%	D1287	HCl	ppm	ppm	ppm	ppm	%	mg/Kg			
-53	55	6.5	2	550	920	350	300	1.2	2250	NOAT	5,000	5,000
-52	53	8	3.7	1150	835	220	310	1.5	2450	NOAT	3,000	3,000
-40	50	8.2	4	1240	840	180	325	1.7	2670	NOAT	1000	1,000

CORROSION METALS				WATER HARDNESS					VISUAL APPEARANCE			
Pb	Fe	Al	Cu	Ca	Chlorides	Sulfates	Total Hardness	TDS	Color	Clarity	Oil Layer	Sediment
ppm	ppm	ppm	ppm	ppm	ppm	ppm	ppm	ppm				
18	125	25	25	NA	14	125	97	870	Green	Hazy	None	None
3	16	15	18	NA	10	90	75	540	Green	Clear	None	None
5	15	11	12	NA	9	80	80	35	Green	Clear	None	None

COOLANT ANALYSIS	Production crawler with NOAT coolant and high iron readings

FIGURE 12-34 Coolant analysis report for a production crawler.

Oil Analysis Cross-Check

As in the preceding case, an oil analysis is needed to confirm whether the increased iron reading could be coming from liner pitting. The results from this crawler's engine are interesting, showing a considerable amount of iron together with the coolant constituents in at least the last two samples. This suggests that the liner cavitation issues started some time ago. Together with coolant, the highlighted numbers indicate abnormal or critical values, such as high viscosity, low total base number, and high total acid number, which are direct consequences of coolant contamination (see Figure 12-35).

METALS							ADDITIVES						
Iron (Fe)	Chromium Cr	Lead (Pb)	Copper (Cu)	Tin (Sn)	Aluminum (Al)	Nickel (Ni)	Magnesium (Mg)	Calcium (Ca)	Barium Ba	Phosphorus (P)	Zinc (Zn)	Molybdenum (Mo)	Boron (B)
182	12	8	13	3	25	<1	65	3400	<1	1495	1685	132	145
135	8	6	10	2	23	<1	57	3507	<1	1578	1634	145	140
8	<1	1	12	<1	2	<1	55	3905	<1	1502	1652	120	132

CONTAMINANTS										PHYSICAL PROPERTIES			
Aluminum (Al)	Silicon (Si)	Sodium (Na)	Potassium (K)	Water (%)	Glycol	Fuel (%)	Soot (%) FTR	Sulphation	Nitration	Viscosity (cSt @ 100 C)	Oxidation	TBN	TAN
25	32	123	95	0.9	YES	<1	0.8	NA	NA	17.2	NA	3.5	4
2	21	45	65	<0.05	Trace	<1	0.5	NA	NA	15.5	NA	7	NA
3	8	15	5	<0.05	NO	<1	0.4	NA	NA	15.3	NA	6	NA

ENGINE OIL ANALYSIS	Production crawler with 5000 hours and 500 hours on the oil. 15W40 oil. High Iron and coolant constituents present. Low TBN high TAN

FIGURE 12-35   Oil analysis high-iron cross-check.

Conclusion

By cross-checking analyses, it is much easier to pinpoint the leaking area and the likely cause of the failure. In this case, an overextended coolant change interval and the lack of nitrite point directly to liner cavitation.

11. Hazy Brown and High TDSs in a Navistar Truck

The operator of a Navistar truck reports a hazy brown coolant with high TDSs.

Coolant Analysis

The coolant analysis results in Figure 12-36 show that at one point this cooling system had a high pH level, which kept aluminum readings high. There is a history of high TDSs, and the organic acid level exceeds that of any known coolant.

PHYSICAL/CHEMICAL										INFORMATION		
Freeze Point D3321	Glycol Content	pH	Reserve Alkalinity	Nitrites D5827	Nitrates	Molybdate	Silicates	OA UPLC	Sodium	Coolant Type	Engine Hours/Miles	Coolant Hours /Miles
°F	%	D1287	HCl	ppm	ppm	ppm	ppm	%	mg/Kg			
-52	57	8.5	8.1	1030	650	220	300	2.5	3900	HOAT	205,000	19,000
-42	52	8.8	4.5	774	466	145	280	1.2	3280	HOAT	186,000	8,150
-27	46	11.5	5	780	476	135	255	1.3	3100	HOAT	177850	15,000

CORROSION METALS				WATER HARDNESS					VISUAL APPEARANCE			
Pb	Fe	Al	Cu	Ca	Chlorides	Sulfates	Total Hardness	TDS	Color	Clarity	Oil Layer	Sediment
ppm	ppm	ppm	ppm	ppm	ppm	ppm	ppm	ppm				
2	12	25	3	6	3	3	3	1810	Brown	Hazy	None	None
3	15	50	15	10	5	4	7	1490	Green	Hazy	None	None
5	14	45	25	12	6	5	8	2060	Green	Hazy	None	Yes

COOLANT ANALYSIS	Navistar truck engine with NOAT

FIGURE 12-36   Coolant analysis report for a Navistar truck.

Conclusion

Given the current and former coolant signatures, it appears that this engine has a mixture of coolants. The increased inhibitor numbers suggest that this coolant has excessive SCAs. The observation about the brown color confirms that this cooling system is using more than one type of coolant, possibly a green NOAT and an orange HOAT. The best option for a cooling system in this condition is to flush and refill with the proper coolant as specified by the manufacturer.

12. Corrosion in a Detroit Diesel V16 149 Two-Stroke Diesel Engine

A Detroit diesel V16 engine powering a generator set for an aggregate plant has started suffering from aluminum connector corrosion. The operator uses water from the desalinization plant to make coolant. The coolant consists of water from the plant plus an SCA, but no glycol because it is in a warm climate. The engine is believed to have been running fine for many years with this coolant. A laboratory coolant report is not available. Field test kits are used to check for coolant condition and water quality (see Figure 12-37).

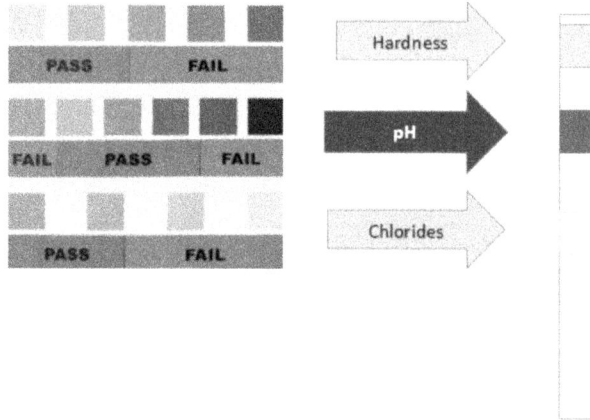

FIGURE 12-37   Hardness, pH, and chlorides result from the coolant field test.

Field Test Results

The field test results for this coolant in Figure 12-38 show that the coolant passed the hardness test but failed the chloride and pH tests. With regard to SCAs, the coolant also failed the organic acid, nitrite, and molybdate tests, which are additives to control corrosion. The charts and colors on the sticks tell the story.

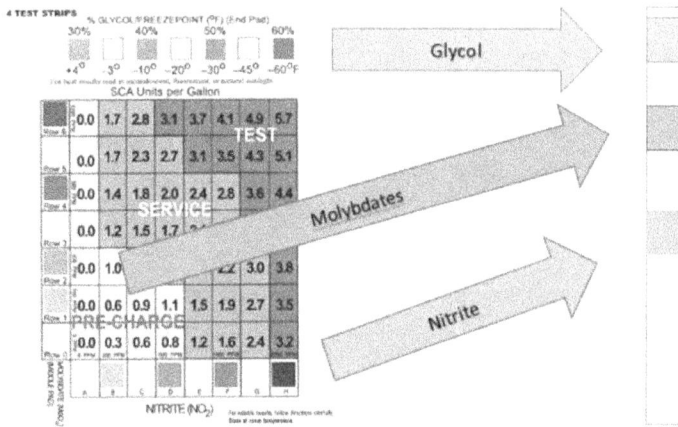

FIGURE 12-38   SCA charge field test.

Conclusion

Although a lab test can provide complete information about coolant condition, field tests are a great option as well. The test results from these field test methods suggest that the coolant for this engine lacks the basic conditions and additives to protect the engine and connectors from corrosion.

CONCLUSION

In this chapter, we went through the origins of coolants and why we need heavy-duty coolants for modern diesel engines. We also discussed the reason for different additive formulations and the importance of water quality. We discussed the different SCA packages and how to test for them. Most important, we discussed real cases and recommendations to solve coolant-related problems. Let us now move the discussion to the next challenging topic—fuels.

CHAPTER 13

FUELS AND FUEL ANALYSIS

Fuels are at the center of energy utilization. Without fuels, there is no mechanical work unless the machines are driven electrically. As we get into the emissions control era, fuels have become more important because we want very efficient engines, and we want engines that are environmentally friendly.

In this era, we confront several challenging goals, especially with diesel engines. The need for emissions compliance requires sophisticated engine manufacturing and electronics to support them. Fuel-related issues and maintenance of the injection systems become as important as lubrication itself.

FUELS

The development of the internal combustion engine began in the second half of the eighteenth century. Progress was slow but advanced over the next 100 years. Around 1892, Rudolf Diesel (Figure 13-1) filed a patent for a reciprocating compression ignition engine that used coal dust as fuel. However, his invention did not work.

Thirty-three years earlier, in 1859, crude oil became available in Pennsylvania, and the first product refined from crude was kerosene for lamps. Because only a fraction of the oil served as lamp oil, refiners had to figure out what to do with the rest of the oil left in the barrel. This was when Rudolf Diesel realized that liquid oil could be a better fuel than coal dust, so he started experimenting with it. The use of petroleum derivatives plus some adjustments to its design resulted in the production of the first successful diesel engine in 1895. Diesel experimented with many compression ratios and in some cases made wrong assumptions, but by trial and error he managed to produce an engine that worked. This is how today both the compression engine and the fuel bear his name. The first engines were large and operated at very low speeds. Today, diesel engines power ships, trains, machines, industrial plants, trucks, and automobiles.

FIGURE 13-1   Rudolf Christian Karl Diesel, 1858–1913.

By 1930, diesel engines powered buses and trucks. During World War II, diesel engines powered submarines and some airplanes (although without great success). After the war, diesel engines powered vehicles and achieved some success in Europe, but not in the United States. Today, diesel engines power transportation, manufacturing, power generation, construction machinery, and agriculture.

There is a wide variety of engines that are as numerous as their applications, from small, high-revving engines with indirect injection to low-speed direct-injection engines that are gigantic with pistons up to 3 feet in diameter. The diesel engine's success is due to its efficiency, economy, and reliability.

Diesel fuel is refined from petroleum, although there is also diesel fuel derived from coal that we will discuss later. Petroleum contains paraffinic, naphthenic, olefinic, and aromatic hydrocarbons. Each class contains a very wide range of molecular weights. As it comes out of the ground, petroleum can be as light and clear as apple cider or as thick and dark as liquid tar.

Oil has a low density and therefore has a very high American Petroleum Institute (API) gravity. In the United States, light crudes are called *high-density crudes* because of their high API gravity. In comparison, heavy crudes with high specific gravities and low API gravities are called *low-density crudes* (see Figure 13-2). Outside the United States, the terminology *light crude* refers to low-density crude and *heavy crude* to high-density crude.

The task of converting crude oil into products of commercial value is called *refining*. The most important products are gasoline, jet fuel, and diesel fuel. Other products are liquid petroleum gas (LPG), heating oil, lubricating oils, waxes, asphalt, and raw materials for the chemical industry. Light crudes contain more of such products as gasoline and generally have less sulfur and nitrogen, which makes them easier to refine. However, the most modern refining processes make it possible to convert heavy crudes into products of high commercial value.

FIGURE 13-2 Naphthenic and paraffinic crude oil samples.

Refineries capable of processing heavy crudes are more expensive, the process is more complex and requires more steps, and they consume more energy, so the cost is higher. The price difference between high- and low-density crudes ultimately affects the cost efficiency of the final refined products (see Figure 13-3).

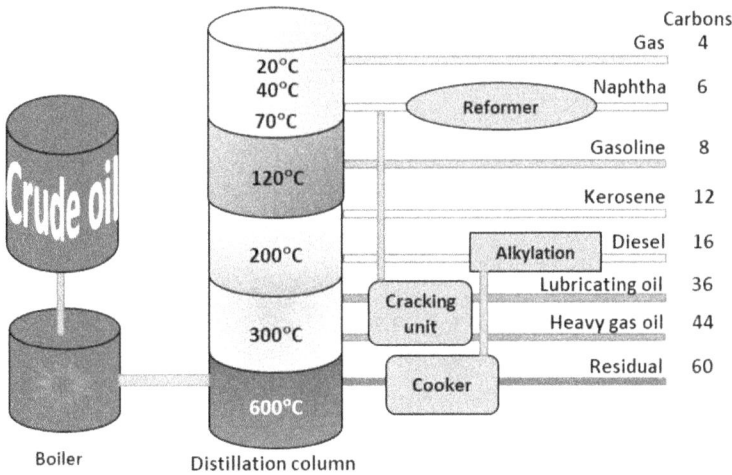

FIGURE 13-3 Crude oil distillation process.

Refining Process

Today's refinery is a complex combination of interdependent processes that are the result of a fascinating intertwining of advances in chemistry, engineering, and metallurgy. There are three basic processes:

- **Separation processes.** The feed to these processes splits into two or more components based on a physical property, usually boiling point. These processes do not otherwise change the feedstock. The most common separation process in a refinery is distillation.
- **Upgrading processes.** These processes improve the quality of a material by using chemical reactions to remove compounds present in trace amounts that give the material an

undesirable quality. Otherwise, the bulk properties of the feedstock are not changed. The most common upgrading process for diesel fuel is hydrotreating to remove sulfur.

- **Conversion processes.** These processes fundamentally change the molecular structure of the feedstock, usually by "cracking" large molecules into small ones (e.g., catalytic cracking and hydrocracking).

FUELS FROM COAL

One of the main methods of direct conversion of coal to liquid by a hydrogenation process is the *Bergius process*, developed by Friedrich Bergius (Figure 13-4). In this process, dry coal is mixed with heavy oil that is recycled from the reactor, and a catalyst is added to the mixture to complete the reaction. The reaction occurs at between 400°C (752°F) and 500°C (932°F) and under 20–70 MPa of hydrogen pressure. The reaction equation is as follows:

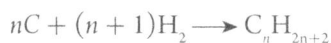

$$nC + (n + 1)H_2 \longrightarrow C_nH_{2n+2}$$

FIGURE 13-4   Fredrick Bergius (1884–1949), Nobel Prize in Chemistry, 1931.

Fischer–Tropsch Process

The Fischer–Tropsch process is a collection of chemical reactions that converts a mixture of carbon monoxide and hydrogen into liquid hydrocarbons. These reactions occur in the presence of metal catalysts, typically at temperatures of 150–300°C (302–572°F) and pressures of one to several tens of atmospheres. The process was first developed by Franz Fischer and Hans Tropsch at the Kaiser-Wilhelm Institute for Coal Research in Mülheim an der Ruhr, Germany, in 1925.

Worldwide production of fuels from coal liquefaction is respectable, and China accounts for most of the production, followed by Australia, South Africa, and some other minor producers in Africa. Although the final products burn much cleaner than petroleum-distilled products, the main concern with coal liquefaction is the pollution created by the production process. There are many projects in the United States that convert coal to fuels, but because of the costs and pollution controls required, they have never come to fruition.

Types of Diesel Fuels

In the American Society for Testing and Materials (ASTM) D975 standard diesel fuels, the following grades are covered:

- Light no. 1 diesel
- Middle no. 2 diesel (most common)
- Fuel oil, heavy no. 4 diesel
- Biodiesel
- Fuel oils, no. 5 and no. 6 marine use (need preheating)

This specification allows up to 5% biodiesel if the blended fuel continues to meet the requirements outlined in ASTM D975.

Diesel Fuel Viscosity

Although viscosity is not a common test for fuels, it becomes important when checking batches of no. 1 or no. 2 diesel fuel during seasonal changes. It is also a valuable tool when calculating oil dilution in sumps, especially in large, dry sumps in the marine industry. Table 13-1 lists the viscosity specifications for fuels. For safety reasons, viscosity tests for fuels are taken only at 40°C.

TABLE 13-1 Viscosity Specifications for Fuels

Fuel	No. 1	No. 2	No. 4	Biodiesel 100%
Viscosity at 40°C (cSt)	1.3–2.4	1.9–4.1	5.5–24	4.6

Four major classes of hydrocarbons are present in fuels—paraffinic, naphthenic, olefinic, and aromatic. Each has different chemical and physical properties, and they vary in the number of carbon atoms. Diesel fuel is a complex mixture of various compounds.

BIODIESEL

Biodiesel is a fuel comprising mono-alkyl esters of long-chain fatty acids derived from vegetable oils and animal fats. Biodiesels are typically blended with petroleum diesel (see Figure 13-5).

The standard specification for biodiesel fuel blends for fuels containing 6%–20% biodiesel (B6 to B20) is ASTM D7467. Biodiesel blends are referred to by their percentage of biodiesel content, and the notation begins with "B." Thus, a B20 blend indicates that 20% of the blend is biodiesel, and pure biodiesel is B100.

Biodiesel has a higher oxygen content (usually 10%–12% higher) and lower energy content (93% of the energy or no. 2 diesel) than petroleum diesel. Biodiesel is much less toxic than petroleum diesel. It also has greater solvency than petroleum diesel. As a result, it can be more aggressive toward some materials, particularly elastomers, which are safe for diesel fuel.

Solvency is usually not much of an issue with low biodiesel blends such as B2 and B5. However, it can clean out residues in storage, delivery, and vehicle tanks and plug filters. Biodiesel is much less toxic than petroleum diesel, and it is biodegradable. Its biodegradability makes it much more susceptible to microbial growth if it becomes contaminated with water.

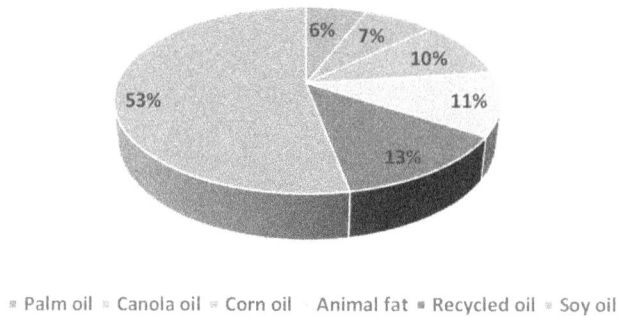

■ Palm oil ■ Canola oil ■ Corn oil ■ Animal fat ■ Recycled oil ■ Soy oil

FIGURE 13-5 Biodiesel USA.

Biodiesel Production

Biodiesel production is the process of producing the biofuel biodiesel through the chemical reactions of transesterification and esterification (see Figure 13-6). This involves vegetable or animal fats and oils being reacted with short-chain alcohols (typically methanol or ethanol). The alcohols used should be of low molecular weight.

Although the transesterification reaction can use either acidic or alkaline catalysts, base-catalyzed reactions are more common. This path has lower reaction times and catalyst costs than those using acidic catalysis. However, alkaline catalysis has the disadvantage of high sensitivity to both water and free fatty acids present in the oils.

It is important that biodiesel (B100) conform to ASTM D6751. Pure vegetable oil contains glycerin, which is sometimes called *sugar alcohol*. Burning vegetable oil as a fuel will quickly produce injection system deposits, sticking, and failure. When converting vegetable oils to biodiesel, the glycerin is removed. Biodiesel should come from a conscientious suppler that completely removes any unreacted oil, fat, alcohol, glycerin, and catalyst. Biodiesel quality has improved greatly since its introduction, but some quality issues still arise.

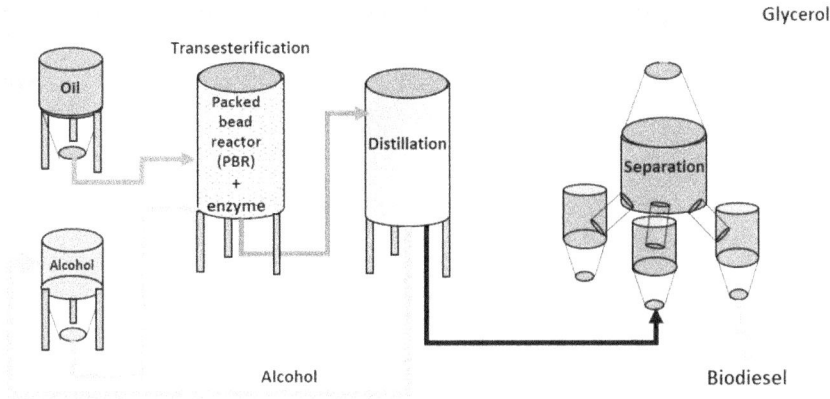

FIGURE 13-6 Transesterification.

Energetic Values of Biodiesel Blends

Biodiesel use has pros and cons. If blended at high rates with diesel, it can affect power, but at 5% or even 10%, the impact could go unnoticed. The real negative is its lack of ability to flow in winter, even though some additives can lessen the problem somewhat. On the positive side, biodiesel improves lubrication that otherwise would need to be provided with additives. Table 13-2 shows the energy content of various fuels.

TABLE 13-2 Energy Content of Various Fuels

ASTM 240	No. 2 diesel	B5	B20	B100
Energy content, BTU/gal	131,000	130,000	128,000	117,000

Biodiesel's Impact on Emissions

Biodiesel is a great fuel in general terms. As Figure 13-7 suggests, it burns cleaner than standard fuel in terms of release of hydrocarbons, particulate matter, and carbon monoxide (CO). However, as the rate of the blend increases, the release of nitrogen oxides goes up.

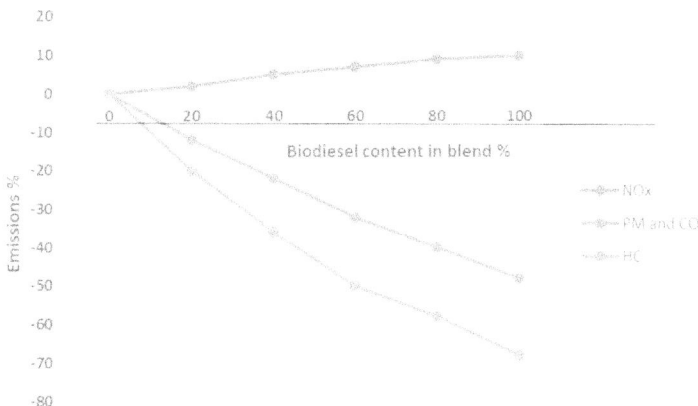

FIGURE 13-7 Emissions versus biodiesel content.

Biodiesel at the Gas Station

Blends of biodiesel are available at gas stations in the United States, with blends ranging as shown in Figure 13-8. Knowing the blend rate is important during cold weather so that users can plan ahead for the proper winter treatment of the fuel. Here fuel analysis is essential.

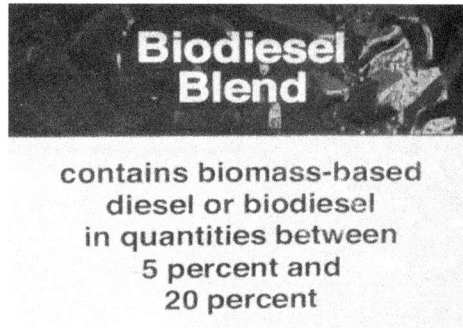

FIGURE 13-8   Biodiesel disclaimer at a gas station.

GASOLINE

The *Oxford English Dictionary* dates its first recorded use of the word *gasolene* in 1863. The term *gasoline* appeared in North America in 1864. The word is a derivation from the word *gas* and the chemical suffixes *-ol*, *-ine*, and *-ene*.

The name *gasoline* is not common outside the United States, and the name *gasolina* is only common in Latin America, with the exception of Argentina, Uruguay, and Paraguay, which call it *nafta*, which comes from the chemical naphtha. In many languages, the name of the product comes from benzene, such as *benzin* in Persian and German, *benzina* in Italian, and *bensin* in Indonesian.

The first internal combustion engines suitable for use in transportation, the so-called Otto engines, were built in Germany during the last quarter of the nineteenth century. The fuel for these early engines was a relatively volatile hydrocarbon obtained from coal gas. With a boiling point near 85°C (185°F; octane boils about 40°C or higher), it worked well for early carburetors (evaporators). The development of a spray-nozzle carburetor enabled the use of less volatile fuels. Further improvements in engine efficiency at higher compression ratios were difficult because of the premature explosion of fuel known as *knocking*.

In 1891, the Shukhov cracking process became the world's first commercial method to break down the heavier hydrocarbons in crude oil to increase the percentage of lighter products compared with simple distillation. In the early 1900s, mandated by high-knocking gasolines available at the time, the first automotive engines had very low compression ratios at around 4.7:1. To get higher power outputs, engine compression ratios needed to go up, particularly in the aviation industry, but this required better gasolines.

In 1909, a group of engineers working for Standard Oil in Indiana discovered that by using high temperatures and pressures, they could thermally crack larger molecules of petroleum into smaller ones and increase the yield of gasoline from <20% to as much as 45%. These engineers requested funds to build stills and commercialize the process, but the risky venture was denied by the main office. Standard Oil's monopoly was broken up in 1911, and the new Standard Oil of Indiana (later to become Amoco) granted funding for the project.

The process not only more than doubled the gasoline yield, but the gasoline also had much better antiknock qualities. With a superior product at a lower cost, Standard Oil of Indiana was able to take market share from the other "baby" Standard Oils. In 1914, Standard Oil of Indiana began to license the thermal cracking technology, which revolutionized the industry by giving refiners flexibility in the product streams they produced.

During World War I, the aviation industry forced an emerging market for high-quality gasoline suitable for high-performance engines, which resulted in the first specifications for gasolines. After the war, between 1917 and 1919, the amount of thermally cracked gasoline used almost doubled. In addition, the use of natural gasoline increased greatly. During this period, many U.S. states established specifications for motor gasoline, but none of the specifications agreed, and they were unsatisfactory from one standpoint or another. Larger oil refiners began to specify the percentage of unsaturated material. Thermally cracked products caused gumming in both use and storage, and unsaturated hydrocarbons are more reactive, tending to combine with impurities, leading to gumming. This controversy found a solution when the Society of Automotive Engineers (SAE), the American Petroleum Institute (API), and the U.S. Bureau of Standards decided to address the problems more scientifically.

In 1921, with the increased use of thermally cracked gasolines came increased concern about its effects on abnormal combustion, and this led to research for antiknock additives. Ultimately, this effort led to the discovery of tetraethyl lead (TEL), which helped with engine knocking and prolonged the life of exhaust valve seats, but at the high cost of human and ambient poisoning.

In 1937, the aviation industry needed gasolines with high octane that were capable of operating at high altitudes without freezing. Catalytic cracking was developed, and high-octane gasolines became available (see Figure 13-9).

In the 1950s, oil refineries started to focus on high-octane fuels and the use of detergents in gasoline to clean the jets in carburetors. The 1970s witnessed greater attention to the environmental consequences of burning gasoline. These considerations led to the phasing out of TEL and its replacement with other antiknock compounds. Subsequently, low-sulfur gasoline was mandated in part to preserve the catalysts in modern exhaust systems.

Commercial gasoline is a mixture of a large number of different hydrocarbons. Gasoline meets a host of engine performance specifications, and many different compositions are possible. Hence the exact chemical composition of gasoline is undefined. The performance specification also varies with season and region, with more volatile blends (due to added butane) during winter to facilitate starting cold engines. At refineries, the composition varies according to the crude source, the type of processing units present, how those units are operated, and which hydrocarbon streams (blend stocks) the refinery opts to use when blending the final product.

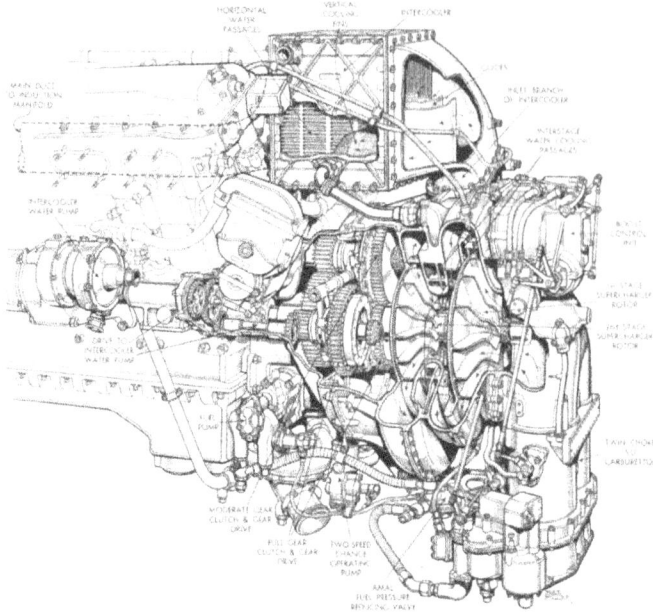

FIGURE 13-9 Two-stage supercharged aviation engine.

Detonation (Knocking) in Gasoline Engines

Detonation (knocking) is the spontaneous combustion of the end gas (remaining fuel–air mixture) in the chamber. It always occurs after normal combustion is initiated by the spark plug. The initial combustion at the spark plug is followed by an abnormal combustion burn. Factors in detonation are heat, too much spark advance, high boost pressure, low-octane gasoline, a lean mixture, and even engine design. The end gas in the chamber does not combust in a progressive and smooth expansion but rather in a spontaneous and quick explosion. Detonation occurs after the commencement of normal combustion with the spark plug.

Modern fuel-injected gasoline engines have knock sensors to retard spark advance in a knock event until the root cause for the knock is no longer present. The knocking vibration is in the 6400-Hz range, and this is why sensors are tuned to this frequency.

To obtain the maximum power from an engine, the location of peak pressure (LPP) in the cylinder needs to be at about 14 degrees after top dead center (TDC). If detonation occurs, the peak pressure will occur much sooner, thus reducing the power of the engine (see Figure 13-10).

Detonation can cause damage to pistons and piston rings, but it usually takes long exposure to this phenomenon to cause serious damage to an engine. Detonation can cause hot ember spots within the combustion chamber that could develop into preignition. Because the detonation combustion occurs so quickly, exhaust temperature goes down, and this phenomenon is used as an indication of detonation in engine tests.

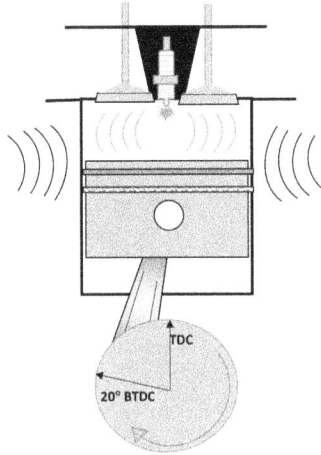

FIGURE 13-10   Gasoline engine detonation (knocking).

Preignition in Gasoline Engines

Preignition is the firing of the fuel–air charge prior to the spark plug spark. Preignition is caused by some other ignition source, such as an overheated spark plug tip, glowing hot carbon deposits in the combustion chamber, or a burned exhaust valve edge acting as a glow plug to ignite the charge. With pre-ignition, the ignition of the charge occurs far in advance of the spark plug firing, near the beginning of the compression stroke. There is no very rapid pressure spike as with detonation. Instead, high pressure is present for the entire compression stroke, acting against the normal rotation of the crankshaft and producing high loads on pistons, connecting rods, and bearings. There is no sharp pressure spike to resonate the block and the cylinder head to cause any detectable noise.

Preignition can cause serious engine damage very rapidly, and the failure mode is going to be different from that of detonation, typically involving damage to the top of the pistons. Preignition is equivalent to advancing the spark by more than 100 degrees (see Figure 13-11).

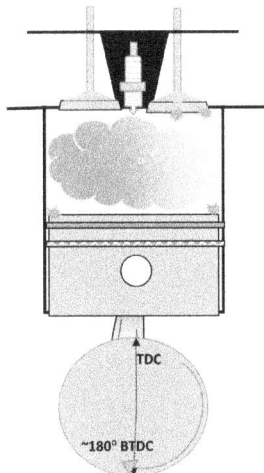

FIGURE 13-11   Engine preignition.

Small turbocharged gasoline direct injection (TGDI) engines have a greater tendency toward preignition than naturally aspirated or port injection engines, particularly under high load and low RPM use. The theory is that a little liquid gasoline gets caught in the crevice above the top piston ring and mixes with engine oil. The hot oil–fuel mixture is thrown up into the cylinder on the compression stroke and ignites the fuel vapors before the spark plug fires. The reason it is mentioned here is that there is a positive correlation between high-calcium detergent levels in engine oil and preignition. It is best to use an oil meeting the manufacturer's recommendations, particularly in these engines. TGDI engines are the wrong place to substitute a diesel engine oil.

Octane Rating

The octane rating of gasoline is measured in a test engine and is defined by comparison with the mixture of 2,2,4-trimethylpentane (isooctane) and heptane that would have the same antiknocking capacity as the fuel under test: the percentage, by volume, of 2,2,4-trimethylpentane in that mixture is the octane number of the fuel. For example, gasoline with the same knocking characteristics as a mixture of 90% isooctane and 10% heptane would have an octane rating of 90. A rating of 90 does not mean that the gasoline contains just isooctane and heptane in these proportions but that it has the same detonation resistance properties; it is a mixture of many hydrocarbons and other additives (see Figure 13-12).

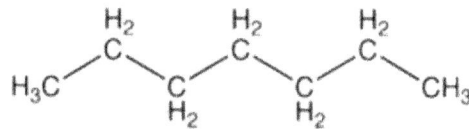

FIGURE 13-12 Heptane.

Engine Compression Ratios

Most engines are of some compression type, but diesel engines produce their own heat source to ignite the fuel. They accomplish this by means of the high compression ratio, which brings air temperature to around 500°C at the end of the compression stroke. Gasoline and gas engines, because of their lower compression ratios, need an ignition source, which is typically a high-energy spark plug. For gasoline engines, compression ratios range from 8:1 to 14.0:1, whereas diesel engines can range from 16:1 to 23:1. Figure 13-13 illustrates the calculation of compression ratios.

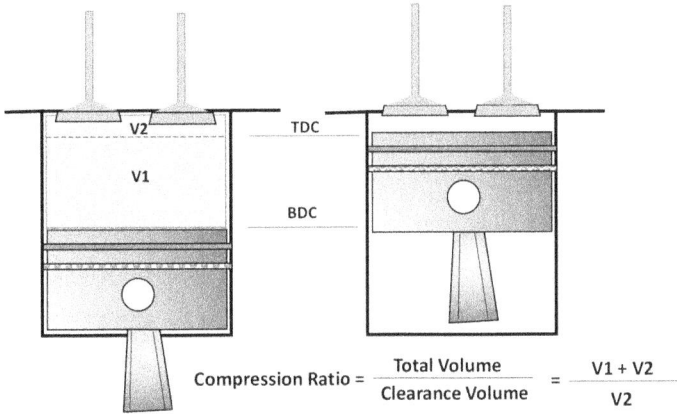

$$\text{Compression Ratio} = \frac{\text{Total Volume}}{\text{Clearance Volume}} = \frac{V1 + V2}{V2}$$

FIGURE 13-13 Compression ratios.

ETHANOL

Ethanol comes from the fermentation of sugars by yeasts or via petrochemical processes such as ethylene hydration. It has medical applications as an antiseptic and disinfectant. It also has uses as a chemical solvent and in the synthesis of organic compounds. It is the alcohol in adult beverages. Ethanol is also an alternative fuel source.

The largest single use of ethanol is as an engine fuel and fuel additive. Brazil in particular relies heavily on the use of ethanol as an engine fuel due in part to its role as one of the globe's leading producers of ethanol. Gasoline sold in Brazil contains at least 25% anhydrous ethanol. Hydrous ethanol (~95% ethanol and 5% water) is also used as a fuel in >90% of new gasoline-fueled cars sold in the country. Brazilian ethanol comes from sugar cane and is noted for high carbon sequestration. The United States and many other countries primarily use E10 (10% ethanol, sometimes known as *gasohol*) and E85 (85% ethanol) ethanol–gasoline mixtures.

Sweet sorghum is another potential source of ethanol that is suitable for growing in dryland conditions. The International Crops Research Institute for the Semi-Arid Tropics (ICRISAT) is investigating the possibility of growing sorghum as a source of fuel, food, and animal feed in arid parts of Asia and Africa. Sweet sorghum has one-third the water requirement of sugarcane over the same time period. It also requires about 22% less water than corn (also known as *maize*). The world's first sweet sorghum ethanol distillery began commercial production in 2007 in Andhra Pradesh, India. According to an industry advocacy group, ethanol as a fuel reduces harmful tailpipe emissions of carbon monoxide, particulate matter, oxides of nitrogen, and other ozone-forming pollutants.

Biodiesel–petrodiesel blend (B20) shows a reduction of 8% in pollutants, conventional E85 ethanol blend shows a reduction of 17%, and cellulosic ethanol shows a reduction of 64% compared with pure gasoline. However, ethanol combustion in an internal combustion engine yields significantly larger amounts of *formaldehyde* and related species such as acetaldehyde. This leads to a significantly larger photochemical reactivity and more ground-level ozone.

Ethanol Blends and Disclaimers

Ethanol is readily available in many states in the United States, and some disclaimers apply for E15 gasoline, especially for vehicles manufactured before 2001 and some manufactured up to 2012. Although E15 has a better octane rating than E10 in the same base gasoline, issues with E15 are generally associated with corrosion in injection systems and impact on seals of vehicles that are not approved for use with ethanol. In some places, fuel is available with 30% or from 51% to 82% ethanol content (see Figure 13-14).

FIGURE 13-14   Ethanol blend disclaimer at a gas station.

Propane

Also known as *liquefied petroleum gas* (LPG) or *propane autogas*, propane is a clean-burning alternative to power light-, medium-, and heavy-duty vehicles. Propane is a three-carbon alkane gas (C_3H_8). It is stored under pressure inside a tank as a colorless, odorless liquid. When released from pressure, the liquid propane vaporizes and turns into the gas used in combustion. An odorant additive, ethyl mercaptan, helps in leak detection.

Propane has a high octane rating, making it an excellent choice for spark-ignited internal combustion engines. If released from a vehicle, it presents no threat to soil, surface water, or groundwater. Propane is a by-product of natural gas processing and crude oil refining. It accounts for about 2% of the energy used in the United States. Of that, <3% goes for transportation. Its main uses include home and water heating, cooking and refrigeration, clothes drying, and powering farm and industrial equipment.

Natural Gas

Natural gas, a fossil fuel composed of mostly methane, is one of the cleanest-burning alternative fuels. It burns in the form of compressed natural gas (CNG) or liquefied natural gas (LNG) to fuel cars and trucks.

Dedicated natural gas vehicles run on natural gas only, whereas bi-fuel vehicles also can run on gasoline or diesel. Bi-fuel vehicles allow users to take advantage of the widespread availability of gasoline or diesel fuel but use a cleaner, more economical alternative when natural gas is available. Because natural gas is stored in high-pressure fuel tanks, bi-fuel vehicles require two separate fueling systems, which take up passenger/cargo space.

Natural gas vehicles are not available on a large scale in the United States—only a few models are available for sale. However, conventional gasoline and diesel vehicles can be equipped to run on CNG.

Fuel Energy Values

The different fuels consumed by industry have different energy values and are used according to the application and economic factors. Because of energy value, economy, safety, auto lubrication, and ease of handling, diesel fuel is very popular. Engines that consume this fuel are much stronger, give more hours per gallon, and last longer. However, they also contribute to certain contaminants that are harmful to the environment and humans.

Biodiesel is popular because it is a renewable product with comparable energy values to diesel fuel and gasoline, and it lubricates the fuel system. However, it also has some tradeoffs, especially in terms of corrosion, nitrogen oxide emissions, and cost. Ethanol is also popular as a blended product in gasoline engines because it is a renewable resource and a natural octane booster, and it burns very cleanly. LPG and CNG are becoming more important because of their abundance on the planet and the discovery of new deposits. They have less energy content but burn very cleanly (see Table 13-3).

TABLE 13-3   Energy Values, Octane Ratings, and Air–Fuel Ratios

Fuel type	Megajoules/ liter	Megajoules/ kilogram	Octane rating	Air-fuel ratio
Diesel	38.6	45.4	25	14.6:1
Gasoline	34.8	44.4	Min 91	14.6:1
Aviation gasoline	33.5	46.8	100/130	14.7:1
Ethanol	21.2	26.8	129	9:01
Methanol	17.9	19.9	123	6.47:1
E85 ethanol	25.2	33.2	105	9.75:1
LPG	26.8	50	108	15.7:1
Gasohol (10% ethanol)	33.7	47.1	93/94	14.04:1
Propane	25.3	46.35	97	15.7:1

Emission Control and Stoichiometric Engines

To reduce the NO_x and CO emissions of their predecessors, the industry needs stoichiometric engines. Stoichiometric engines are capable of a nearly complete reaction, where all the fuel and all the oxygen are consumed during combustion. To match the amount of fuel molecules sprayed by the injector, the appropriate amount of air needs to be available (by means of turbocharging) and condensed (by means of a charge air cooler) to completely burn the fuel without having oxygen left over.

U.S. Tier 4 engines and European Stage IV engines are stoichiometric engines. Injection retardation and high-pressure electronic injection help in achieving stoichiometric combustion. To facilitate the stoichiometric conditions, the electronic injector provides ramped-up injections at the ends of compression and power cycles to achieve complete combustion. At the same time, much more air comes from high-pressure in-series intelligent turbochargers. Together with the engine controller—which tailors fuel delivery based on many parameters, including temperatures, NOx, and pressures—to achieve stoichiometric combustion (see Figure 13-15).

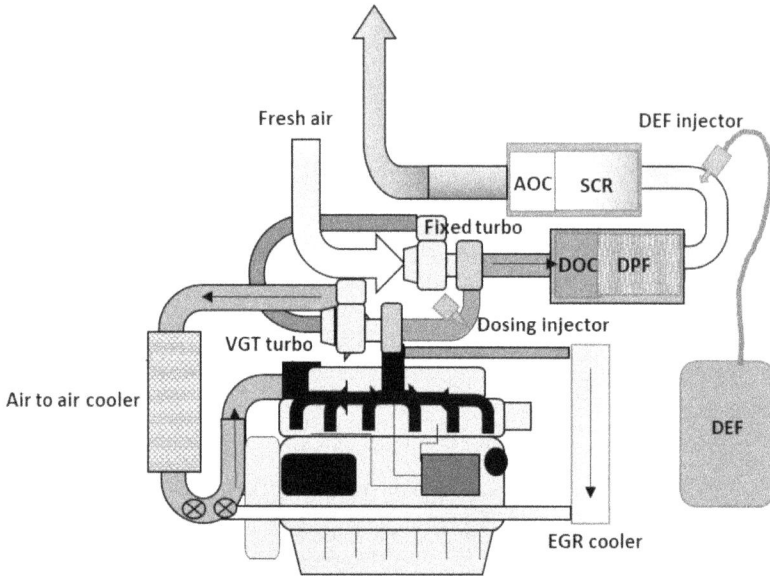

FIGURE 13-15   Tier 4 engine typical configuration.

On the emissions side, final Tier 4 and European Stage IV engines may use exhaust gas recirculation (EGR) to reduce NO$_x$ emissions and are equipped with four stages of filters/catalytic converters. A diesel oxidation catalyst (DOC) and a diesel particulate filter (DPF) accomplish the first stage of cleaning. This is followed by selective catalyst reduction (SCR) and an ammonia oxidation catalyst (AOC) to handle excess NO$_x$ with the help of ammonia injection that converts the remaining NO$_x$ to nitrogen and water.

On the design side, engines have evolved to handle the extra requirements by adding capacity to coolant and oil flows and by adding extra valves to the combustion chamber, oil-cooled piston heads, and in some cases switching over to two-piece steel/aluminum pistons.

HIGH-PRESSURE INJECTION SYSTEMS

Current injection systems have evolved from the in-line or rotary fuel pumps of the past to high-pressure common rail (HPCR) or high-pressure unit injector systems to better atomize fuel inside the combustion chamber (see Figure 13-16). New systems inject fuel at 10 times higher pressures than 20 years ago, achieving higher levels of atomization for better combustion characteristics. New materials that are more durable are required for injectors and pumps, and electronics are essential to map the injection profile that can play with ramp-up injection and delivery for the desired result. Hydrocracking is used to remove sulfur in making ultralow-sulfur diesel (ULSD), but the process also destroys the fuel's natural ability to lubricate fuel systems.

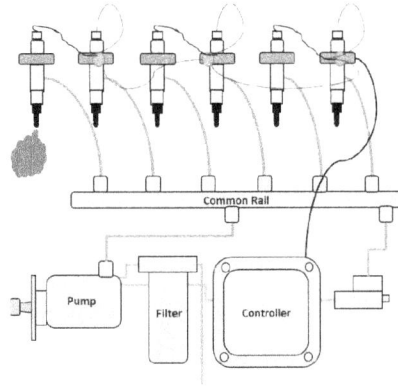

FIGURE 13-16   High-pressure common rail (HPCR) system.

Fuel manufactures use lubricity additives to bring the fuel into compliance with the 520-μm maximum requirement in the high-frequency reciprocating-rig (HFRR) test (ASTM D6079), as specified by ASTM D975 in the United States or the more stringent 460-μm maximum as required by the EN590 specification in Europe. Biodiesel is an excellent lubricity additive and more than adequate at 2% (B2).

Fuel Filtration

To accommodate high-pressure fuel injection, manufactures equip their engines with filtration that typically has two or three stages. The filters go from 30 μm down to 10 μm with a final filtration of 2 or 3 μm. Some manufactures go from 10 to 2 μm or from 30 to 10 μm and then to 5 μm. However, what feeds the engine depends on the user and the care taken while practicing regular sampling and filtration of fuel farms (see Figures 13-17 and 13-18).

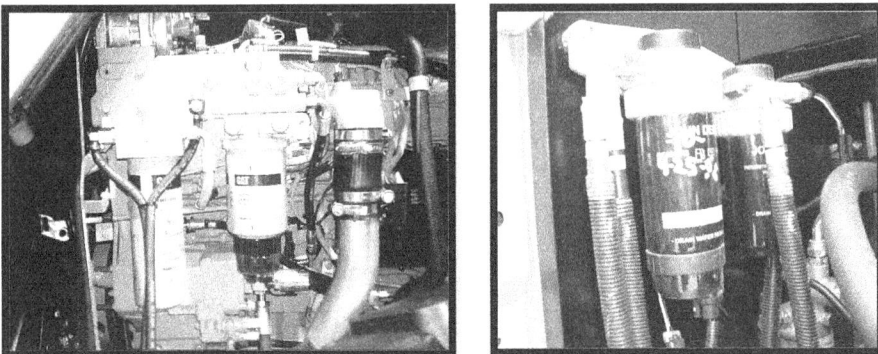

FIGURE 13-17   Different manufacturers' filter arrangements

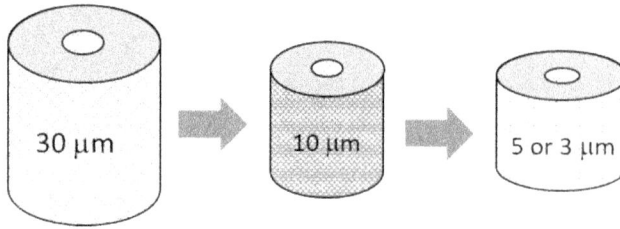

FIGURE 13-18 Fuel filter sizes.

Sulfur in Fuels Around the World

Sulfur is a compound found in all petroleum-based diesel fuels. However, when sulfur is oxidized during combustion, it forms sulfur dioxide (SO_2) and sulfur trioxide (SO_3), which, when combined with water, create sulfuric acid (H_2SO_4). Diesel fuel comes with a variety of acceptable sulfur concentration levels. Most standard diesel fuels have a sulfur content of no more than 0.1% or 1,000 ppm.

Low-sulfur diesel (LSD) has a sulfur content of no greater than 0.05% or 500 ppm and was required for on-road engines in the United States beginning in 1993. In an effort to put tighter controls on emissions, ULSD was mandated for on-road engines in 2006 to enable the new emissions equipment of the 2007 model year. ULSD has a sulfur content no greater than 0.015% or 15 ppm. The emissions requirements and sulfur limits were phased into the off-road market a few years after the on-road requirements for both LSD and ULSD.

High levels of sulfur limited the opportunities to use new and emerging emissions technologies such as EGR found on Tier 3 engines and DPFs found on interim Tier 4, final Tier 4, and Stage IV engines. Use of high-sulfur fuels with engines equipped with an EGR can result in failure of the EGR valve. On Tier 4 engines with exhaust filters, high-sulfur fuel can poison the oxidation catalyst and plug the DPF to the point of failure. With ULSD, the number of pollutants released during operation goes down, and this enables the use of emissions-reduction technologies. The adoption of ULSD is not consistent around the world, and many countries still lag behind in its adoption. In some countries, ULSD is available only in large metropolitan areas.

DIESEL FUEL ADDITIVES

A wide variety of fuel additives and solutions is available to meet the new maintenance requirements for fuel. These products include diesel fuel conditioners, fuel detergents, and cleaning solvents. The most complete fuel conditioners offer the following:

- Dispersants
- Lubricity additives
- Cetane improvers
- Stability improvers/oxidation inhibitors
- Rust inhibitors, demulsifiers
- Metal deactivators

- Cold-flow improvers
- Wax antisettling agents
- Anti-icing additives

Some aftermarket additives offer reduced or more targeted treatment packages designed to address specific issues, such as lubrication and gelling. While many aftermarket additives are beneficial, some are better at marketing than performance (see Table 13-4).

TABLE 13-4 Fuel Additives

Function	Purpose	Benefit
Biocide	To treat and prevent microbes, fungi, and algae	Kills microbial growth in fuel tanks
Cetane improver	To raise cetane number	Improves cold and high-altitude starts
Cold-flow improver	Improves cold filter plugging	Allows fuel flow at lower temperatures
Conditioning	Storage	Improves stability and inhibits oxidation
Dispersant	Strong cleaning	Cleans and prevents deposits in injectors
DW10B detergent	For mild cleaning	Inhibits buildup on injectors
Solvent	Strong cleaning	To flush systems that already have deposits
Demulsifier	Helps shedding water	Sheds water in summer and prevents icing in winter
Metal deactivator	Inhibits fuel degradation by catalysis from copper	Oxidation stability
Wax antisettling agent	Suspends cold-weather wax cloud	Prevents filter plugging during warm-up

We are not going to discuss all the additive components in detail, but a little understanding of cetane improvers and cold-flow additives can help avoid some of the pitfalls and get the most out of these additives.

Cetane Improver

Alkyl nitrate cetane improvers are an effective way to increase cetane. In engines where the timing was retarded to reduce nitrogen oxide emissions, using a higher-cetane fuel, which lights more quickly after the beginning of injection, can give back some of the timing. These engines respond with more power and efficiency. The engine does not care if the higher cetane came naturally or from an additive, but cetane improvers do not lower the fuel's energy content, whereas a lighter fuel blend does.

Cetane improvers work by making the fuel unstable, exploding in the heat of the cylinder and igniting the fuel around it. Unfortunately, this instability can lead to injector deposits when there is a hot soak of the injector tip, and perhaps a worst-case example is Jake Brake operation. In contrast, dispersant additives can prevent deposits. When using cetane-improving additives, they should always be coupled with injector cleaners. Fuels with naturally high cetane will respond to cetane improvers much more than low-cetane fuels. The fuels that need cetane help the most, benefit the least.

Finally, there is a tendency for additive marketers to exaggerate the cetane increase that their product provides. 2-Ethylhexyl nitrate is currently the "gold standard" for cetane improver, and its concentration shows up on Safety Data Sheets because it is hazardous. The SDS may be more effective in choosing a cetane additive than the product literature.

Diesel Winter Additives

Cold-flow additives can add 10°F (5.6°C) or more to the operating range of diesel equipment in winter. They coat the wax crystals as they fall out of the fuel, preventing them from agglomerating. The wax still falls out, so there is little or no effect on cloud point, but the crystals remain small enough to pass the filter. The fuel must be treated at temperatures above the cloud point for cold-flow improvers to reach their optimal performance. Their performance can be measured with before and after cold filter plugging point (CFPP) tests.

Wax antisettling agents help. If the wax cloud settles overnight, that concentrated wax is in the first fuel hitting the filter. Often the engine will start and idle okay. As the machine is put into use, fuel flow increases, and the wax that is concentrated at the bottom of the tank packs the filter. Usually, the machine stalls when it has traveled just far enough to be extremely inconvenient. Antisettling agents suspend or disperse the cloud to prevent the machine from drawing in that concentrated wax. After 15 minutes of use or so, enough warm fuel has returned to the tank that filter plugging is no longer a threat (see Figure 13-19).

There are emergency additives on the market designed to be added directly to a plugged filter to dissolve fuel wax or dissolve wax in the tank. They are generally aggressive solvents, such as xylene or toluene, and while they can be effective, they are risky for the injection system.

Icing can plug a filter as quickly as gelling. However, it is a different problem with a different solution. As a general rule of thumb, if cold-weather plugging occurs at temperatures above 0°F (−18°C), the problem is icing. If it occurs below 0°F (−18°C), it is usually gelling.

Anti-icing agents are generally alcohols or glycols. No additive can remove water, so it becomes a matter of what the additive does with it. Typically, anti-icing agents work by contaminating the water, lowering its freezing point. Once the water has frozen to form ice, anti-icing agents will not thaw it out. Alcohol should never be used in a diesel engine because it can suspend relatively large amounts of water. The last thing you want in a high-pressure injection system is to defeat the water separator.

Pure biodiesel (B100) has a pour point at around 32°F (0°C). Keep in mind that cold-flow additives work in diesel fuels and have very little effect on biodiesel. They are effective in lean biodiesel blends such as B2 or B5, but CFPP would still be lower in the base fuel before the addition of biodiesel.

Diesel fuel at rest overnight at temperatures below the cloud point. All are the same, but the two on the right are treated at two levels with antisettling additive. All have the same amount of wax, but the untreated sample has all the wax concentrated at the bottom, which is more likely to pack the filter during warmup.

FIGURE 13-19 Fuel samples treated with Cen-Pe-Co wax antisettling agent.

Blends of B20 or higher are unworkable in most cold-weather climates without massive doses of kerosene (no. 1 diesel). At that point, the fuel has a much lower energy value, and fuel economy suffers, making it hard to be green. It is always important that the biodiesel used in fuel blends conforms to ASTM D6571, but it becomes critical in the cold. If the reaction in making biodiesel is incomplete, glycerin, oils, or fats can separate in cold temperatures and plug filters.

ADDITIVES FOR GASOLINE

Detergents

Combustion engines consistently fall victim to carbon buildup, notably on the fuel injectors, because they make the most contact with the gasoline. As fuel injection became more common than carburetors in the 1980s, carbon buildup on fuel injectors became an increased concern. One of the first and most used intake valve deposit cleaners is polyether amine (PEA), released under the name Techron in 1995 by Chevron. In 1996, the U.S. Environmental Protection Agency (EPA) created the set lowest additive concentration (LAC) standard to enforce a minimum amount of fuel detergent in gasoline to control intake valve deposits.

As injection systems continued to evolve, the LAC became insufficient for some engines. This inspired Toyota, Honda, General Motors, Audi, and BMW to establish the top tier gasoline standard for detergent additives in 2004. The top tier designation is a significantly stricter deposit control standard than the LAC that requires fuel to use larger amounts of certified detergent additives. Retailers labeling their gasoline pumps with "Top Tier" must meet the standard in every grade they sell because the designation is not related to octane.

Friction Modifiers

Approximately 25% of the burned gasoline lubricates the upper cylinder wall, which by design is neglected by the lubrication system. Friction modifiers adhere to the wall of the cylinder, forming a microlayer while the rest of the gasoline burns.

Corrosion Inhibitors, Demuslsifiers, and Solvents

Most of the gasolines sold in the United States contains corrosion inhibitors, demulsifiers, and solvents. Solvents and demulsifiers also work to keep additives in suspension while helping to separate out water. Some antiadhesion additives prevent the fuel additives from forming unwanted films on surfaces.

Octane Boosters

Since tetraethyl lead and later methyl *tert*-butyl ether (MTBE) were phased out, the industry started using benzene, toluene and ethylbenzene, and xylene (BETX) as octane boosters. However, starting in 2007, the EPA mandated that gasoline could not contain >0.62% benzene; thus, ethanol became the option.

FUEL SAMPLING

Fuel sampling and testing should be an essential part of any equipment operation. At present, 85% of injection system downtime comes from water and particulates in the fuel. However, only 1% of fluid sampling is about fuel. It appears that most of the attention is paid to oils and coolants.

Fuel testing is expensive compared with oil or coolant testing. Therefore, it is reasonable to plan a good strategy to practice well-informed fuel sampling and analysis. A good strategy could be, for example, to test fuel farms regularly because fuel farms feed entire fleets. Occasional machine testing to understand isolated cases is necessary but should be the exception rather than the rule.

However, sampling from fuel farms requires understanding some basic guidelines. Otherwise, the results will be so inconsistent that finding value in the sampling activity becomes elusive. Here the ASTM standards play an important role by establishing the rules of the game so that the results are meaningful, consistent, and conducive to good maintenance decisions.

First, you need to define what you want to test—the bulk tank, the fuel provider, or the machine. Each one is going to produce different results, and those results are going to be different depending on the location and time at which the sampling occurred.

Fuel Farm Sampling

The ASTM 4057 standard provides guidelines for fuel sampling from fuel tanks. The nomenclature in Figure 13-21 helps with the discussion. The use of a fuel bacon bomb to collect a representative sample is essential in obtaining consistent fuel samples for testing (Figure 13-20).

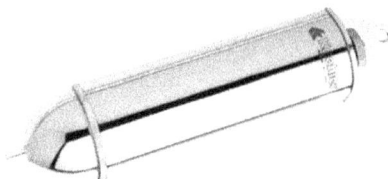

FIGURE 13-20   Bacon bomb sampler.

In the past, fuel samples were taken without regard to location (depth) or time of extraction or even defined purpose. Here are some specific recommendations for successful fuel farm sampling:

- If you want to test the supplier, take a sample from the middle of the tank after refill.
- If you want to test the fuel that feeds the equipment, you need to take a sample from a lower point in the tank. It needs to be from the lower third of the tank, but not at the outlet point.
- If you want to test tank cleanliness, take a sample from the bottom.
- Combined samples from middle and top locations make a composite sample that could be appropriate for fuel quality analysis (see Figure 13-21).

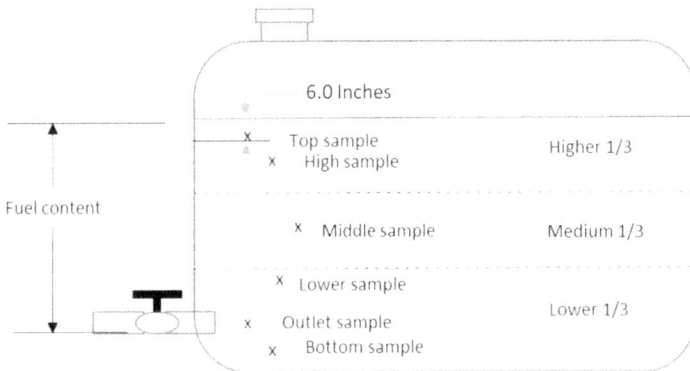

FIGURE 13-21   Fuel sampling.

Machine Fuel Sampling

With regard to sampling a machine's tank, use these steps:

- The machine's fuel gets cleaner as the machine accumulates hours during the day. This happens because ~60% of the fuel sent to the injectors returns to the tank in a closed-loop path. It works this way because the fuel is used to cool the injectors after the injection phase, which happens at very high pressures in common-rail systems these days.
- If the machine is suffering from short fuel filter life, it is in the best interest of the user to sample the machine shortly after refueling. This will uncover issues with the main fuel supply.
- Do not sample a machine that has been at rest overnight because the sample will not be able to collect debris or water.
- Do routine fuel testing on the machine early in the workday when the machine is already active. This will allow some signals to get into the sample bottle; otherwise, they will be lost in the filters.
- Do not sample at the end of a journey unless you are looking for fuel specifications and not for fuel contaminants.

- In a machine that is active, it does not matter at what height in the tank you take the sample. If the machine is equipped with a fuel sampling valve, make sure that the valve is not located after the filters. Dispose of a certain amount of fuel that remains stagnant in the line before collecting the sample.
- Use only approved fuel bottles.
- Make sure that the packaging for shipment complies with the rules of hazardous materials, which typically include absorbent material, double containers, and a sealed bag.
- Fill in the sample identification form (SIF) with information about the fuel.

Fuel Farm Tank Maintenance

Fuel farms and fuel trucks suffer from similar issues in terms of contaminants. The difference is that fuel farms are static and can drain water and debris more efficiently, whereas fuel trucks can do this only in the morning before the truck starts moving.

In one year, 1 million gallons of fuel at 0.05% by volume contamination can produce 500 gallons of water and 250 pounds of dirt, enough to plug 7200 fuel filters with 4 oz of dirt each. This is more than enough to produce frequent downtime from filter plugging, corrosion, and abrasion.

Typical types of contaminants found in fuel tanks include water, inorganic waste (e.g., sand, rust, dust, and metal particles from the surfaces of the tanks), and organic (carbon-containing) waste such as sludge, asphaltenes, polymers, wax, and microbial contamination. More than 50% of tank contamination is generally organic. The task of the maintenance crew is to prevent these contaminants from entering fuel farms so that the cleanest possible fuel feeds the fleet in the field (see Figure 13-22).

FIGURE 13-22 Fuel tanks.

Water Condensation

Condensation occurs because of temperature changes. In a half-full fuel tank, the air at the top of the tank contains humidity that condenses when it comes in contact with the tank's walls. This is more notorious at night when the temperature drops below the dew point. The moisture also leads to the formation of rust in the upper spaces of the fuel tank. Contamination occurs through refueling, filler caps, breather points, and defective seals (see Figure 13-23). Thus:

- To diminish the effect of temperature changes on water condensation, painting external tanks white has its merits.
- Adding desiccant breather filters to fuel tanks can bring additional benefits in controlling water condensation, rust, and bacteria.
- It is true that water can come with new fuel deliveries, but most of the water in fuel tanks is usually the by-product of humid air.
- Drain water from tanks weekly or as required.

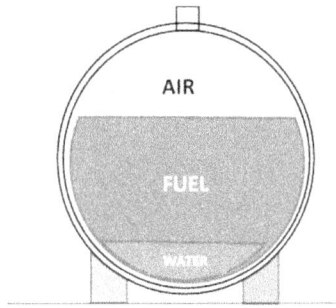

FIGURE 13-23 Fuel tank water condensation.

Large operations that confront harsh environments such as high humidity, dust, and several fuel transfer logistics could benefit from the use of a fuel centrifugal apparatus like those used by the marine industry (see Figure 13-24). Send a sample to the lab to check for basic parameters. Such a test is not expensive.

FIGURE 13-24 Alfa Laval water separator (*left*); Des Case fuel desiccant filter (*right*).

Long-Term Fuel Storage Package

Complete the task of maintaining the fuel farm tanks with a long term fuel storage analysis as indicated in the following tests.

- Fuel stability by ASTM D6468
- Color by ASTM D1500 and D156
- Free water, volumetric
- Suspended water, ppm
- Particulate contamination by weight, mg/L
- Biodiesel concentration

Send a fuel sample to the lab for a comprehensive analysis every three months.

- Calculated cetane index (incudes API gravity and distillation)
- Free water, volumetric
- Suspended water, ppm
- Particulate contamination by weight, mg/L
- Appearance, by numbering system (new change)
- Biodiesel concentration
- Sulfur content
- Bacteria (summer)
- Cold filter plugging point (winter)

Complete the SIF appropriately.

Field Test: Water-Finding Paste

Although laboratory tests are much better at determining the amount of water contamination, a water-finding paste can reveal whether there is water at the bottom of your tank. It can be used in vehicle tanks, storage tanks, and intermediate tanks. Water-finding paste also can be used to check whether there is water after every delivery (see Figure 13-25).

FIGURE 13-25   Kolor Kut water-finding paste.

To test, simply smear some water-finding paste along the bottom of a clean stick, rod, or tape measure, and lower it to the lowest point of the tank. If water has settled to the bottom of the tank, the paste turns red. If not, it remains brown. How far up the rod the paste turns red indicates the depth of the water. The same paste can be used in diesel fuel or gasoline. Ensure that a compatible paste is being used when testing fuels containing biodiesel or alcohol.

Machine Fuel Tank Maintenance

Machine tanks have some of the same issues as fuel farm tanks with a difference that the fuel in machines gets cleaner during operation, thanks to machine closed-loop fuel filtration. Still, there are points that can help reduce machine downtime if put into practice:

- Fill the vehicle fuel tank at the end of use every day, if possible. This will keep humid air out of the tank.
- Drain some fuel from the fuel tank drain valve every day before operation. This is no longer feasible in some machines because manufacturers have eliminated those valves.
- Drain water and sediment from filter bowls when they are visible.
- Make sure that the fuel tanker has filtration.
- Clean the machine's fuel tank strainer at least twice a year or more often when water and bacteria dictate shorter service intervals (see Figure 13-26).

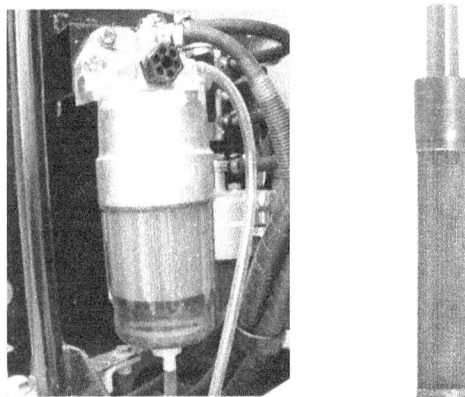

FIGURE 13-26   Fuel filter bowl (left); fuel tank strainer (*right*).

Sample Information Form (SIF)

For labs, it is essential to count on some basic information to be able to return sound recommendations and sometimes to make sense of abnormal readings. If you send a sample, it is in your best interest to complete the information on the SIF as completely as possible based on your knowledge of the operation. Help the lab do its job so that it returns usable results to you. The information on a SIF may contain some of the elements in the sample SIF in Figure 13-27. This SIF only shows information about the fuel, but all the SIFs from labs will contain spaces to fill information from the user and the machine in question, if any.

DIESEL FUEL SAMPLE INFORMATION FORM (SIF)							
	S15	S500	S5000	BIODIESEL %	#1	#2	#4
TYPE OF FUEL	☐	☐	☐	☐	☐	☐	☐
	FARM TANK	TANKER TRUCK	MACHINE	OTHER	TIME OF SAMPLING	AM	PM
PLACE AND TIME OF SAMPLING	☐	☐	☐	☐		☐	☐
	AT 1/3 OF BOTTOM	AT MIDDLE	AT 1/3 OF TOP	OTHER	COMMENTS		
SAMPLING HEIGHT	☐	☐	☐	☐			
	BIOCIDE	CETANE IMPROVER	STABILIZER	LUBRICANT	DETERGENT	EMULSIFIER	OTHER
ADDITIVES USE	☐	☐	☐	☐	☐	☐	☐

FIGURE 13-27   Sample information form.

Fuel Sampling Bottles

To collect and ship fuel samples to the lab, a dedicated and a safe bottle/package are necessary. The bottles are generally provided by the lab on a prepaid analysis basis and contain the SIF label and absorbent material in case of a leak. Bottles for fuel are generally more resistant and seal better than oil bottles.

DIESEL FUEL TESTING

Table 13-5 provides a comprehensive list of the tests available for diesel fuel. Highlighted in the top two sections of the table are the tests that any good fuel analysis service should provide on a regular basis. The additional tests are available to test the diesel fuel for specific issues that the user may find with his or her fleet. The tests highlighted in the bottom three lines are seasonal and are related to cold-weather operation. Others are specific for high-altitude operation or simply for quality checks of the final product. Some of the tests are suitable for the fuel farm at the user's end to ensure that the product the user stores and the product he or she buys meet the minimum requirements of the fleet.

TABLE 13-5 Diesel Fuel Tests

Test	Method	No. 1 Diesel	No. 2 Diesel	No. 4 Diesel	Units
Flash point	ASTM D93	100 Min	125 Min	130 Min	Degrees F
Water and sediment	ASTM D2709	0.05 Max	0.05 Max	Not specified	Volume %
Water by Karl Fischer	ASTM D6304	<200	<200	Not specified	ppm
Particulate matter	ASTM D2276/ D5452 ISO	10 Max <18/16/13	10 Max <18/16/13	Not specified	ISO
Distillation temp 90% recovered	ASTM D86	500 Max	640 Max	Not specified	Volume %
Cetane number	ASTM D613	40 Min	40 Min	Not specified	ASTM number
Cetan index	ASTM D976	40 Min	40 Min	30 Min	ASTM number
Biodiesel content	FTIR	0–20	0–20	Not specified	Volume %
Sulfur	ASTM D5453/ D2622	15, 500, 5000	15, 500, 5000	2.0 Max	ppm
TAN	ASTM D664	0.15 Man	0.15 Max	Not specified	MGk OH/gr
Cold filer plugging point avg. NA	ASTM D6371	−12		Not specified	Degrees F
Cloud point avg. NA	ASTM D2500	−1		Not specified	Degrees F
Pour point avg. NA	ASTM D97/ D5949	−25		Not specified	Degrees F
Viscosity	ASTM D445	1.3 Min, 2.4 Max	1.9 Min, 4.1 Max	5.5 Min, 24.0 Max	cSt
Ash content	ASTM D482	0.01	0.01	0.1	Mass %
Corrosion Cu strip	ASTM D130	No. 3 Max	No. 3 Max	Not specified	ASTM
Carbon residue	ASTM 524	0.15 Max	0.35 Max	Not specified	Mass %
Lubricity, HFRR	ASTEM D6079	520 Max	520 Max	Not specified	Microns
Long-term storage ASTM 4625	ASTM 2625				
High temperature stability	ASTM 6468				
Conductivity	ASTM D2624/ D4308	25 Max	25 Max	Not specified	ps/m Pico Siemens/m
Bacteria growth	ASTM D6469	Culture	Culture	Not specified	Reported

1. Flash Point (ASTM D93)

This test needs to be in the basic group of fuel tests. The flash point is the lowest temperature at which an ignition source causes the fuel vapors to ignite. It can indicate contamination by lighter- or heavier-grade fuels or solvents, and it is a complementary test for the distillation test. The fact that no. 1 diesel's flash point is lower than no. 2 diesel's is one more justification for the use of no. 1 diesel during the cold season (see Figure 13-28).

	Diesel No 1	Diesel No 2
Flash Point	38 ° C min. (100.4 °F)	52 °C min. (125.6 °F)

FIGURE 13-28   Anton-Paar Wiki flash point tester.

2. Water and Sediment (ASTM D2709)

This test method provides an indication of free water and sediment suspended as haze, cloudiness, or droplets in middle distillate fuels such as no. 1 and no. 2 fuel oils. It uses centrifugal force in a graduated tube to separate the amount of water and sediment. The capillary tip of the tube is capable of measuring down to 0.01 ml and estimating to 0.005% of these contaminants (see Figure 13-29).

Fuel	Diesel No 1	Diesel No 2
Particles and Water % Vol	0.05 Max	0.05 Max

FIGURE 13-29   Capillary tip centrifuge tube.

3. Water (ASTM 6304)

One of the most critical fuel tests is the precision water test by Karl Fischer using ASTM 6304. Water, as shown earlier, causes injector failure if not kept in check. The test uses iodine to measure the reaction with water, which is proportional to the amount of water. The reagent is added until an electrometric endpoint is reached. Water concentration in parts per million or percent is calculated from the amount of reagent used to reach the endpoint (see Figure 13-30).

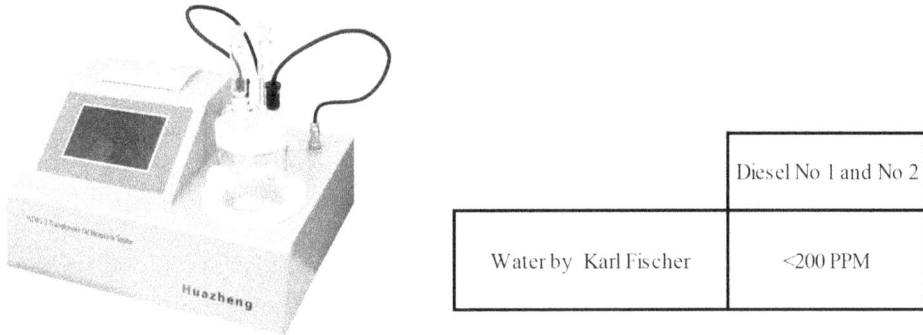

	Diesel No 1 and No 2
Water by Karl Fischer	<200 PPM

FIGURE 13-30 Huazheng titrator.

Impact of Water on Injector Life

The life of injectors depends very much on fuel cleanliness, and water is a known killer. Figure 13-31 illustrates the impact of water on the life of injectors. The figure suggests that the life of injectors can reach 100% of their design life when water content is between 300 and 500 ppm. Astonishingly, the life of injectors can double if water content is below 100 ppm. Keeping fuels free of water is the number-one task on which a maintenance manager needs to focus.

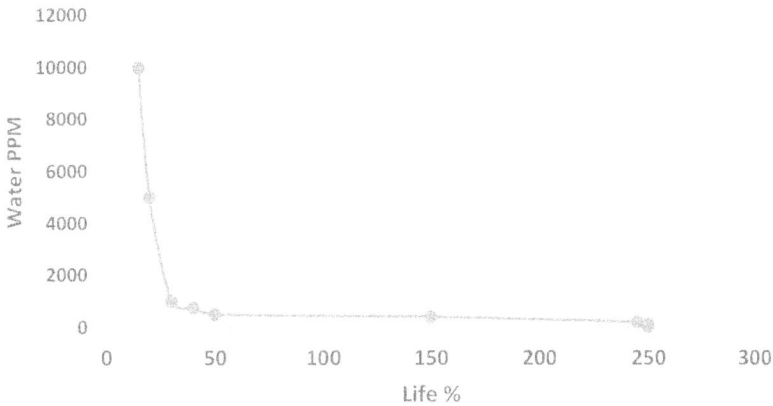

FIGURE 13-31 Impact of water on injector life.

4. Fuel Cleanliness, Particulate Matter (ASTM D2276 or D5452)

Equally important to fuel is cleanliness. Whatever fuel goes into a machine needs to be clean enough that the machine's filters can take it to an acceptable level of cleanliness while in operation. Although the ASTM standard for fuel cleanliness provides the guidelines discussed below, the requirements of current engine technologies demand much cleaner fuel. The reason is injection pressures. In the past, injectors operated in the range of 2,000–3,000 psi. Current final Tier 4 and Stage IV engines often operate at pressures 15–20 times higher. Any trace of hard particles or water becomes very abrasive and shortens injector life.

Current standards for fuel cleanliness have not kept pace with real-life requirements in the field. The cleanliness specification of 18/16/13 works well in hydraulic systems working at 6,000 psi. Current injection systems work at four times that pressure at least, so having a fuel cleanliness of 16/14/11 would be more appropriate.

Current filtration systems on machines and heavy-duty vehicles allow reaching this cleanliness level because final filtration is typically around 3 μm, and 60% of the fuel recycles through the filters over and over again during the work journey (see Figure 13-32). Measuring particles counts on diesel fuel is difficult with a particle counter, so the preferred method is with a graduated patch.

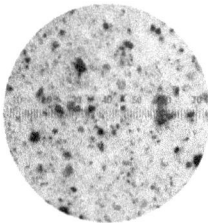

Fuel	Diesel No 1	Diesel No 2
Particles mg/Liter	10	10
Particle Counts	<18/16/13	<18/16/13

FIGURE 13-32 Microscopic patch.

5. Distillation (ASTM D86)

Distillation is a prime test to be included in any fuel analysis. The distillation measurements allow for determining whether contamination with lighter or heavier fuels is present in the sample. The most typical mistakes with construction equipment involve mixes with gasoline, which increase the cetane index to high levels, causing the diesel engine to knock out of control with serious consequences for pistons and piston rings (see Figure 13-33).

	Diesel No 1	Diesel No 2
Distillation	90% Rec. 288 °C (550 °F)	90% Rec. 338 °C (640 °F)

FIGURE 13-33 PAC Opti PMD Microdistillation analyzer.

6. Cetane Number (ASTM D613-18a)

Cetane number is the time between the start of injection and the actual ignition of the fuel. The higher the number, the shorter the delay interval.

Cetane

Cetane ($C_{16}H_{34}$, hexadecane or isocetane) is a collection of unbranched open-chain alkane molecules that ignites very easily under compression:

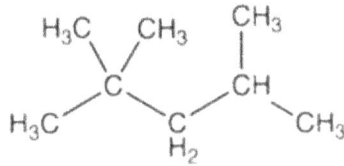

Measurement requires burning the fuel in a special diesel engine called a *Cooperative Fuel Research* (CFR) engine (see Figure 13-34). This engine (ASTM D613) is a single-cylinder indirect injection engine with variable compression that compares the combustibility of the test fuel with that of reference fuels. The scale for cetane number goes from 0 to 100. A low number may cause slow cold starting, excessive smoking, and erratic operation.

A good cetane number is also important for high-altitude operation. A low-cetane fuel will cause difficulty starting a cold engine at high altitude because the temperature to ignite the fuel is harder to achieve. A low cetane number also makes it more difficult to start engines at extreme cold temperatures and prolongs the time to smooth running. In contrast, a high cetane number tends to produce more smoke and reduce engine output.

FIGURE 13-34   CFR test engine.

7. Cetane Index (ASTM D976/D4737)

Needless to say, engine-run tests to determine cetane number are rather expensive, so an estimate based on distillation, called the *cetane index*, is much more popular. The calculated cetane index by four-variable equation (ASTM D4737) provides a reasonably reliable estimate of the fuel's cetane number. This is generally faster and much less costly than measuring the actual cetane, making it more practical for fleet managers. It can be a quick check to ensure that the cetane quality is appropriate for the equipment for which it is intended.

Within the range of cetane numbers from 32.5 to 56.5, the expected error of prediction of the cetane index (ASTM D4737 Procedure A) will be less than ±2 cetane numbers for 65% of the distillate fuels evaluated. Errors may be greater for fuels whose properties fall outside the recommended range of application. Obviously, cetane-improving additives do not affect the way a fuel was distilled. Therefore, the cetane index does not include the effect of cetane-improving additives.

Test method ASTM D6890 works to obtain a derived cetane number (DCN). This automated laboratory test measures ignition delay in a combustion chamber filled with compressed air. Because it measures ignition delay, it provides a better estimate of cetane number than the cetane index. It provides a superior result when the fuel's combustibility does not fit the distillation curve, such as alternative fuels like biodiesel or those treated with a cetane improver.

8. API Gravity (ASTM287/D1298)

API gravity is a scale comparing the density of petroleum or a petroleum product with the density of water. It is used primarily to measure the energy content of crude oil, but it is also applied to diesel fuel. An API gravity of 10 is equal to the density of water, and numbers below 10 sink in water, and numbers above 10 float on water. Thus, products with higher numbers have lower density and vice versa. In this respect, API gravity is the opposite of specific gravity. API gravity is expressed in degrees with the same notation as temperature.

Formulas can be used to convert from specific gravity to API gravity or the other way around. Table 13-6 compares API gravities and specific gravities. Below is the formula to convert specific gravity (SG) to API gravity and an example equalizing the density of the petroleum sample with that of water:

$$\frac{141.5}{1.0} - 131.5 = 10° \text{ API}$$

$$\text{API gravity} = \frac{141.5}{\text{SG}} - 131.5$$

TABLE 13-6 API Gravity versus Specific Gravity

API gravity	Specific gravity	Weight	
		lb/U.S. gal	kg/m³
8	1.014	8.448	1012
9	1.007	8.388	1005
10	1	8.328	998
15	0.966	8.044	964
20	0.934	7.778	932
25	0.904	7.529	902
30	0.976	7.296	874
35	0.85	7.076	848
40	0.825	6.87	823
45	0.802	6.675	800
50	0.78	6.49	778
55	0.759	6.316	757
58	0.747	6.216	745

9. Biodiesel Content by Fourier Transform Infrared (FTIR) Spectroscopy (ASTM D7371)

Biodiesel content can be determined by using FTIR spectroscopy to measure the ester peak that is not present in conventional distillate fuel (see Figure 13-35). As established by engine manufacturers, the maximum blend allowed without any consequences in terms of corrosion, seal failure, and/or engine power loss is 20%. There are applications running on 100% biodiesel, but most equipment manufacturers have set 20% as the permissible limit without noticeable consequences.

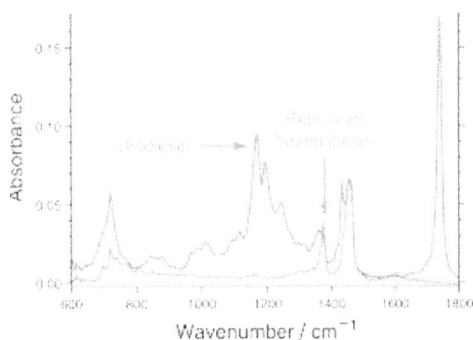

FIGURE 13-35 Jasco Global FTIR 8,000 apparatus.

10. Sulfur (ASTM 2622/5453)

Sulfur tests were mandatory when high-sulfur fuels were the norm. They are still mandatory in markets where the adoption of ULSD is still awaiting implementation. This test measures the sulfur content for operational concerns and for compliance with EPA regulations (see Figure 13-36).

Reduced sulfur content is necessary with Tier 3, interim Tier 4, and Tier 4 engines to protect several elements from the emissions control accessories. The use of fuels containing sulfur in these engines results in catalyst poisoning and/or DPF plugging.

High-sulfur fuels create acids that deplete the total base number (TBN) in engine oil much faster. Sulfur in fuels is also responsible for corrosion that is evident through metal generation, specifically iron from liners and lead from bearings. Even in countries where ULSD is available, there could be batches of high-sulfur fuel that end up in users' hands for reasons that are difficult to explain, but labs get high-sulfur fuel samples from time to time.

	ULSD Diesel No 1	ULSD Diesel No 2	Low Sufur Diesel No 1 and No 2
Sulfur Content	< 15 PPM	< 15 PPM	< 500 PPM

FIGURE 13-36   Parker Kittiwake fuel sulfur test apparatus.

11. Acid Number (ASTM D644)

Total acid number (TAN), which is more common with hydraulic fluids, also works for fuels. The test measures acidic constituents of fuel for corrosion potential. It also measures the level of free fatty acids present in biodiesel and biodiesel blends (see Figure 13-37).

	Diesel No 1 and No 2
TAN measured by mg/KOH/gm	0.15 mg KOH/gm max

FIGURE 13-37   OelCheck acid number test apparatus.

12. Cold Filter Plugging Point (ASTM D6371)

The cold filter clogging point is the temperature at which 20 mL of fuel will not pass through a 45-μm wire-mesh screen in 60 seconds under vacuum. This test helps in determining the uptime capabilities of equipment when extreme low ambient temperatures are present (see Figure 13-38).

Traditionally, the two main considerations for cold-weather operability of diesel fuel have been cloud point (CP) and cold filter plug point (CFPP). The CP determines the point where wax becomes visible in a fuel sample. This wax first appears as a floating cloudiness in a transparent fuel and is the first indicator of the cold-temperature limit at which the fuel is flowing. Most of the time, there is a spread between CP and CFPP of 0–4°C in fuels that are not treated with cold-flow additives. By determining the CP and the CFPP of a fuel, the cold-weather operability limits can be estimated.

Antigelling additives help to disrupt the agglomeration of wax crystals to improve cold flow. Because they work by keeping wax particles separated, they must be added before the fuel reaches its CP to achieve their full potential. Flow improvers have little or no effect on CP, but they lower the CFPP. Fuels with more than 4°C between the CP and CFPP likely have been treated with cold-flow additives.

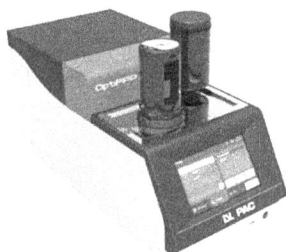

	Diesel No 1 and 2
Cold Filter Plugging Avg. NA °F	-12

FIGURE 13-38   PAC OptiFPP cold filter plugging point apparatus.

13. Cloud Point (ASTM D2500)

The CP test is a seasonal test that allows users to take measures to prevent machine stoppage caused by blocked fuel lines or filters. The test is about finding the temperature at which the first formation of haze (i.e., wax cloud) occurs and the temperature where low-temperature operability issues may occur (see Figure 13-39).

	Diesel No 1 and 2
Cloud point Avg. NA ° F	-1

FIGURE 13-39   Tanaka Cloud Point Tester.

14. Pour Point (ASTM D97 and D5949)

The pour point is the temperature at which the fuel becomes solid. This test is a seasonal test or an artic test that is essential under extreme cold temperatures to determine whether the fuel can be pumped. It is well below the CFPP or the fuel temperature where diesel equipment will run.

The D5949 test method determines the pour point in a shorter period than manual method D97. Less operator time is required to run the test using this automatic method. Additionally, it does not use an external chiller bath or refrigeration unit. The range goes from −57°C to +51°C. Results come at 1°C or 3°C testing intervals. This test method has better repeatability and reproducibility than the manual method D97 (see Figure 13-40).

	Diesel No 1 and 2
Pour Point Avg. NA °F	-25

FIGURE 13-40   PAC 70Xi pour point apparatus.

15. Viscosity (ASTM 445 at 40°C)

Although this is not the typical test in every sample, knowing fuel viscosity when issues arise is a good idea (1) because it allows the user to check that the supplier is delivering the correct fuels for a given season, and (2) because low viscosity increases the potential for fuel injection pump and injector leakage, which can lead to power loss and/or oil dilution. Also, the correct viscosity allows the injection system to properly atomize the fuel, and it lubricates the pump and injectors. Viscosity that is too high or too low also can alter fuel delivery metering, and users need to be aware that a certain injection retardation occurs with higher viscosities, which also applies to the use of biodiesel (see Figure 13-41).

	No 1	No 2	No 4	B100
Viscosity @ 40 °C cSt	1.3 to 2.4	1.9 to 4.1	5.5 to 24	4.6

FIGURE 13-41   U.S. Solid viscometer.

16. Ash Content (ASTM D482)

Ash content is not a regular test, but it could be included if needed, but only if some issues are occurring. This test allows users to confirm whether abrasive solids (ash) are contributing to wear and deposits. In addition, soluble metal soaps contribute to engine deposits.

This test covers the determination of ash in the range 0.001–0.180 mass% from distillate and residual fuels, gas turbine fuels, crude oils, lubricating oils, waxes, and other petroleum products, in which any ash-forming materials are normally undesirable impurities or contaminants. The test method is limited to petroleum products, which are free from metallic (ash-forming) additives, including detergents, antiwear additives, and certain phosphorus compounds (see Figure 13-42).

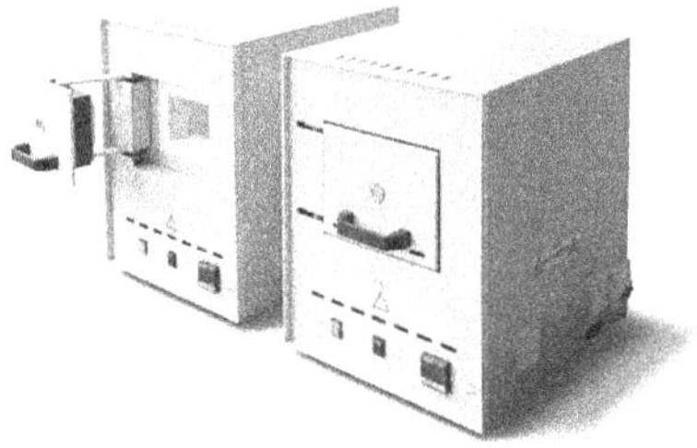

	Diesel No. 1	Diesel No. 2
Ash Content	0.01% max.	0.01% max.

FIGURE 13-42 Lin Tech ash content oven.

17. Copper Strip Corrosion (ASTM D130)

The copper strip corrosion test became popular with the introduction of biodiesel. The test determines corrosive potential on copper, brass, and bronze components of the fuel system (see Figure 13-43).

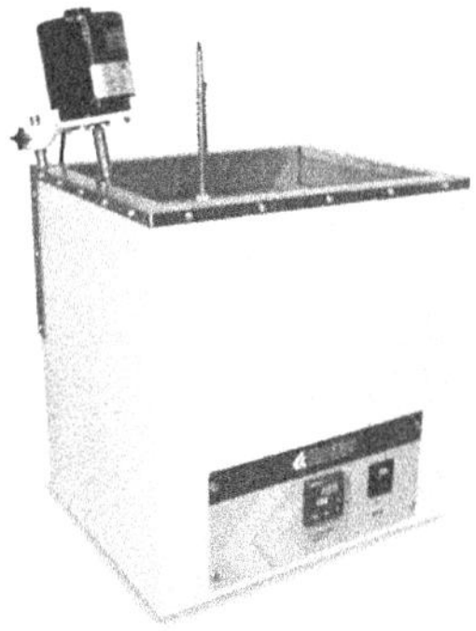

	Class	Designation	Description
1a	1	Slight Tarnish	1a Light orange, almost the same as a freshly polished strip
1b			1b Dark orange
2a	2	Moderate Tarnish	2a Claret red
2b			2b Lavender
2c			2c Multi colored with lavender blue and/or silver overlaid or claret red
2d			2d Silvery
2e			2e Brassy or gold
3a	3	Dark Tarnish	3a Magenta overcast on brassy strip
3b			3b Multicolored with red and green showing (peacock) but not gray
4a	4	Corrosion	4a Transparent black, dark gray or brown with peacock green barely showing
4b			4b Graphite or lusterless black
4c			4c Glassy or jet black

FIGURE 13-43 Koehler K25339 copper corrosion test bath (*left*); copper strip corrosion chart (*right*).

The copper test strip is immersed in a fuel sample and cooked for 19 hours in a closed stainless-steel vessel at 140°C. At the end of the test, the lab personnel compare the copper strip against the corrosion comparator scale and assign a class.

18. Carbon Residue (ASTM 524)

This test is not in the normal core fuel-testing program but is available if required. The test measures a fuel's tendency to form carbon, which is an estimate of the potential to form deposits in an engine. The carbon residue value of burned fuel serves as a rough approximation of the tendency of the fuel to form deposits, and assuming that there are no alkyl nitrates (cetane improver) or additives, the carbon residue of diesel fuel correlates approximately with combustion chamber deposits (see Figure 13-44).

FIGURE 13-44   Conradson carbon residue burner.

19. Lubricity (ASTM D6079)

The HFRR lubricity test is not done regularly by end users, but only when lack of fuel lubricity is suspected. Fuel functions as a lubricant in most components of the fuel injection system, and proper lubricity is important to reduce wear of fuel injection pumps and injectors. The severe processing used to reduce sulfur and aromatics in making ULSD may decrease the amount of surface protecting agents in the fuel. Essentially, a steel ball is rubbed against a metal disc that is flooded with the test fuel. On completion, the wear scar on the steel ball is measured and reported in microns (see Figure 13-45). The test can be expensive and suffers from poor repeatability. It is not necessary as a regular test.

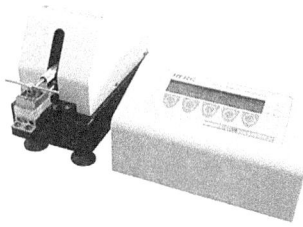

	Diesel No 1 and 2
Lubricity	460 microns

FIGURE 13-45 PCS Instruments high-frequency reciprocating rig.

20. Long-Term Storage (ASTM D4625)

For operations that depend on fuel storage of some length, that is, that the fuel consumption does not exhaust inventory on a weekly basis, fuel stability becomes important, especially for fuel in storage that contains biodiesel. Fuel oxidation and other degradative reactions leading to the formation of sediment (and change in color) accelerate mildly by the test conditions of ASTM D4625 compared with typical storage conditions. The results of this test can predict storage stability more reliably than other more accelerated tests (see Figure 13-46).

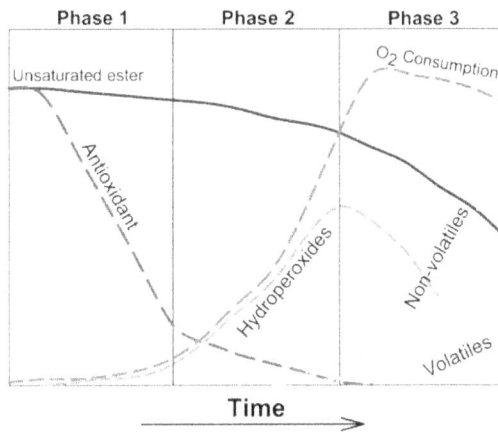

FIGURE 13-46 Phases of oxidation of polyunsaturated fatty acid esters (ASTM D4625).

21. High-Temperature Stability (ASTM D6468)

The high-temperature stability test provides an indication of thermal oxidative stability of distillate fuels when heated to high temperatures that simulate those that occur in some types of fuel systems that return a lot of warm fuel to the fuel tank. Results do not correlate with the engine or the operation. The test can be useful for investigation of operational problems related to fuel thermal stability (see Figure 13-47).

The test method uses a filter paper with a nominal porosity of 11 μm, which will not capture all the sediment formed during aging but allows differentiation over a broad range. The color of filterable insolubles affects reflectance ratings, which may not correlate to the mass of the material filtered from the aged fuel sample. Therefore, no quantitative relationship exists between the pad rating and the gravimetric mass of filterable insolubles.

Before Stability Test ASTM D6468   After Stability Test ASTM D6468

FIGURE 13-47   TC 40 Analytic high-temperature fuel stability oven (*left*); ASTM D6468 (*center* and *right*).

22. Conductivity (ASTM D2624/4308)

There are growing operational safety concerns regarding the handling of modern ULSD and kerosene-based automotive fuels. While the use of additives may assist in reducing potential conductivity problems, fuel retailers must test these fuel types for conductivity to ensure that they are within safe limits for storage and transportation.

Conductivity testing is required by fuel retailers and is not normally part of fuel testing by end users. However, fleet managers should be aware that hydrocarbons can develop a static charge during pumping. Fuel tanks, fuel trucks, and vehicles need to remain properly grounded, and dispensing equipment should bond the tanks to each other.

The ASTM D975 diesel fuel specification mandates the measurement of conductivity for no. 1 and no. 2 diesel fuels and states a 25 pS/m maximum conductivity requirement, especially in instances of high-velocity handling such as fuel transfer/loading (see Figure 13-48). Hydrocarbons circulating at high speeds generate static current, which could be a hazard if the fuel also had high conductivity.

	Diesel No 1 and 2	BioDiesel
Conductivity	25 ps/m Max	25 ps/m

FIGURE 13-48   Fuel conductivity apparatus.

23. Bacteria (ASTM D6469)

Bacterial growth is very prevalent in diesel fuel, especially in machines or tanks that remain half-full at night or during resting periods, thus allowing water to condense and work as an ideal medium for bacteria growth. One of the advantages of biodiesel is its biodegradability, which means that bacteria "eat" it. This attraction of microbes makes biodiesel blends much more susceptible to microbial growth if they become contaminated with moisture (see Figure 13-49).

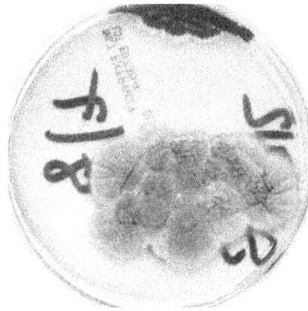

FIGURE 13-49   Lab culture for bacteria analysis.

The most common symptoms associated with microbial growth contamination are filter plugging and corrosion. Even after treatment with biocide, problems can persist if the fuel system is not clean of the waste by-products or if excessive water contamination is still present. Dead bacteria show up as hard particles, and if not caught by the filters, they can be detrimental to injection pumps or injectors.

The confirmation of bacteria by lab analysis dictates the use of diesel fuel biocides that can blend with the diesel fuel without causing detrimental effects to fuel lubrication and/or engine power. Check tanks for moisture to prevent its return because bacteria will not grow in the absence of water.

Fuel Analysis

For fuel analysis cases, we will use the format of preceding case discussions but with tests that are specially designed for fuel analysis. The task here is to assess whether the fuel in question represents a product that will produce the expected power from the engine, to establish how clean it is, and ultimately to determine whether the fuel represents the product that the user is buying (see Figure 13-50).

When collecting the sample, it is important to specify the location and the timing of the sample collection. As discussed earlier under "Fuel Sampling," this is going to make a big difference in results; otherwise, the ghost will never be caught.

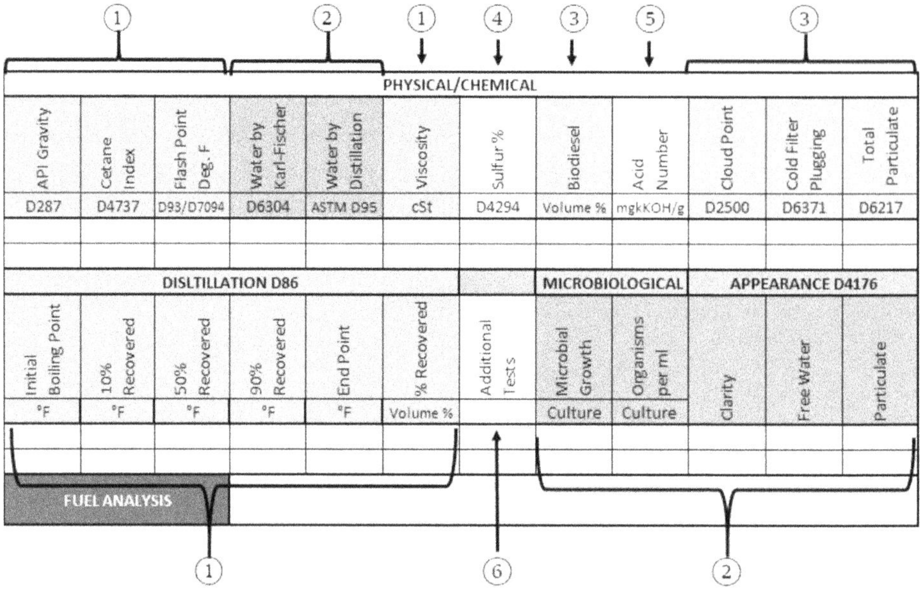

FIGURE 13-50 Fuel analysis format.

With regard to things to look for, everything depends on the reason for the analysis. If users want to check the amount of water and bacteria from the fuel farm tank, they need to focus on zone 2 tests. However, if they want to check their winter readiness, then they need to check the zone 1 tests and zone 3 tests. The same concept applies for a machine fuel tank test. If users want a test because the machine has difficulties starting in the morning, they want to know the cetane index from this fuel under zone 1 tests, but if the problem is short fuel filter life, then they need to pay attention to microbes, appearance, and water content under zone 2 tests. Keep in mind that time and point of sample collection are important if you really want your money back from this activity. In some markets, knowing sulfur content and acid number could be a regular requirement (zones 4 and 5), whereas in others, knowing fuel storage stability or high-temperature stability could be more important.

CASE DISCUSSIONS

The following case discussions are real-life occurrences that bring light to the subject of fuel condition and contamination. They should represent most of the cases experienced by mobile equipment users and should be of great help when diagnosing situations involving fuel-related problems.

1. Water-Excavator with Fuel Pump Failures

A construction excavator equipped with a Tier 2 engine was operating in harsh conditions in gold mines located someplace in South America. The machine was having issues with fuel pump failures.

Fuel Analysis

The fuel analysis results in Figure 13-51 shows high water content with the Karl Fischer titration test and by distillation. It also shows free water and particulates. Sulfur content is 2,000 ppm, and the fuel looks opaque.

PHYSICAL/CHEMICAL											
API Gravity	Cetane Index	Flash Point Deg. F	Water by Karl-Fischer	Water by Distillation	Viscosity	Sulfur %	Biodiesel	Acid Number	Cloud Point	Cold Filter Plugging	Total Particulate
D287	D4737	D93/D7094	D6304	ASTM D95	cSt	D4294	Volume %	mgkKOH/g	D2500	D6371	D6217
37.1	48.9	145	1695	0.9	NA	2000	<0.1	0.07	NA	NA	10

DISLTILLATION D86							MICROBIOLOGICAL		APPEARANCE D4176		
Initial Boiling Point	10% Recovered	50% Recovered	90% Recovered	End Point	% Recovered	Additional Test	Microbial Growth	Organisms per ml	Clarity	Free Water	Particulate
°F	°F	°F	°F	°F	Volume %		Culture	Culture			
365	417	512	621	664	98.1		Neg	0	Opaque	YES	69

FUEL ANALYSIS	Excavator working in a gold mine in South America

FIGURE 13-51 Fuel analysis report for an excavator with fuel pump failures.

Conclusion

Fuel retrieved from the machine shows high contamination with water and particulates, but bacteria did not show up in culture. Ideally, both the fuel supply and storage are investigated. Also, filling procedures for the machine and whether filtration was used during refueling should be reviewed. Refueling the machine at the end of daily use could avoid water condensation and ultimately prolong injector/pump life.

2. Hard Cold Starting in a Production 4WD Loader

A production four-wheel-drive (4WD) loader working in a copper mine in high-altitude operation was experiencing hard cold starting and some smoking. The machine was relatively new and in mechanically good operating condition.

Fuel Analysis

The fuel analysis results in Figure 13-52 show a diesel fuel that is free of water and particulates. However, the fuel has a cetane index of 38.2, and its API gravity is at 32.7. The fuel is a low-sulfur fuel with 510 ppm sulfur.

PHYSICAL/CHEMICAL											
API Gravity	Cetane Index	Flash Point Deg. F	Water by Karl-Fischer	Water by Distillation	Viscosity	Sulfur %	Biodiesel	Acid Number	Cloud Point	Cold Filter Plugging	Total Particulate
D287	D4737	D93/D7094	D6304	ASTM D95	cSt	D4294	Volume %	mgkKOH/g	D2500	D6371	D6217
32.7	38.2	141	167	0.03	NA	510	3.6	0.06	NA	NA	2

DISLTILLATION D86							MICROBIOLOGICAL		APPEARANCE D4176		
Initial Boiling Point	10% Recovered	50% Recovered	90% Recovered	End Point	% Recovered	Additional Test	Microbial Growth	Organisms per ml	Clarity	Free Water	Particulate
°F	°F	°F	°F	°F	Volume %		Culture	Culture			
364	406	500	604	668	97.6		Neg	0	Opaque	NO	NO

FUEL ANALYSIS	Production 4WD loader working in high altitude in a copper mine is having issues with starting in cold temperatures.

FIGURE 13-52   Fuel analysis report for a production 4WD loader.

Conclusion

This problem could have a fix with a cetane-improving additive, or fuel quality could be discussed with the fuel supplier. It is also good practice to test batches of fuels supplied to the job site to confirm conformance to specifications. Given that this machine operates at high altitude, it would make sense to test for fuel viscosity and establish fuel grade. In addition, it would be good practice to test for fuel flash point and determine whether this fuel complies with specifications for high-altitude, cold-weather operation.

3. Injector Failures in a Motor Grader

A motor grader with a Tier III engine has had injector failures. This is the second time in 3,500 hours that the machine has suffered problems with injectors.

Fuel Analysis

The fuel analysis report in Figure 13-53 shows some water, which is over the limit of <200 ppm. Additionally, sulfur at 1,800 ppm is pushing TAN up to 1.2. The limit for TAN is 0.07. Particulate contamination exceeds specifications. A filter patch accompanies this report, showing metal and carbon particulates (see Figure 13-54).

					PHYSICAL/CHEMICAL						
API Gravity	Cetane Index	Flash Point Deg. F	Water by Karl-Fischer	Water by Distillation	Viscosity	Sulfur %	Biodiesel	Acid Number	Cloud Point	Cold Filter Plugging	Total Particulate
D287	D4737	D93/D7094	D6304	ASTM D95	cSt	D4294	Volume %	mgkKOH/g	D2500	D6371	D6217
35	41	144	350	1.2	NA	1800	<0.1	0.07	NA	NA	12

	DISLTILLATION D86						MICROBIOLOGICAL		APPEARANCE D4176		
Initial Boiling Point	10% Recovered	50% Recovered	90% Recovered	End Point	% Recovered	Additional Test	Microbial Growth	Organisms per ml	Clarity	Free Water	Particulate
°F	°F	°F	°F	°F	Volume %		Culture	Culture			
366	410	511	615	665	96.5		Neg	0	Cloudy	Yes	Yes

FUEL ANALYSIS	Motor Grader with Tier III experiencing injector failure. Patch report attached

FIGURE 13-53   Fuel analysis report for a motor grader.

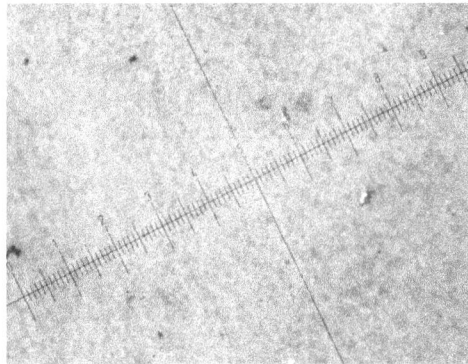

FIGURE 13-54   Fuel patch.

Conclusion

The presence of high sulfur and water is triggering the TAN increase that could be contributing to the production of metal particles from the injection pump. Particles could be contributing to the injector failure because these particles originate past the filtration point. The recommendation starts with reducing water content. By controlling water, the opportunity for metal corrosion is also under control. Controlling sulfur is out of the question unless the local fuel supply also offers low-sulfur fuel.

4. Heavy Knocking and Smoke in a Backhoe

A backhoe loader was experiencing abnormal engine operation with heavy knocking, smoke, and lack of power.

Fuel Analysis

The fuel analysis report in Figure 13-55 shows high API gravity and high cetane, both of which exceed those of standard diesel fuel. The distillation results show recovery at temperatures lower than specified for diesel fuels. Aside from that, the fuel is clean and does not show other out-of-specification numbers.

PHYSICAL/CHEMICAL											
API Gravity	Cetane Index	Flash Point Deg. F	Water by Karl-Fischer	Water by Distillation	Viscosity	Sulfur ppm	Biodiesel	Acid Number	Cloud Point	Cold Filter Plugging	Total Particulate
D287	D4737	D93/D7094	D6304	ASTM D95	cSt	D4294	Volume %	mgkKOH/g	D2500	D6371	D6217
53.8	69	144	17	<0.1	NA	16	<0.1	0.06	NA	NA	<1

DISLTILLATION D86							MICROBIOLOGICAL		APPEARANCE D4176		
Initial Boiling Point	10% Recovered	50% Recovered	90% Recovered	End Point	% Recovered	Additional Test	Microbial Growth	Organisms per ml	Clarity	Free Water	Particulate
°F	°F	°F	°F	°F	Volume %		Culture	Culture			
102	192	400	555	619	98		Neg	0	Cloudy	Yes	Yes

FUEL ANALYSIS	Backhoe Loader operating in Central America. High cetane number and low distillation points.

FIGURE 13-55 Fuel analysis report for a backhoe.

Conclusion

High API gravity, high cetane index, and low distillation temperature recoveries are an indication of diesel fuel contaminated with gasoline. Operating a diesel engine with a mixed fuel showing an excessively high cetane index could be highly detrimental. The machine needs a flushing of the fuel system. Additionally, fuel samples from the fuel farm, if any, need to go to the lab for evaluation. The user needs to determine the point of contamination with gasoline.

5. Fuel Starvation During Winter in a 4WD Loader

A construction-size 4WD loader operating in Canada was having issues with fuel starvation during the winter months. The operator was using fuel additives to lower the fuel pour point without success.

Fuel Analysis

The fuel analysis results in Figure 13-56 show that the machine is using a high percentage of bio-diesel and that the viscosity puts the fuel in the no. 2 grade. It also shows that the cloud point is –5°F and that the cold filter plugging point is at –10°F.

PHYSICAL/CHEMICAL											
API Gravity	Cetane Index	Flash Point Deg. F	Water by Karl-Fischer	Water by Distillation	Viscosity	Sulfur %	Biodiesel	Acid Number	Cloud Point	Cold Filter Plugging	Total Particulate
D287	D4737	D93/D7094	D6304	ASTM D95	cSt	D4294	Volume %	mgkKOH/g	D2500	D6371	D6217
36.9	47.3	145	64	<0.1	No 2	8	15	0.07	-5	-10	<1

DISLTILLATION D86							MICROBIOLOGICAL		APPEARANCE D4176		
Initial Boiling Point	10% Recovered	50% Recovered	90% Recovered	End Point	% Recovered	Fuel Grade	Microbial Growth	Organisms per ml	Clarity	Free Water	Particulate
°F	°F	°F	°F	°F	Volume %	EPA	Culture	Culture			
360	390	488	620	648	97		NA	NA	Clear	Yes	Yes

FUEL ANALYSIS	4WD Loader operating in Canada is having fuel starvation problem in cold temperature.

FIGURE 13-56   Fuel analysis report for a 4WD loader.

Conclusion

With biodiesel content at 15%, cloud point at –5°F, and cold filter plugging point at –10°F, it would be difficult for the machine to operate satisfactorily in winter. Additionally, the machine is operating with what appears to be a no. 2 summer diesel fuel. The operator needs to reduce the amount of biodiesel in the mix and blend with no. 1 diesel or switch over to no. 1 diesel for the duration of the winter. Cold-flow-improving additives also give a larger response to blends with no. 1 diesel fuel.

6. Turbocharger Failures in a Production Bulldozer

A production bulldozer equipped with a Tier III engine and operating in South America had been suffered turbocharger failures. The failure mode is corrosion of the knuckles of the variable-geometry turbocharger mechanism, locking the vanes in one position.

Fuel Analysis

The fuel analysis results in Figure 13-57 show a fuel sample with high sulfur content and bacteria.

PHYSICAL/CHEMICAL											
API Gravity	Cetane Index	Flash Point Deg. F	Water by Karl-Fischer	Water by Distillation	Viscosity	Sulfur ppm	Biodiesel	Acid Number	Cloud Point	Cold Filter Plugging	Total Particulate
D287	D4737	D93/D7094	D6304	ASTM D95	cSt	D4294	Volume %	mgkKOH/g	D2500	D6371	D6217
39.6	58.6	143	50	<0.1	NA	8000	<0.1	0.15	NA	NA	<1

DISLTILLATION D86						OTHER	MICROBIOLOGICAL		APPEARANCE D4176		
Initial Boiling Point	10% Recovered	50% Recovered	90% Recovered	End Point	% Recovered	Fuel Grade	Microbial Growth	Organisms per ml	Clarity	Free Water	Particulate
°F	°F	°F	°F	°F	Volume %	EPA	Culture	Culture			
390	449	520	598	634	99	No2	POS	1000	Cloudy	No	Yes

FUEL ANALYSIS	Production crawler operating in South America. Turbocharger operating mechanism failure

FIGURE 13-57   Fuel analysis report for a production bulldozer.

Conclusion

Operating a machine with 8,000 ppm sulfur fuel will create acids and corrosion that lock up the turbocharger and could be detrimental to the life of the engine, especially the bearings. Sulfur from high-sulfur fuels combines with water from compression to create sulfuric acid, thus causing corrosion of the turbocharger operating mechanism. There is no fix through additives for sulfur contents. The only alternative to repeated failures is to get diesel fuel with a sulfur content that is within the operable range for this engine.

7. Filter Plugging in a Production Bulldozer

A production bulldozer operating in Colorado was experiencing continuous fuel filter plugging. The plugging occurred at the rate of once every 50 hours. The operator expected fuel filters to last at least 500 hours.

Fuel Analysis

The fuel analysis results in Figure 13-58 show a high water content, a high particulate content, and the presence of bacteria.

API Gravity	Cetane Index	Flash Point Deg. F	Water by Karl-Fischer	Water by Distillation	Viscosity	Sulfur %	Biodiesel	Acid Number	Cloud Point	Cold Filter Plugging	Total Particulate
D287	D4737	D93/D7094	D6304	ASTM D95	cSt	D4294	Volume %	mgkKOH/g	D2500	D6371	D6217
38	43	142	1500	0.08	#1	15	<0.1	0.14	NA	NA	35

Initial Boiling Point	10% Recovered	50% Recovered	90% Recovered	End Point	% Recovered	Fuel Grade	Microbial Growth	Organisms per ml	Clarity	Free Water	Particulate
°F	°F	°F	°F	°F	Volume %	EPA	Culture	Culture			
385	438	510	584	625	97	No2	POS	1000	Cloudy	POS	POS

FUEL ANALYSIS — Production crawler operating in Colorado. Filter plugging

FIGURE 13-58 Fuel analysis report for a production bulldozer.

Conclusion

The presence of water and bacteria suggests that the cause for filter plugging is the continuous growth of bacteria that is thriving in ideal conditions for growth. The first recommendation would be to use a biocide to kill the bacteria and then to flush the fuel tank. If the tank is not flushed, filter plugging will continue, at least for a while, because dead bacteria still plug filters. The possibility of having repeated occurrences of this issue is high unless the fuel supply is also free of water and bacteria. This is a case where both the machine and the fuel supply/storage need biocide treatment. As a final recommendation, refueling with filtration and refueling at the end of each day will keep water condensation in check. Recheck for diesel fuel bacteria and water content regularly.

8. Engine Smoking in a Tracked Tree Harvester

A tracked tree harvester machine operating in the Amazon jungle had been smoking. The operator worried that the smoke was an indication that the engine was developing mechanical problems with the injection system. The machine operated normally with a slight power loss.

Fuel Analysis

The fuel analysis results in Figure 13-59 show several abnormal indicators. To start with, the flash point at 145°F is high, and the API gravity is low. The cetane index is also low. The results do not show issues with bacteria, although water is at the top limit of 200 ppm. The sample also fails the distillation recovery path, as highlighted in the report.

PHYSICAL/CHEMICAL											
API Gravity	Cetane Index	Flash Point Deg. F	Water by Karl-Fischer	Water by Distillation	Viscosity	Sulfur %	Biodiesel	Acid Number	Cloud Point	Cold Filter Plugging	Total Particulate
D287	D4737	D93/D7094	D6304	ASTM D95	cSt	D4294	Volume %	mgkKOH/g	D2500	D6371	D6217
35	38	154	**200**	0.01	4.5	800	<0.1	**0.05**	NA	NA	<10

DISLTILLATION D86						OTHER	MICROBIOLOGICAL		APPEARANCE D4176		
Initial Boiling Point	10% Recovered	50% Recovered	90% Recovered	End Point	% Recovered	Fuel Grade	Microbial Growth	Organisms per ml	Clarity	Free Water	Particulate
°F	°F	°F	°F	°F	Volume %	EPA	Culture	Culture			
402	455	550	615	674	96	No2	NEG	NO	Opaque	NO	NO

FUEL ANALYSIS	Tracked tree harvester machine working in the Amazon jungle. Smoking engine.

FIGURE 13-59   Fuel analysis report for a tracked tree harvester.

Conclusion

The distillation recovery suggests that this fuel is blended with something heavier than no. 2 diesel fuel. The operator indicated that he burns used engine oil from the operation mixed with fuel in a ratio of 5%–10% oil to fuel. The machine has an onboard filtration system that meters used oil into the fuel. Although this procedure is no longer in use in most countries around the world due to environmental concerns and was never recommended in Tier II or higher engines, there are areas where the system is still in place. The issue with burning waste oil is that it introduces engine oil type additives and wear particles into the high-pressure injection system. These additives and particles pose a risk to the high-pressure injection system and the emissions control system and constitute an environmental hazard.

9. Injection Pump Failure in a Grapple Skidder

A grapple skidder operating in the jungles of eastern Peru had a failed injection pump. The machine had a Tier III engine equipped with a high-pressure common-rail system. Because of the remote location and logistics, the operator had been using jet fuel A to simplify the operation. The operator was aware of the possible consequences of using jet fuel in a diesel engine and had been adding some additives to the fuel. The operator wanted to know whether the fuel may have caused the failure he was experiencing.

Fuel Report

The fuel analysis results in Figure 13-60 show a fuel that is very close to diesel fuel. However, the jet fuel shows 3,000 ppm sulfur. The TAN is higher than for a diesel fuel, but in general, the fuel appears clean and free of water and particulates. Nevertheless, based on the type of failure, the lab also provided an HFRR lubricity test that came in at 625.

PHYSICAL/CHEMICAL											
API Gravity	Cetane Index	Flash Point Deg. F	Water by Karl-Fischer	Water by Distillation	Viscosity	Sulfur %	Biodiesel	Acid Number	Cloud Point	Cold Filter Plugging	Total Particulate
D287	D4737	D93/D7094	D6304	ASTM D95	cSt	D4294	Volume %	mgkKOH/g	D2500	D6371	D6217
43	45	110	150	0.01	1.1	3000	<0.1	0.03	NA	NA	<10

DISLTILLATION D86						OTHER	MICROBIOLOGICAL		APPEARANCE D4176		
Initial Boiling Point	10% Recovered	50% Recovered	90% Recovered	End Point	% Recovered	HFRR	Microbial Growth	Organisms per ml	Clarity	Free Water	Particulate
°F	°F	°F	°F	°F	Volume %	ASTMD6079	Culture	Culture			
360	425	502	576	603	98	625	NEG	NO	CLEAR	NO	NO

FUEL ANALYSIS	Grapple Skidder operating in Canada. Injection pump failure

FIGURE 13-60 Fuel analysis report for a grapple skidder.

Conclusion

Jet fuel A lacks the antiscuffing additives used in diesel fuel to protect moving parts within the injection system. Although the operator depends on fuel additives, the fuel sample taken from the machine did not pass the high-frequency reciprocating rig (HFRR) test, and this explains the failure of the injection pump.

10. Quality Check—Fleet Operation

A large forestry fleet operation submitted its fuel farm samples regularly to check for fuel quality and contaminants and whether the supplier provides good-quality fuels. The company had specified 5% biodiesel in its purchases.

Fuel Analysis

The fuel analysis results in Figure 13-61 show a clean fuel with an API gravity of 36.9, a flash point of 135°F, and a cetane index of 47. 3. It also shows final recovery of 90% at 620°F from the distillation test. The tested fuel also confirms 15 ppm sulfur and 7% biodiesel. The sample is free of water, particulates, and bacteria.

PHYSICAL/CHEMICAL											
API Gravity	Cetane Index	Flash Point Deg. F	Water by Karl-Fischer	Water by Distillation	Fuel Type	Sulfur %	Biodiesel	Acid Number	Cloud Point	Cold Filter Plugging	Total Particulate
D287	D4737	D93/D7094	D6304	ASTM D95	D975	D4294	Volume %	mgkKOH/g	D2500	D6371	D6217
36.9	47.3	135	100	0.01	#2	15	7	0.03	-10	-33	<10

DISLTILLATION D86						OTHER	MICROBIOLOGICAL		APPEARANCE D4176		
Initial Boiling Point	10% Recovered	50% Recovered	90% Recovered	End Point	% Recovered	HFRR	Microbial Growth	Organisms per ml	Clarity	Free Water	Particulate
°F	°F	°F	°F	°F	Volume %	ASTMD6079	Culture	Culture			
360	390	488	620	648	97	NA	NEG	NO	CLEAR	NO	NO

FUEL ANALYSIS	Fuel quality check from a large forestry company. Fuel farm sample

FIGURE 13-61 Fuel analysis report for a fleet.

Conclusion

The flash point and distillation confirm that this is a no. 2 diesel fuel. The biodiesel content at 7% is slightly over the specification requested by the customer. However, the general results indicate that the fuel is of good quality and is clean.

CONCLUSION

Up to this point, we have covered most of the fluids used in mobile equipment, the sampling procedures, testing, and case discussions. We cannot leave exhaust fluid untouched because it is already being used by mobile equipment. This is the last topic we need to discuss and is the subject of Chapter 14.

CHAPTER 14

DIESEL EXHAUST
FLUID ANALYSIS

The race to reach the emissions mandates for diesel engines started early in 1996 with Tier 1 engines in the United States and Stage I engines in Europe. The schedule and limits for emissions were stringent, and the industry had no option but to comply through the different stages. The changes in engine design were drastic, and electronics were part of the solution. Some manufacturers managed to comply with the target emissions limits by playing with camshaft and valve design, whereas others needed extra help from cooled exhaust gas recirculation (EGR) and variable-geometry turbochargers (VGTs). All of them needed the help of high-pressure injection, electronics, and injection retardation. Reaching the Tier 4 and Stage IV levels was a great accomplishment, and some engine categories required the use of diesel exhaust fluid (DEF), the subject of this chapter.

The emissions reduction mandates required reductions in hydrocarbons (HCs), particulate matter (PM), and nitrogen oxides (NO_x), as shown in Figure 14-1.

In conjunction with the introduction of Tier 4 and Stage IV engines, a blend of urea and water became the option to reduce nitrogen oxides at the ends of the exhaust filters. The exhaust fluid that burns in the final stages of the exhaust system needs to comply with certain specifications to ensure that the selective catalytic converter (SCR) is not poisoned with other elements that do not belong to the exhaust fluid and to guarantee that the machine meets the emission standards regarding nitrogen oxides (see Figure 14-2).

DEF is a blended aqueous urea solution of 32.5% high-purity urea and 67.5% distilled water that converts excess NO_x into inert nitrogen and water. The reaction of ammonia (NH_3) and oxygen with the help of a catalyst forms water (H_2O) and nitrogen (N_2). DEF injection occurs into the exhaust gas at the inlet of a selective catalytic reducer (SCR) in on-highway truck engines and, more recently, in off-road mining and construction equipment.

EPA and EU non road emisions regulations (50-750 HP)

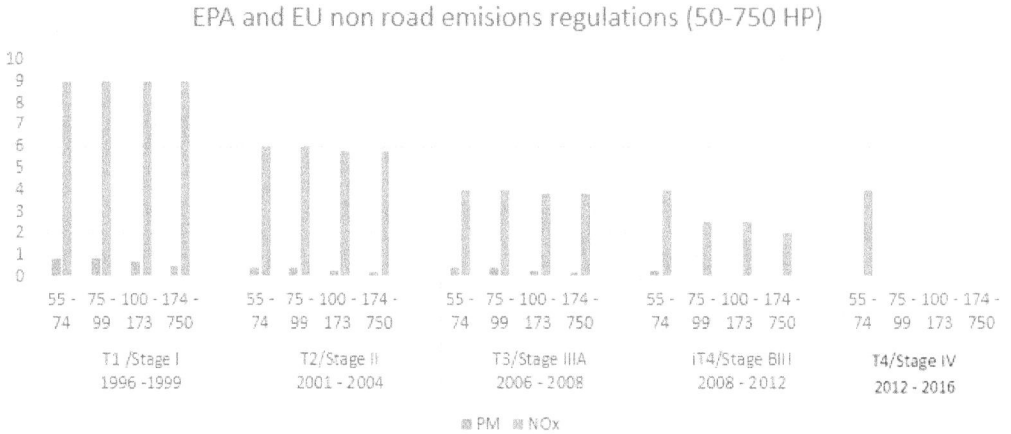

FIGURE 14-1 Particulate matter (PM) and nitrogen oxide (NO_x) emissions from Tier 1–4 off-road vehicle engines.

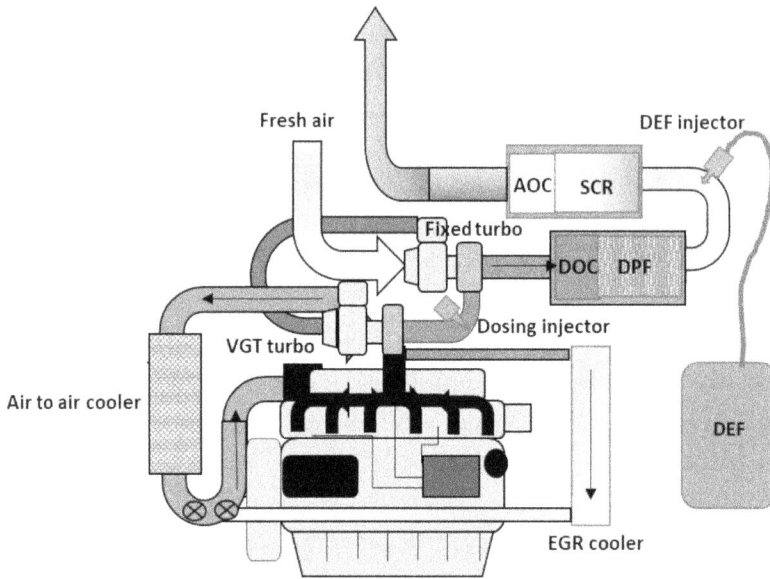

FIGURE 14-2 Tier 4 engine.

The reaction in the SCR looks like this:

$$2NH_3 + NO + NO_2 \rightarrow 2N_2 + 3H_2O$$

Why the 32.5% DEF-to-Water Ratios?

This percentage is the best approach for the mix because the combined elements freeze at −11°C (12.2°F). In this way, if freezing conditions prevail, the DEF is not diluted or superconcentrated.

The need for quality control is critical for optimal operation of the equipment when it comes to DEF. Prevention and proactive monitoring of the DEF can help to mitigate issues. Typical issues with poor handling of DEF include:

- Overdosing. Too much DEF may not fully hydrolyze, leading to crystalized deposits in the exhaust or injector nozzle.
- Contamination. Improper storage or handling of the DEF can result in dirt entry, rust, or tank scale.
- Dilution. Topping up the DEF with water will dilute the solution with a resulting drop in NO_x neutralization effectiveness.
- Freezing. DEF may gel or freeze in colder climates, which may prevent engine operation.
- Other fluids. Contamination from other fluids introduced to the DEF tank (e.g., diesel fuel, coolant, or engine oil) can cause damage to the SCR system and engine shutdown.

Contaminants or impurities in your DEF could cause the SCR system in your vehicle/machine to malfunction or fail. Should a fault code for your DEF system occur, testing can detect the cause of the problem. There are two primary methods of testing DEF: field testing and laboratory testing.

Field Testing

Visually inspect the DEF for color and clarity. DEF is naturally clear. If it appears cloudy or colored, you may have an issue with age or contamination. If engine or hydraulic fluid is present, it will form a separate layer in the sample taken. You can also visually inspect the DEF filter and storage containers for dirt or metal debris.

Measure urea concentration with a lens or an electronic refractometer or a gravity tester to ensure that 32.5% of the fluid is urea (see Figure 14-3).

FIGURE 14-3   DEF electronic refractometer (*left*); DEF gravity meter (*right*).

Use test strips to check for hydrocarbon contamination of the DEF fluid (see Figure 14-4).

Inspect exhaust piping and DEF injectors for the presence of crystallization. Crystallization builds up over time, and testing of DEF currently in the tank may or may not indicate issues that could cause crystallization.

FIGURE 14-4 Test strips.

Lab Testing

Labs refer to ISO 22241-2 2019 as the reference for the different tests required for DEF. Table 14-1 summarizes the typical tests for DEF. ASTM 7821-12 is an alternative method to measure urea content.

TABLE 14-1 Tests for Diesel Exhaust Fluid (DEF)

Test	Unit	Typical	Min	Max
Urea content	% (m/m)	32.5	31.8	33.2
Density at 20°C	kg/L	1.0899	1.087	1.093
Refractivity index at 20°C	—	1.383	1.3814	1.3843
Alkalinity as ammonia	% (m/m)	<0.1	—	0.2
Biuret	% (m/m)	0.2	—	0.3
Aldehydes	ppm	<0.5	—	5
Insoluble matter	ppm	<0.1	–	20
Phosphates	ppm	<0.05	—	0.5
Calcium	ppm	<0.1	—	0.5
Iron	ppm	<0.1	—	0.5
Copper	ppm	<0.1	—	0.2
Zinc	ppm	<0.1	—	0.2
Chrome	ppm	<0.1	—	0.2
Nickel	ppm	<0.1	—	0.2
Aluminum	ppm	<0.1	—	0.5
Magnesium	ppm	<0.1	—	0.5
Sodium	ppm	<0.1	—	0.5
Potassium	ppm	<0.1	—	0.5
Identity (FTIR spectroscopy)	ppm	Identical to reference	—	—

Why Those Tests?

Table 14-2 shows the need for certain tests and the impact in the DEF emissions equipment. Of special interest are biuret and aldehydes.

TABLE 14-2 DEF Tests

Shelf life	Alkalinity as NH_3
Clogs spray nozzle	Insolubles Calcium Magnesium
Gummy deposits in exhaust pipe	Aldehyde
Damages SRC catalyst	Sodium Potassium Phosphate PO_4 Biuret Aluminim Iron Copper Zinc Chromium Nickel

Biuret

One of the most important tests is the test for biuret. Biuret forms deposits above 200°C in the SCR that later solidify, damaging the SCR or making it inefficient. Biuret is a chemical compound with the chemical formula $[H_2NC(O)]_2NH$. It is a white solid that is soluble in hot water. It is an undesirable impurity in urea-based fertilizers. Because biuret is toxic to plants, its percentage in fertilizers needs to be low. For our discussion, biuret in DEF also needs to be low. Biuret will cause gummy deposits in the exhaust system of a Tier 4 engine (see Figure 14-5).

FIGURE 14-5 Biuret.

Aldehydes

Aldehydes (HCO) are nonregulated pollutants that cause harmful effects in humans. Aldehydes can participate in complex reactions in the atmosphere, generating other gaseous oxidants such as ozone, which causes respiratory problems.

DEF Analysis

The same as we have been doing with other fluids, DEF analysis follows a similar methodology, as illustrated in Figure 14-6.

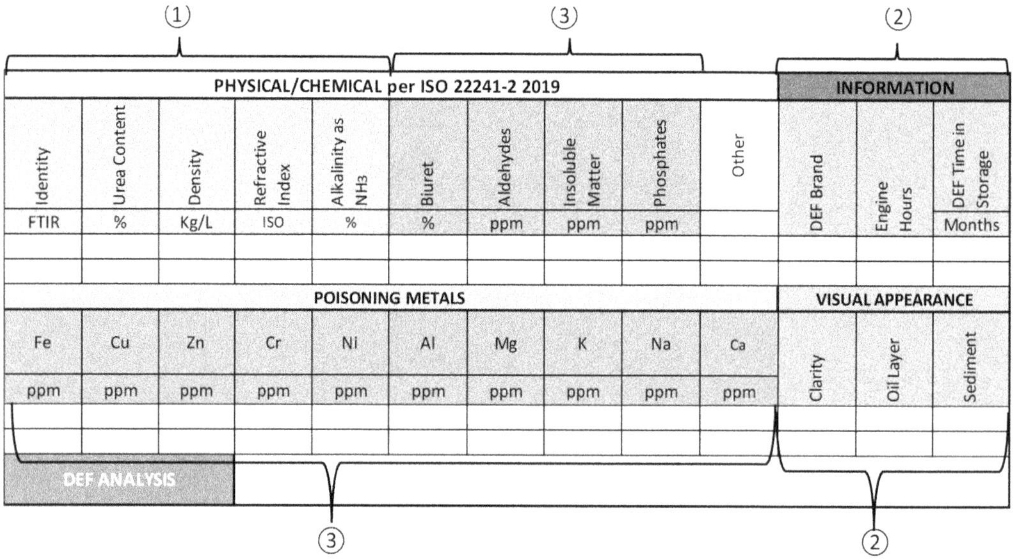

FIGURE 14-6 DEF analysis report format.

Although the format from every lab is going to be different, the one we use here keeps pace with the other reports discussed within this book, so we can have an educated discussion via several cases.

There are three specific groups of indicators that we need to keep an eye on with regard to DEF quality and condition: (1) product identity and urea content, (2) appearance and time in inventory, and (3) contaminants (e.g., metals, insoluble matter, biuret, aldehydes, and phosphates). It all depends on what the user wants to check. If testing is done only for checking supplier compliance with product specifications, the focus is on indicators under point (1). If it is the result of machine malfunction or fault codes, then indicators points (1) and (2) need attention, but if the issue is with component damage or a DEF plugged nozzle, points (1) and (3) could lead to the correct answer. Sampling DEF follows the same rule as with other fluids. It is in the interest of users to collect a representative DEF sample so that the results are not skewed by improper sampling techniques.

CASE DISCUSSIONS

1. Engine Deration in a Production 4WD Loader

A production-size 4WD loader equipped with a Tier 4 engine had been derating to the point that the machine was not operable anymore. Fault codes suggested a malfunction in the SRC. On disassembling the SRC and related tubing, some degree of urea crystallization was seen to obstruct the injector.

DEF Analysis

The DEF analysis report in Figure 14-7 shows several indicators in gray. To start with, the urea concentration is at 40% and aldehydes are at the top limit of guidelines. Several metals, although small in number by any standard, also show over the allowable limits. The identity of the product is also in question because the figures do not match the new fluid.

PHYSICAL/CHEMICAL per ISO 22241-2 2019										INFORMATION		
Identity	Urea Content	Density	Refractive Index	Alkalinity as NH3	Biuret	Aldehydes	Insoluble Matter	Phosphates	Other	DEF Brand	Engine Hours	DEF Time in Storage
FTIR	%	Kg/L	ISO	%	%	ppm	ppm	ppm				Months
No Match	45	1.1	1.42	0.15	0.32	5	9	0.3		Blue	3,580	5

POISONING METALS										VISUAL APPEARANCE		
Fe	Cu	Zn	Cr	Ni	Al	Mg	K	Na	Ca	Clarity	Oil Layer	Sediment
ppm	ppm	ppm	ppm	ppm	ppm	ppm	ppm	ppm	ppm			
1	<0.5	0.3	0.4	0.1	3	1	2	0.5	0.5	Clear	No	No

DEF ANALYSIS	Production 4WD loader with Tier IV engine. Engine derating and fault codes.

FIGURE 14-7 DEF analysis report for a production 4WD loader.

Conclusion

The major issue in this case is the urea concentration at 45% that inflates other indicators such as biuret and aldehydes. Some metals are also over the guideline limits. The use of highly concentrated DEF is the direct cause for injector crystallization (biuret deposits) that cause fault codes and engine deration. This DEF deposit needs flushing, and a sample from the stock DEF needs to go to the lab for analysis to understand the differences between the DEF in inventory and the DEF in the machine.

2. Engine Derating Intermittently in a Utility Bulldozer

A utility bulldozer in California had been having intermittent engine deration. The company bought DEF for the fleet in bulk from a local supplier in town. Other machines in the fleet also had been having issues with deration.

DEF Analysis

The DEF analysis report in Figure 14-8 shows a DEF that matches the signature of the product. However, the results also show high contamination with calcium, magnesium, and zinc.

PHYSICAL/CHEMICAL per ISO 22241-2 2019										INFORMATION		
Identity	Urea Content	Density	Refractive Index	Alkalinity as NH3	Biuret	Aldehydes	Insoluble Matter	Phosphates	Other	DEF Brand	Engine Hours	DEF Time in Storage
FTIR	%	Kg/L	ISO	%	%	ppm	ppm	ppm				Months
Match	31	1.09	1.38	0.15	0.23	0.7	10	0.05		Blue	4,580	3

POISONING METALS										VISUAL APPEARANCE		
Fe	Cu	Zn	Cr	Ni	Al	Mg	K	Na	Ca	Clarity	Oil Layer	Sediment
ppm	ppm	ppm	ppm	ppm	ppm	ppm	ppm	ppm	ppm			
1	0.2	12	0.4	0.1	0.3	8	0.2	0.5	25	Opaque	POS	No

DEF ANALYSIS	Utility bulldozer with Tier IV engine, frequent engine deration.

FIGURE 14-8   DEF analysis report for a utility bulldozer.

Conclusion

The results are abnormal with high rates of calcium, magnesium, and zinc. These three elements are typically from lubricants. Further investigation indicated that the company's service personnel were filling empty oil containers with DEF for field service purposes. Small amounts of oil remaining in the lubricant pails contaminated the DEF enough to cause DEF injectors to plug and engine deration.

3. SCR Failure in a Production Excavator

A production excavator had recently had an SCR failure, and the user was interested in knowing the reasons behind the failure at such low hours. The SCR could not keep up with NO_x production, triggering fault codes and deration.

DEF Analysis

The DEF analysis results in Figure 14-9 show a DEF fluid that appears to match that of the product brand reported. However, the report shows 25 ppm of aluminum.

PHYSICAL/CHEMICAL per ISO 22241-2 2019										INFORMATION		
Identity	Urea Content	Density	Refractive Index	Alkalinity as NH3	Biuret	Aldehydes	Insoluble Matter	Phosphates	Other	DEF Brand	Engine Hours	DEF Time in Storage
FTIR	%	Kg/L	ISO	%	%	ppm	ppm	ppm				Months
Match	32	1.089	1.38	0.14	0.24	2.7	9	0.03		AC Delco	5,670	6

POISONING METALS										VISUAL APPEARANCE		
Fe	Cu	Zn	Cr	Ni	Al	Mg	K	Na	Ca			
ppm	ppm	ppm	ppm	ppm	ppm	ppm	ppm	ppm	ppm	Clarity	Oil Layer	Sediment
0.2	0.2	**0.1**	0.4	0.1	25	0.4	0.2	0.5	0.3	Clear	No	No

DEF ANALYSIS	Production size excavator with Tier IV engine, SCR failure

FIGURE 14-9 DEF analysis report for a production excavator.

Conclusion

The limit for aluminum should not be higher than 0.5 ppm. The question now is: How did the aluminum get into the DEF fluid? Information provided by the user indicates that bulk DEF is now stored in all-aluminum tanks that came from a dairy farm, which the company thought were stainless steel. The fluid is six months old, and it is the supply for the whole fleet. This explains the presence of aluminum and the reason for the SCR failure.

4. Fault Codes in a Dump Truck

A dump truck from a construction company had been giving fault codes related to the emissions control system. The truck has 37,540 miles, and this was the first time the problem had shown up. A DEF sample was sent to the lab.

DEF Analysis

The DEF analysis results in Figure 14-10 show a diluted DEF fluid with only 11% of urea content. The product does not match the brand for this reason.

PHYSICAL/CHEMICAL per ISO 22241-2 2019										INFORMATION		
Identity	Urea Content	Density	Refractive Index	Alkalinity as NH3	Biuret	Aldehydes	Insoluble Matter	Phosphates	Other	DEF Brand	Engine Hours/Miles	DEF Time in Storage
FTIR	%	Kg/L	ISO	%	%	ppm	ppm	ppm				Months
NO Match	11%	1.06	1.25	0.08	0.06	0.8	14	0.12		Armor Blue	37,540	3

POISONING METALS										VISUAL APPEARANCE		
Fe	Cu	Zn	Cr	Ni	Al	Mg	K	Na	Ca	Clarity	Oil Layer	Sediment
ppm	ppm	ppm	ppm	ppm	ppm	ppm	ppm	ppm	ppm			
0.1	0.1	0.1	0.1	0.1	0.3	0.4	0.2	0.5	0.3	Clear	No	No

DEF ANALYSIS	Dump truck with Tier IV engine, emission related fault codes

FIGURE 14-10 DEF analysis report for a dump truck.

Conclusion

What we have here is a case of DEF dilution. Another sample from the inventory showed that the product complies with specifications. This leads to the conclusion that the DEF tank on the vehicle received water and not DEF fluid during service. Although this situation does not cause damage to the SCR, it produces fault codes, and the engine is not compliant with emissions standards.

5. Quality Check: Fleet Inventory

A large construction company wanted to check the quality of the DEF fluid it was buying in bulk quantities. The company was not having issues with the current product, but it wanted to switch to a cheaper product for cost-reduction purposes.

DEF Analysis

The DEF analysis results in Figure 14-11 show a product that is marginal in several parameters that are highlighted. To start with, the urea content is low; the biuret, although within acceptable levels, is at the upper limit. The same applies to aldehydes, and finally, the calcium content is at the upper limit.

PHYSICAL/CHEMICAL per ISO 22241-2 2019										INFORMATION		
Identity	Urea Content	Density	Refractive Index	Alkalinity as NH3	Biuret	Aldehydes	Insoluble Matter	Phosphates	Other	DEF Brand	Engine Hours/Miles	DEF Time in Storage
FTIR	%	Kg/L	ISO	%	%	ppm	ppm	ppm				Months
Match	30	1.088	1.3814	0.15	0.3	4	<0.1	0.01		ACME Blue	0	0

POISONING METALS										VISUAL APPEARANCE		
Fe	Cu	Zn	Cr	Ni	Al	Mg	K	Na	Ca	Clarity	Oil Layer	Sediment
ppm	ppm	ppm	ppm	ppm	ppm	ppm	ppm	ppm	ppm			
<0.1	<0.2	<0.1	<0.1	<0.1	0.2	0.2	0.1	<0.1	0.5	Clear	No	No

DEF ANALYSIS	Construction company. DEF Inventory quality check

FIGURE 14-11   DEF analysis report for a fleet.

Conclusion

The product submitted for analysis is compliant in all but one specification, which is urea content. This company's exercise allowed it to decide whether the product in question is worth buying or at least use these results to request a revision of the specifications with the supplier. There are going to be differences between suppliers, and checking specifications is a healthy practice. Even if the supplier guarantees its product and has lab results handy, it is always good to cross-check product specifications.

CONCLUSION

With this chapter, we have reached the end of the fluid analysis for mobile equipment theme, and we hope that readers find the information usable and pertinent to their needs. As in any technical material, the information may become less useful over the years, and we will try to keep this publication current as long as we have the energy and motivation to continue with this endeavor.

ACRONYMS

ACEA	European Auto Manufacturers Association
AOC	Ammonia oxidation catalyst
API	American Petroleum Institute
ASTM	American Society for Testing and Materials
AW	Antiwear
BTU	British thermal unit
CAN	Controller area network
C3	Ttransmission fluid specification by Allison
C4	Updated transmission fluid specification by Allison
$CaCl_2$	Calcium chloride
CAT	Caterpillar, a construction and mining equipment manufacturer
CBM	Condition-based maintenance
CE	API service for high-performance supercharged diesel engines
CF	API service category for indirect diesel fuel–injected engines (1994)
CF-2	API service category for two-stroke diesel engines (1994)
CF-4	API service category for four-stroke high-speed diesel engines (1990)
CFPP	Cold filter plugging point
CG-4	API service category for four-stroke high-speed diesel engines (1995) for use with off-highway diesel fuel with up to 5000 ppm sulfur
CH-4	API service category for four-stroke high-speed diesel engines (1998) to meet certain emissions standards
CI-4	API service category for four-stroke high-speed diesel engines (2002) to meet new emissions standards of 2004

CJ-4	API service category for four-stroke high-speed diesel engines (2007) to meet new emissions standards and diesel fuel with less than 500 ppm sulfur
CK-4	API service category for four-stroke high-speed diesel engines (2017) to meet new emissions standards and diesel fuel with less than 15 ppm sulfur
CO	Carbon monoxide
CO_2	Carbon dioxide
cP	Centipoise, unit of absolute viscosity
cSt	Centistoke, unit of kinematic viscosity
DEF	Diesel exhaust fluid
DEO	Diesel engine oil
DH-1	Japanese oil specification for heavy-duty diesel engines
DOC	Diesel oxidation catalyst
DPF	Diesel particulate filter
EGR	Exhaust gas recirculation
ELC	Extended-life coolant
EP	Extreme pressure
GC	Grease specification for bearing grease, covers GA, GB, GC, LA
GF6	International Lubricant Standards Approval Committee current oil classification for passenger car gasoline engines
GL4/5	API service characteristics for gear oil
HFRR	High-frequency reciprocating rig
HOAT	Hybrid organic acid technology
HPCR	High-pressure common rail
ICP	Inductivity couple plasma
ILSAC	International Lubricant Standards Approval Committee
IP	Institute of Petroleum (United Kingdom)
IR	Infrared
ISO	International Standards Organization
J20-C	John Deere specification for universal tractor fluid
J20-D	John Deere specification for winter universal tractor fluid
JASO	Japan Automotive Standards Organization
JD	John Deere, a farm and construction equipment company

LNG	Liquefied natural gas
LPG	Liquid petroleum gas
MERCON	Ford Motor Company trademark for ATF specification
NGEO	Natural gas engine oil
NH_3	Ammonia
NH_4OH	Hydroxide ammonia—amonium hydroxide
NO_2	Nitrite
NO_3	Nitrate
NOAT	Nitrite organic acid technology
NOx	Nitrogen oxides
OAT	Organic acid technology
OEM	Original equipment manufacturer
PAG	Polyalkene glycols
PM	Particulate matter
RA	Reserve alkalinity
RBOT	Rotating bomb oxidation test
RON	Research octane number
SAE	Society of Automotive Engineers
SCA	Supplemental coolant additive
SCR	Selective catalyst reduction
SJ	API service category for gasoline engines from 2001 and older
SL	API service category for gasoline engines from 2004 and older
SM	API service category for gasoline engines from 2010 and older
SN	API service category for gasoline engines from 2020 and older
SP	API service category for gasoline engines introduced in May 2020 to provide protection from low-speed preignition timing-chain wear protection
STLE	Society of Tribologists and Lubrication Engineers
TAN	Total acid number
TBN	Total base number
TCP	Tricresyl phosphate
TDS	Total dissolved solids
TO-4	Caterpillar specification for transmission and drivetrain fluid requirements

ULSD   Ultralow-sulfur diesel

UPCL   Ultrapressure liquid chromatography

ZDDP   Zinc dialkyl dithiophosphate

ZF   A German transmission manufacturer

MACHINE PROFILES AND APPLICATION

PROFILE	CLASS	APPLICATION	OPERATION	LOOK FOR
1	4WD Loader or articulated front end loader.	Loose material handling, truck loading, hoper feeding, feedlot material mixing	Constant directional changes, long idling periods, varied engine RPM	Engine oil dilution, transmission fluid viscosity loss, brake wear, differential contamination
2	Compact hydrostatic loader	Light material handling, truck loading, hoper feeding, chemical plant, feedlot material handling	Constant directional changes, varied engine RPM	Hydraulic/hydrostatic fluid and axle contamination
3	Backhoe loader	Trenching, loose material handling, pipe laying, truck loading, sewer work	As a backhoe, constant engine RPM. As a pipe layer long idling periods. As a loader, constant directional changes. Long hauling distances over road	As pipe layer loader look for engine oil dilution or soot. As a loader alone, look for transmission fluid viscosity loss and brake wear.
4	Skid steer loader	Tight quarters loader, loose material handling, multi use loader, farm application	Constant RPM medium load application	Look for hydraulic/hydrostatic fluid contamination
5	Hydrostatic bulldozer	Slot trenching, road building, short distance dirt push, scraper push, ripping ground, leveling	Constant use of power and spaced directional changes	Engine dirt contamination and final drive dirt check
6	Elevated track bulldozer	Slot trenching, road building, short distance dirt push, scraper push, ripping ground, leveling	Constant use of power and spaced directional changes	Engine dirt contamination. Transmission fluid viscosity check and metal generation. Final drive dirt check
7	Utility bulldozer (Hydrostatic)	Housing development, light dirt removal, leveling, tight area dozing	Medium power use and close turning with load	Engine dirt check and hydraulic/hydrostatic dirt and water check
8	Hydrostatic crawler loader	Slot trenching, digging, cutting, truck loading, trenching, ripping, hoper filling,	High power utilization, constant change of direction	Hydraulic/hydrostatic fluid physical properties and water and dirt check. Final drive contamination check.
9	Wheel bulldozer	Loose material spreading, dirt push, leveling	Continuous directional changes and medium power utilization. Low use of brakes	Axle and transmission fluid contamination and physical properties checks

427

10

| Trash compactor | Sanitary land fills, trash compaction, dirt dozing, spreading | Constant directional change, heavy wheels and heavy loads on axles | Look for wear in axles, dirt in engine |

11

| Construction Excavator | Excavation, material handling, pipe lying, truck loading, trench digging, dredging | Intermittent use of power, long idling periods, continuous house swing, limited propelling | Look for engine oil fuel dilution. If dredging look for hydraulic water content. Final drive dirt and water check. Check for hydraulic fluid mixing |

12

| Utility excavator | Housing development, utility services, trenching, pipe laying, ditch cleaning, sewer work | Intermittent use of power, long idling periods | Check for engine oil dilution and hydraulic fluid mixing and contamination |

13

| Wheel excavator | Excavation, material handling, pipe lying, truck loading, sewer | Intermittent use of power, long idling periods, continuous house swing, long distance travel | Look for engine oil fuel dilution. Check for hydraulic fluid mixing |

14

| Excavator with hydraulic hammer | Demolition, concrete/asphalt breaking, boulder splitting | Constant use of medium power, hydraulic pulsation in one pump, continuous intermittent propelling | Look for hydraulic dirt and metal contamination, check for hydraulic fluid mixing |

15

| Demolition excavator | Demolition, concrete breaking, construction steel cutting | Constant use of medium power to operate tool, short distance propelling | Look for hydraulic fluid mixing |

16

| Elevated cab scrap handler | Metal and scrap material handling | Intermittent use of full power, continuous house swing, limited propelling | Look for hydraulic metal contamination, check for hydraulic fluid mixing |

17

| Extended arm Excavator | Dredging, ditch cleaning | Constant RPM, full power utilization | Look for water/dirt in hydraulics and fluid mixing. |

18

| Hydrostatic Compactor | Surface and asphalt compaction | Low power utilization, constant directional changes, constant engine RPM | Look for hydraulic fluid mixing |

#		Machine	Application	Operating Conditions	Inspection
19		Motor grader	Precise ground cut, land leveling, high speed dirt or asphalt spreading, ditch cutting, road slope combing, ripping	Constant use of power, long directional changes, long roading	Look for transmission fluid physical properties
20		Self Propelled Scraper	Landfill, short distant dirt transport and spreading	Constant engine RPM and full power utilization	Look for engine dirt contamination and transmission and axle oil physical properties.
21		Tracked pull scraper (Hydrostatic)	Landfill, short distance dirt transport and spreading	Constant engine RPM and full power utilization	Look for cross contamination from scraper hydraulics, check for dirt in hydraulics, check for final drive dirt contamination
22		Quad track pull scraper	Landfill, short distance dirt transport and spreading	Constant engine RPM and full power utilization, Short travel distances	Look for cross contamination from scraper hydraulics, check for dirt in hydraulics. Look for axle wear. Check for transmission fluid physical properties and contamination
23		Articulated wheel pull scraper	Landfill, short distance dirt transport and spreading	Constant engine RPM and full power utilization. Short travel distances	Look for cross contamination from scraper hydraulics, check for dirt in hydraulics. Look for axle and brake wear. Check for transmission fluid physical properties and contamination
24		Trencher	Trenching	Constant RPM and load, continuous low speed propelling	Check for hydraulic/hydrostatic fluid contamination with dirt and final drive contamination
25		Grapple/Cable Skidder	Log pulling	Varied engine RPM, medium power utilization, short distance travel	Look for transmission metal generation and water and dirt in axles. Check for brake wear. Check for water in fuel
26		Wheel Tree harvester/process or/ delimbing	Tree harvesting and processing, delimbing and cut to length	Constant RPM and medium load	Check for hydraulic fluid contamination due to hose repairs in the field
27		Forwarder	Loading and carrying	Medium power utilization, variable engine speeds medium hauling distance	Check for bogie oil's contamination and wear. Hydraulic/Hydrostatic fluid condition.
28		Tracked tree harvester/ delimber with leveling cab	Tree harvesting and processing in side hill operation	Constant engine RPM and medium power utilization. Constant propelling. Constant swing	Check for hydraulic fluid contamination due to hose repairs in the field, check for final drive contamination and wear

29		Tracked feller buncher	Three harvesting	Constant engine RPM and medium power utilization. Constant propelling	Check for hydraulic fluid contamination due to hose repairs in the field, check for final drive contamination and wear
30		Hydrostatic Wheel Feller Buncher	Tree harvesting	Constant directional changes, medium power utilization, little brake utilization	Check for engine fuel dilution
31		4WD Log Loader	Carrying, loading, accommodating and piling logs	Constant directional changes, medium power utilization, high brake utilization, varied engine RPM	Check for engine fuel dilution, differential and brake wear
32		Dump Truck	Long distance loose material transport	Constant speed, medium power utilization, short idling periods	Check for engine fuel dilution
33		Articulated off-road dump truck	Off-road dirt/rock hauling	Medium distance hauling with heavy loads. Uphill and down hill constant use of retarder and converter lock up	Look for transmission fluid physical properties and metal generation. Check for differential dirt/water contamination
34		Mining Truck	Heavy off-road dirt/rock hauling	Medium distance hauling with heavy loads. Uphill and down hill constant use of retarder and converter lock up	Look for transmission fluid physical properties and metal generation. Check for differential wear
35		Mining Shovel	Open pit mining, high output production, off highway truck loading	Full hydraulic power utilization at constant engine RPM. Constant swing	Look for hydraulic fluid physical properties and contamination. Check for engines dirt contamination. Test grease from swing bearing
36		Farm tractor	Plowing, fertilizing, planting, pulling	Varied power utilization depending on the application. Short utilization during the year	Look for engine dirt contamination and fuel specifications and condition
37		Rubber track farm tractor	Heavy duty low ground pressure plowing, fertilizing, planting, pulling	Constant engine RPM, seasonal utilization	Look for engine dirt contamination and fuel specifications and condition
38		Articulated quad track farm tractor	Heavy duty low ground pressure plowing, fertilizing, planting, pulling	Constant engine RPM, seasonal utilization	Look for engine dirt contamination and fuel specifications and condition

39

Articulated wheel farm tractor	Heavy duty plowing, fertilizing, planting, pulling	Constant engine RPM, seasonal utilization	Look for engine dirt contamination and fuel specifications and condition

40

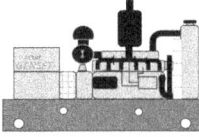

Genset	Stand-by or full utilization	Constant engine RPM and load	In stand-by use look for fuel condition and contamination and for coolant pH and additives.

INDEX

Absolute viscosity, 32
Absorption inhibitors, for coolant, 329
Acid number (AN). *See also* Total acid
    number
    of API Group I base oil group, 31
    for diesel fuel, 392
    lubricant testing for, 138
    oil breakdown and, 40
Additives. *See also specific types*
    for AW, 49, 52, 145
    in axles, 187
    for coolant, 327–328
    decline of, 59
    for diesel fuel, 374–377
    EP and, 49
    in final drive, 187
    for gasoline, 377–378
    in gear oils, 56
    in hydraulic fluid, 54
    load and, 47–48
    for lubrication, 44–53
    in oil analysis, 52, 178
    storage of, 68
    temperature and, 48
Advanced prognosis, in CBM, 6
Air
    filtration with, 96–98
    particle count and, 106
    powertrain oil analysis for, 276
Aldehydes, in DEF, 414, 415, 417
Alkyl naphthalene (AN), 30
Aluminum, 90, 91, 103. *See also* Dirt
    in axles, 291, 293, 294, 296, 297, 299, 301

in brakes, 291, 293, 294, 296, 297, 299, 301
cavitation and, 21
coolant analysis for, 8
coolant and, 348
in DEF, 414, 419
engine oil analysis for, 258
in generator set, 353
grease, 70, 79
hydraulic fluid analysis for, 235–236
lubricant testing for, 139
oil analysis for, 177
in powershift transmission, 278–279,
    282–284, 290–291
powertrain oil analysis for, 278–279,
    282–284, 290–291, 293, 294, 296, 297,
    299, 301
wear tables for, 208
American Petroleum Institute (API)
    base oil groups of, 29–41
    on copper, 256
    diesel fuel categories, 37–38
    gasoline engine oil classifications, 39–40
    gasoline standards of, 365
Ammonia oxidation catalyst (AOC), 372
AN. *See* Acid number; Alkyl naphthalene
Anodic inhibitors, for coolant, 329
Anti-icing agents, for diesel fuel, 376
Antioxidants, 46
Antiwear (AW)
    additives for, 49, 52, 145
    boundary lubrication for, 43
    grease, 71, 81
    grease testing for, 162

Antiwear (AW) (*continued*):
    hydraulic fluid, 36, 54, 106, 123, 140, 155,
        182, 211, 226
    lubrication for, 41–42
    lubricity and, 47
AOC. *See* Ammonia oxidation catalyst
API. *See* American Petroleum Institute
API gravity
    of diesel fuel, 390–391
    heavy knocking and, 404
Arrhenius rule, 40
Ash content, of diesel fuel, 37, 395
ASTM. *See* Automotive Service Greases
    Standard
ATFs. *See* Automatic transmission fluid
Attenuated total reflection (ATR), 145
Automatic lubricators, 83
Automatic transmission fluid (ATFs)
    for powershift transmission, 53
    TAN for, 155
    water in, 184
Automatic transmission fluids (ATFs),
    signature of, 55–56
Automotive Service Greases Standard
    (ASTM)
    coolant standards of, 324–325, 331
    diesel fuel standards of, 361
    for grease, 73–79
    on oil analysis, 167
    on standard deviation cleaning, 206
    viscosity tests, 140–143
    water lubricant testing, 143–144
Aviation
    CBM in, 15, 16
    coolant for, 321
    esters for, 30
    gasoline for, 365, 366
AW. *See* Antiwear
Axles
    additives in, 187
    aluminum in, 291, 293, 294, 296, 297,
        299, 301
    chromium in, 291, 293, 294, 296, 297,
        299, 301, 302, 304, 306, 308
    copper in, 291, 293, 294–296, 297, 299, 301
    crackle test for, 187
    dirt in, 187
    friction modifiers in, 187, 274
    ICPMS for, 190
    iron in, 291–293, 294, 296–298, 299, 301,
        302, 304, 306, 308
    lead in, 291, 293, 294–296, 297, 299, 301,
        302, 304, 306, 308
    lubrication for, 53
    oil analysis for, 186–190
    oil hours for, 190
    overheating of, 96
    oxidation in, 188
    powertrain oil analysis for, 273–274,
        291–308
    PQ index for, 190
    TAN in, 187, 305–306
    viscosity in, 188
    water in, 187, 293–294
    wear metals in, 188–189
    wear tables for, 291, 293, 294, 296, 297,
        299, 301, 302, 304, 306, 308

Bacon bomb sampler, 378
Bacteria
    in diesel fuel, 399
    filter plugging from, 406–407
Balance Charge Agglomeration (BCA), 95
Barium
    in gear oil additives, 56
    lubricant testing for, 139
    in oil analysis, 52
Base number. *See also* Total base number
    lubricant testing for, 138
Base oil groups, of API, 29–41
BCA. *See* Balance Charge Agglomeration
Benzene, toluene and ethylbenzene, and
    xylene (BETX), 378
Bergius, Friedrich, 360
Beta ratio/rating, for filtration, 59, 93–95,
    116, 127, 213
BETX. See Benzene, toluene and
    ethylbenzene, and xylene
Biodegradable hydraulic fluid, 229–230
Biodiesel, 361–364
    cold-flow additives for, 376–377
    emissions impact of, 363

energetic values of, 363, 371
engine oil analysis for, 255
FTIR for, 391
fuel analysis for, 9
at gas stations, 364
production of, 362–363
solvency of, 362
thickening of, 255
Biodiesel-petrodiesel blend, 369
Bittering agent, coolant and, 325
Biuret, in DEF, 414, 415
Bogie
chromium in, 316–317
iron in, 316–317
powertrain oil analysis for, 275, 316–317
Boiling point
of coolant, 325
in crude oil refining separation processes, 360
of gasoline, 364
GC and, 147
Boiling protection, from coolant, 322, 330
Boron
in additives in oil analysis, 52
in gear oil additives, 56
lubricant testing for, 139
Boundary lubrication, 41, 43
Brakes
aluminum in, 291, 293, 294, 296, 297, 299, 301
chromium in, 291, 293, 294, 296, 297, 299, 301
copper in, 291, 293, 294–296, 297, 299, 301
iron in, 291–293, 294, 296–298, 299, 301
lead in, 291, 293, 294–296, 297, 299, 301
powertrain oil analysis for, 273–274, 291–308
sediment on, 292
tractor fluids and, 56
water in, 293–294
wear metals for, 103
wear tables for, 291, 293, 294, 296, 297, 299, 301
Breather filters, 41
powertrain oil analysis for, 276

Bypass filter, 41
beta ratio for, 94
cleanliness with, 95, 107
in hydraulic fluid analysis, 225–226
for varnish, 127
for weak acids, 262

Calcium
in DEF, 414, 418
in gear oil additives, 56
in grease thickeners, 71
lubricant testing for, 139
in oil analysis, 52
TAN and, 59–60
Calcium sulfonate grease, 81
CAN. See Controller area network
Carbon, in petroleum, 28
Carbon monoxide (CO)
from biodiesel, 363
emission control of, 371–372
from ethanol, 369
Carbon residue
in diesel fuel, 396
in SAE J300, 32
Carburetors, 364
Catalysts
in biodiesel, 362
DOC, 372
for filtration, 98
for gasoline, 365
Catastrophic failure
from contaminants, 86, 89
hydraulic fluid analysis for, 237–239
Cathodic inhibitors, for coolant, 329
Cavitation
from air, 97
aluminum and, 21
from contaminants, 85
coolant and, 324, 331, 351–352
coolant leaks from, 174
copper etching and, 51
of filter caddy, 118
CBM. See Condition-based management
Centipoise (cP), of viscosity, 32
Centistokes (cST), of viscosity, 32
Cetane improvers, for diesel fuel, 375–376

Cetane index
 for diesel fuel, 390
 fuel analysis for, 9
 heavy knocking and, 404
Cetane number, for diesel fuel, 389
CFPP. *See* Cold filter plugging point
CFR. *See* Cooperative Fuel Research
CH-4 API diesel engine oil category, 38
Cheese curding, in biodegradable hydraulic
  fluid, 229–230
Chlorides
 in coolant, 8, 337–338, 340
 HPM+CR and, 77
Chromium
 in axles, 291, 293, 294, 296, 297, 299, 301,
  302, 304, 306, 308
 in bogie, 316–317
 in brakes, 291, 293, 294, 296, 297, 299, 301
 as contaminant, 103
 in DEF, 414
 lubricant testing for, 139
 oil analysis for, 177
 in powershift transmission, 281, 284, 287,
  288, 290, 291
 powertrain oil analysis for, 316–317
CI-4 API diesel engine oil category, 38
 engine oil signatures and, 57
CJ4 API diesel engine oil category, 37–38
 engine oil signatures and, 57
 TBN for, 176
CK4 API diesel engine oil category, 37–38
 engine oil signatures and, 57
 optimization of, 65
 TBN for, 176
Cleaning
 of hydraulic system, 113–128
 lubrication for, 27
 for oil sampling, 130
 standard deviation and, 205–207
Cleanliness
 with bypass filter, 95, 107
 component life and, 107–108
 of diesel fuel, 388
 filtration and, 107–108
 hydraulic fluid analysis and, 212
 particle count and, 102, 105, 106, 150–153

Cloud point (CP)
 for diesel fuel, 393
 fuel analysis for, 9
 fuel starvation in winter and, 404
CNG. *See* Compressed natural gas
CO. *See* Carbon monoxide
Coal, diesel fuel from, 358, 360–361
Cold filter plugging point (CFPP)
 for diesel fuel, 393
 fuel analysis for, 9
 fuel starvation in winter and, 404
Cold-flow additives
 for biodiesel, 376–377
 for diesel fuel, 376
Colloid, 47
Colorimetry, for grease testing, 163
Communication
 in CBM, 7, 22
 from lubrication, 28
Component tolerances, contaminants and,
  112
Compressed natural gas (CNG), 370–371
Compression loss, lubrication for, 28
Compression ratios, for gasoline, 368–369
Condition-based management (CBM)
 advanced prognosis in, 6
 basic principles of, 6–7
 communication in, 7, 22
 coolant analysis in, 8–9
 data management in, 20
 diagnostics in, 12–13
 failure modes and, 4, 12–13
 fault codes in, 13
 fluid sensors in, 19–20
 fuel analysis in, 9–10
 hydraulic fluid analysis and, 211
 infrared imaging for, 11
 inspections in, 16–17
 machine health correlations in, 20–22
 maintenance paradigms of, 4–5
 maintenance types in, 5–7
 in micro world, 10–25
 oil analysis in, 6, 7–8, 129, 165
 operating costs and, 15–16
 operator in, 17
 P-F curve in, 14–15

preventive maintenance and, 1, 5–6, 15
RCA in, 6, 22–25
telematics in, 6, 17–21
thought process of, 1
visibility in, 12
wear and, 2–4, 14–15
wear tables and, 204
Conductivity
of diesel fuel, 398
TDSs and, 336
Cone-penetration test, for grease, 161
Contaminants, 85–114. *See also* Filtration; specific types
additive decline and, 59
component tolerances and, 112
in coolant, 345–346
from failure modes, 125–126
fluid sensors for, 19–20
foam inhibitor and, 46
gravimetric measurement for, 105
in hydraulic fluid analysis, 211, 212–213, 241–242
in hydraulic system, 120–128
internally generated, 113–114
lubrication and, 27, 41
oil analysis for, 112, 176
particle count of, 97, 102, 105–107
particle size and visibility of, 85–92, 102, 112
portable oil labs for, 110–111
powertrain oil analysis for, 276
Rossin-Rammler particle distribution for, 90–92
silicon as, 90–92
solvency and, 31
source of, 86
water as, 121–122
wear from, 86
wear metals and, 103–104
Controller area network (CAN), 20
Conversion processes, in crude oil refining, 359–360
Coolant and coolant analysis, 321–355
absorption inhibitors for, 329
additives for, 327–328
aluminum and, 348

anodic inhibitors for, 329
boiling point of, 325
boiling protection from, 322, 330
case discussions on, 342–354
cathodic inhibitors for, 329
cavitation and, 324, 331, 351–352
in CBM, 8–9
chlorides in, 8, 337–338, 340
composition of, 325
contaminants in, 345–346
copper in, 346–347, 349–351
corrosion and, 353–354
corrosion inhibitor and, 322, 329
development of, 321–322
ELCs, 327–328
engine oil analysis for, 246–249
freeze point of, 327, 334, 341
for freeze protection, 330, 342–343
glycol concentration in, 335
glycol degradation in, 333
for heat transfer, 322, 323
heavy-duty, 327–328
ICPMS for, 337
iron in, 351–352
lab and field tests for, 333–338
lead in, 346–347
marine operation for, 323
for new technologies, 324
nitrite depletion with, 332
nitrite in, 335
on-site testing of, 338–342
pH of, 327, 332–333, 338–339, 344
phosphates in, 325, 327
production of, 322
RA in, 327, 334–335, 341
reports for, 341–342
requirements of, 322
sampling of, 325–326
silicate in, 325, 327
specific heat of, 324
standards for, 325–326, 331–332
sulfates in, 337–338, 340
TDSs in, 336, 339–340, 352–353
types of, 322
water in, 331, 337–338, 340
wear metals in, 337

Coolant leaks
    from cavitation, 174
    to generator set, 344
    glycol from, 173–175
Cooperative Fuel Research (CFR), 389
Copper
    in axles, 291, 293, 294–296, 297, 299, 301
    in brakes, 291, 293, 294–296, 297, 299, 301
    as contaminant, 92, 103, 123–124
    in coolant, 346–347, 349–351
    in coolant leak, 247
    in DEF, 414
    engine oil analysis for, 256, 259
    engine oil signatures and, 57
    etching, 51
    as filtration catalyst, 98
    in hydraulic fluid analysis, 219–220, 221–222
    in hydraulic system, 18, 123–124, 193–194
    lubricant testing for, 139
    oil analysis for, 7, 176, 177, 193–194
    passivators, 50
    in powershift transmission, 281–282, 284–285
    powertrain oil analysis for, 281–282, 284–285, 291, 293, 294–296, 297, 299, 301
    strip corrosion test, for diesel fuel, 395–396
Corrosion. *See also specific corrosives*
    contaminants and, 88
    coolant and, 324, 353–354
    from ethanol, 370
    grease for, 70
    ZDDP for, 47
Corrosion inhibitor/inhibition
    coolant and, 322, 325, 329
    copper passivators for, 50
    for gasoline, 378
    glycol contaminants and, 49
    lubrication as, 27
    in oil analysis, 52
    of SAE J300, 32
Coulometric titration, 137, 143, 154
    for hydraulic fluid analysis, 228
    in oil analysis, 181
cP. *See* Centipoise

CP. *See* Cloud point
CR. *See* Saltwater corrosion resistance
Crackle test, 144–145
    for axles, 187
    for final drive, 187
    for water, 100–101, 187
Crude oil, 28
    refining of, 358–360
cST. *See* Centistokes
Cumulative curve, standard deviation and, 201–202

DEF. *See* Diesel exhaust fluid
Demulsifiers, for gasoline, 378
Density, fluid sensors for, 19–20
Desiccant filters, 41, 95–96
    storage of, 67, 69
Detergents, 47
    for contaminant filtration, 41
    for gasoline, 377
    in oil analysis, 52
    soot from, 110
Determinism, in RCA, 23
Detonation (knocking)
    with gasoline, 366–367
    heavy, 403–404
Diagnostics, in CBM, 12–13
Dielectric, fluid sensors for, 19
Diesel, Rudolf Christian Karl, 357
Diesel exhaust fluid (DEF), 411–421
    aldehydes in, 414, 415, 417
    biuret in, 414, 415
    fault codes for, 416–417, 419–420
    field testing of, 413–414
    lab testing of, 414
    quality check for, 420–421
    water and, 412–413
Diesel fuel, 28. *See also* Biodiesel
    AN for, 392
    additives for, 374–377
    anti-icing agents for, 376
    API gravity of, 390–391
    ash content of, 37, 395
    bacteria in, 399
    carbon residue test for, 396
    cetane improvers for, 375–376

cetane index for, 390
cetane number for, 389
CFPP for, 393
cleanliness of, 388
from coal, 358, 360–361
cold-flow additives for, 376
compression ratio for, 368
conductivity of, 398
copper strip corrosion test for, 395–396
CP for, 393
distillation of, 388
energetic values of, 371
filtration of, 388
flash point of, 386
with gasoline, 404
grades of, 361
high-temperature stability of, 397–398
injection pump failure and, 408
lacquer with, 110
long-term storage of, 397
lubricity of, 396–397
natural gas and, 370–371
particulate matter in, 388
from petroleum, 358–360
pour point for, 394
sediment in, 386
specific gravity of, 358, 390–391
sulfur in, 374, 392
TAN for, 392
TBN for, 60, 154
testing of, 384–399
ULSD, 158, 176
viscosity of, 192–193, 361, 394
water in, 386–387
wax antisettling agents for, 376, 377
Diesel oxidation catalyst (DOC), 372
Diesel particulate filter (DPF), 372
sulfur and, 374
Diesters, in Group V API base oil group, 30
Differential
LS, 51, 187
PQ index for, 159
viscosity tests for, 140
Dirt
in axles, 187
as contaminant, 88–112, 120

in final drive, 187
in gearbox, 184
in hydraulic fluid analysis, 213–214, 230–231
in hydraulic system, 120, 181
in hydraulic system analysis, 211
oil analysis for, 171–173, 181, 184
in powershift transmission, 184, 279, 287–288
powertrain oil analysis for, 276, 279, 287–288, 311–312
in pump drives, 311–312
in splitter drive, 311–312
Discovery, in RCA, 23
Dispersants, 47
for soot, 50, 110
Distillation
of diesel fuel, 388
heavy knocking and, 404
Distillation point, fuel analysis for, 9
DOC. See Diesel oxidation catalyst
DPF. See Diesel particulate filter
Dropping point, grease testing for, 160–161

EGR. See Exhaust gas recirculator
Elastohydrodynamic lubrication, 44
Elastomers, biodiesel and, 362
ELCs. See Extended-life coolant
Engine oil
dilution of, 146–148
dirt in, 171
lubricant testing of, 141–142
optimization of, 65
particle count for, 108
signatures of, 57
TAN for, 155
viscosity of, 141–142, 148
Engine oil analysis, 245–270
for aluminum, 258
for biodiesel, 255
for coolant analysis, 246–249
for copper, 256, 259
for extended service intervals, 265–270
for forensics, 263–264
for fuel dilution, 253–255, 262
for iron, 259
for oxidation, 259–260

Engine oil analysis (*continued*)
    for soot, 250–252, 262
    for sulfation, 259–260
    for sulfur, 260–261
    for TAN, 256–257
    for TBN, 256
    for viscosity, 252–253
    for water, 250
    for weak acids, 262–263
EP. *See* Extreme pressure
Esters
    grease, 70
    in Group V API base oil group, 30
Ethanol, 369–371
    corrosion from, 370
    disclaimers for, 370
    energetic values of, 371
    octane rating for, 370
Ether polyoils, in Group V API base oil
    group, 30
Ethylene glycol, 321, 322
    specific heat of, 324
Excel, for standard deviation, 205–210
Exhaust gas recirculator (EGR), 372
    coolant leak through, 248–249
    for diesel fuel exhaust, 411
    sulfur and, 374
Extended service intervals, engine oil
    analysis for, 265–270
Extended-life coolant (ELCs), 327–328
    guidelines for, 340–341
Extreme pressure (EP)
    additives and, 49
    grease, 71, 79, 81
    grease testing for, 162
    lubrication for, 41–42
Extrusion test, for grease testing, 163

FA4 API diesel engine oil category, 37–38
Failure modes
    CBM and, 4, 12–13
    contaminants from, 125–126
    diagnostics for, 12–13
    in hydraulic system, 125–126
    powertrain oil analysis for, 277
    RCA and, 22–25

Fault codes
    in CBM, 13
    for DEF, 416–417, 419–420
FID. *See* Flame ionization detector
FIFO. See First in, first out
Filter caddy
    for hydraulic system, 116–119, 121
    storage of, 67, 69–70
Filter plugging, 406–407
Filtration, 92–114. *See also specific types*
    with air, 96–98
    beta ratio/rating for, 59, 93–95, 116, 127,
        213
    catalysts for, 98
    cleanliness and, 107–108
    of diesel fuel, 388
    of fuel, 373
    for gearbox, 318–319
    with heat, 96–97
    hydraulic fluid analysis and, 212
    inspections of, 98
    of lubrication, 41
    powertrain oil analysis for, 318–319
    with rare earth magnets, 96
    with static current, 99
    of water, 99–102
Final drive
    additives in, 187
    crackle test for, 187
    dirt in, 187
    friction modifiers in, 187
    ICPMS for, 190
    iron in, 309–311
    lubrication for, 53
    oil analysis for, 186–190
    oil hours for, 190
    oxidation in, 188
    powertrain oil analysis for, 274–275, 298–302
    PQ index for, 159, 190
    TAN in, 187
    viscosity in, 188
    water in, 187
    wear metals in, 103, 188–189
    wear tables for, 302, 304
First in, first out (FIFO), for lubricant
    storage, 67, 68

Fischer, Franz, 360–361

Fischer, Karl, 100–101, 121, 137, 143, 181, 228

Fishbone diagram (Ishikawa cause-and-effect diagram), for RCA, 24–25

Flame ionization detector (FID), 147

Flash point
    of diesel fuel, 386
    of SAE J300, 32
    Setaflash for, 147–148

Fluid mixtures
    of gear oils, 67
    in hydraulic fluid analysis, 226–227
    in hydraulic system, 123, 181
    oil analysis for, 181
    particle count and, 106
    in powershift transmission, 285–287
    powertrain oil analysis for, 285–287

Fluid sensors, in CBM, 19–20

Foam inhibitor, 46
    contaminants from, 91
    coolant and, 325
    oil analysis for, 7, 52
    in powershift transmission, 279, 290
    silicon in, 171

Foaming, in hydraulic fluid analysis, 226–227

Forensics, engine oil analysis for, 263–264

Formaldehyde, from ethanol, 369

Fourier transform infrared (FTIR), 100–101
    for biodiesel, 391
    for DEF, 414
    for glycol, 146, 173
    for grease testing, 163
    for hydraulic fluid analysis, 228
    for lubricant testing, 137, 144, 146
    for nitration, 157
    for oil analysis, 181
    for oxidation, 155–157
    for water, 144

Frauler, Hebert, 46–47

Freeze point, of coolant, 327, 334, 341

Freeze protection, coolant for, 330, 342–343

Friction modifiers, 51
    in ATFs, 55
    in axles, 187, 274
    in final drive, 187
    for gasoline, 377

FTIR. *See* Fourier transform infrared

Fuel and fuel analysis, 357–410. *See also* Diesel fuel; Ethanol; Gasoline
    in CBM, 9–10
    filter plugging, 406–407
    filtration of, 373
    fuel starvation in winter, 404–405
    hard cold starting and, 401–402
    heavy knocking, 403–404
    injection pump failure, 408–409
    injector failures, 402–403
    quality check, 409–410
    sulfur in, 374
    turbocharger failure, 405–406

Fuel dilution
    engine oil analysis for, 253–255, 262
    idle time and, 19

Fuel farms
    long-term storage at, 382
    sampling, 378–379
    tank maintenance, 380

Fuel oils, 361

Fuel pump, 372
    failure, 399–400

Fuel sampling, 378–384
    bottles for, 384
    SIF for, 383–384
    water in, 381–383

Fuel starvation in winter, 404–405

Gas chromatography (GC), 147
    in powershift transmission, 279, 281, 282, 284

Gasohol, 369
    energetic values of, 371
    natural gas and, 370–371

Gasoline, 28, 358
    additives for, 377–378
    for aviation, 365, 366
    boiling point of, 364
    carburetors for, 364
    compression ratios for, 368–369
    corrosion inhibitor for, 378
    demulsifiers for, 378
    detergents for, 377
    detonation with, 366–367

Gasoline (*continued*)
    diesel fuel with, 404
    energetic values of, 371
    with ethanol, 369–370
    friction modifiers for, 377
    HCs in, 365
    lead in, 365
    octane boosters for, 378
    octane rating for, 368
    preignition with, 367–368
    solvents for, 378
    top tier, 377
Gasoline engines
    oil classifications, 39–40
    varnish in, 110
GC. *See* Gas chromatography
Gear lube, 171
    oxidation of, 188
    viscosity of, SAE J300 standard for,
        35–36
Gear oils
    LS, 57–60, 67
    lubricant testing of, 142
    optimization of, 67
    phosphorus in, 187
    powertrain oil analysis for, 306–308
    signatures of, 56–57
    TAN for, 155
    viscosity in, 142
    viscosity shear-down of, 59
Gearbox
    filtration for, 318–319
    iron in, 313–314
    lubrication for, 53
    oil analysis for, 183–186
    powertrain oil analysis for, 275, 313–314,
        317–319
    viscosity tests for, 140
    wear metals in, 184, 317–318
    wear tables for, 314, 320
Generator set
    aluminum in, 353
    coolant leaks to, 344
Glycerin, in biodiesel, 362
Glycol. *See also Coolant*
    concentration of, in coolant, 335
    from coolant leaks, 173–175
    corrosion inhibitor and, 49
    degradation of, in coolant, 333
    fluid sensors for, 19
    grease, 70
    lubricant testing for, 137, 138, 146
    oil analysis for, 173–175
    specific heat of, 324
Glycolates, 333
Gravimetric measurement, for contaminants,
    105
Grease
    aluminum, 70, 79
    ASTM for, 73–79
    AW, 81
    classification of, 71
    colors of, 72–73
    compatibility of, 80–81
    CR, 77
    EP, 79, 81
    high load (HL), 78, 82
    HPM, 74–79
    LT, 78
    selection of, 79
    sodium, 79
    temperature and, 71–72
    thickeners for, 71, 80–81
    WR, 77
Grease testing, 159–164
    for AW, 162
    cone-penetration test, 161
    for dropping point, 160–161
    for EP, 162
    oil-separation test for, 161
Grease thief testing, 163
Group I API base oil group, 29
    antioxidants for, 46
    as hydraulic fluid, 54
    solvency of, 32
Group II API base oil group (synthetic
    blends), 29, 30
    antioxidants for, 46
    solvency of, 32
Group III API base oil group, 29
    antioxidants for, 46
    as hydraulic fluid, 54

Group IV API base oil group (synthetic oils), 30
   antioxidants for, 46
   for gasoline engine oil, 39
   as MV hydraulic fluid, 55
Group V API base oil group, 30–31

Hard cold starting, 401–402
HCs. *See* Hydrocarbons
Heat
   filtration with, 96–97
   powertrain oil analysis for, 277
Heat transfer
   coolant for, 322, 323
   lubrication for, 27
Heavy-duty coolant, 327–328
Herztian force, 44
HFRR. *See* High-frequency reciprocating-rig
High load (HL) grease, 78, 82
High-density crudes, 358
High-frequency reciprocating-rig (HFRR), 373
   lubricity test, 396–397, 408
High-performance multiuse (HPM), grease, 74–79
High-pressure injection systems (HPCR), 372–373
High-temperature, high-shear test (HTHS), 34
   FA4 API diesel fuel category, 37
High-temperature stability, of diesel fuel, 385, 397–398
HL. *See* High load
HOAT. *See* Hybrid-organic acid technology
HPCR. *See* High-pressure injection systems
HPM. *See* High-performance multiuse
HTHS. *See* High-temperature, high-shear test
Hybrid-organic acid technology (HOAT) coolant, 327
   on-site testing of, 338
Hydraulic fluid
   AW, 36, 54, 106, 123, 140, 155, 182, 211, 226
   biodegradable, 229–230
   blackening of, 127–128
   degradation of, 182

desiccant filters for, 95
dirt in, 171
flushing of, 67, 121
in hydraulic system cleaning, 119
lubricant testing of, 140–141
MV, 55
optimization of, 65
particle count in, 105, 106, 123–124
R&O, 36, 140
seal-swell agents for, 31
signature of, 53–55
TAN for, 154–155
TCP for, 47
viscosity of, 140–141
viscosity shear-down of, 58
Hydraulic fluid analysis, 211–244
   for aluminum, 235–236
   for biodegradable hydraulic fluid, 229–230
   bypass filter in, 225–226
   for catastrophic failure, 237–239
   CBM and, 211
   cleanliness and, 212
   contaminants in, 211, 212–213, 241–242
   copper in, 219–220, 221–222
   dirt in, 213–214, 230–231
   filtration and, 212
   fluid mixtures in, 226–227
   foaming in, 226–227
   for hydraulic hammer, 230–231
   iron in, 220, 222–223
   for microdieseling, 234–235
   oxidation in, 223–224
   particle count in, 213–214, 218–219, 228–229
   silt and, 215–216
   temperature in, 224–225
   for titanium, 235–236
   of tractor fluid, 219–220, 226–227, 239–241
   varnish in, 232–233
   for viscosity, 236–237
   water in, 216–217, 242–244
   wear metals in, 214, 217, 221, 222
   wear tables for, 213, 226, 228, 236, 241
Hydraulic hammer, 230–231

Hydraulic system and hydraulic system
    analysis, 113–128
    contaminants in, 120–128
    copper in, 18, 193–194
    dirt in, 120, 181, 211
    failure modes in, 125–126
    filter caddy for, 116–119, 121
    fluid mixtures in, 123, 181
    hydraulic fluid blackening in, 127–128
    oil analysis for, 7–8, 123–124, 179–182,
        193–194
    overheating of, 96
    oxidation in, 122
    particle count in, 123–124, 181
    TAN in, 122, 182
    varnish in, 126–127
    water in, 121–122, 181, 211
    wear metals in, 181
Hydrocarbons (HCs)
    from biodiesel, 363
    breakdown of, 40
    in diesel fuel exhaust, 411
    in gasoline, 365
    in Group IV API base oil group, 30
    in petroleum, 28–29, 358
Hydrodynamic lubrication, 41, 42–43
Hydrogen, in petroleum, 28

ICPMS. See Inductively coupled plasma mass
    spectrometry
ICRISAT. See International Crops Research
    Institute for Semi-Arid Tropics
Idle time, 19
Inductively coupled plasma mass
    spectrometry (ICPMS), 102
    for axles, 190
    for coolant, 337
    for final drive, 190
    for lubricant testing, 138–139, 158–159
    for PQ index, 158–159, 190
Infrared imaging, for CBM, 11
Injection pump
    engine oil analysis and, 257
    failure, 408–409
Injector, 11
    failures, 402–403

fuel dilution and, 147
idle time and, 19
Inspections
    in CBM, 16–17
    of filtration, 98
Intelligent filter caddy. See Filter caddy
International Crops Research Institute for
    Semi-Arid Tropics (ICRISAT), 369
International Standards Organization (ISO)
    hydraulic fluid viscosity grades of, 54, 55
    on particle cleanliness, 150–152
    on particle count, 106, 111
    tractor fluid grades of, 56
    viscosity grades of, 36–37
Iron
    in axles, 291–293, 294, 296–298, 299,
        301, 302, 304, 306, 308
    in bogie, 316–317
    in brakes, 291–293, 294, 296–298, 299, 301
    as contaminant, 92, 103
    in coolant, 351–352
    in DEF, 414
    engine oil analysis for, 259
    as filtration catalyst, 98
    in final drive, 309–311
    in gearbox, 313–314
    in hydraulic fluid analysis, 220, 222–223
    lubricant testing for, 139
    in powershift transmission, 279, 281, 282,
        284–285, 287, 288, 290, 291
    powertrain oil analysis for, 284–285,
        291–293, 296–298, 309–311,
        313–314, 316–317
Ishikawa cause-and-effect diagram (fishbone
    diagram), 24–25
ISO. v International Standards Organization

Kerosene, 28
Kinematic viscosity, 32
Knocking (detonation)
    with gasoline, 366–367
    heavy, 403–404

LAC. See Lowest additive concentration
Lacquer, with diesel fuel, 110
LaserNet Fines, 149

Lead
    in axles, 291, 293, 294–296, 297, 299, 301,
        302, 304, 306, 308
    in brakes, 291, 293, 294–296, 297, 299, 301
    as contaminant, 104, 123
    in coolant, 346–347
    in gasoline, 365
    in hydraulic system, 123
    in powershift transmission, 279, 281, 282,
        284, 287, 288, 290, 291
    powertrain oil analysis for, 291, 293,
        294–296, 297, 299, 301, 302, 304, 306,
        308
Light crudes, 358
Light no. 1 diesel, 361
Limited-slip (LS)
    decline of, 59
    differential, 51, 187
    gear oils, 57–60, 67
    powertrain oil analysis for, 277
    TAN and, 59–60
Liquefied natural gas (LNG), 370
Liquefied petroleum gas (LPG, propane),
        358, 370
    energetic values of, 371
Lithium grease, 70, 79
LNG. See Liquefied natural gas
Load
    additives and, 47–48
    HL grease, 78, 82
Logic tree, for RCA, 24
Low temperature (LT) grease, 78
Lowest additive concentration (LAC), 377
Low-sulfur diesel (LSD), 374
LPG. See Liquefied petroleum gas
LS. See Limited-slip
LSD. See Low-sulfur diesel
LT. See Low temperature
Lubricant/lubrication, 27–83
    additives for, 44–53
    API base oil groups, 29–41
    API diesel engine oil categories, 37–38
    automatic lubricators for, 83
    for AW, 41–42
    compatibility of, 61–62
    contaminants in, 41

    degradation of, 89
    for EP, 41–42
    filtration of, 41
    functions of, 27–28
    gasoline engine oil classifications, 39–40
    grease, 70–74
    Hertzian force and, 44
    ISO viscosity grades, 36–37
    oil breakdown, 40–41
    optimization of, 63–67
    oxidation of, 59–60
    petroleum for, 18–19
    physical properties of, 59–60
    SAE J300 standard for, 32–36
    shelf life of, 67–68
    signatures of, 53–67
    storage of, 67–70
    TAN of, 59–60
    TBN of, 59–60
    transportation of, 63
    tribology for, 27
    types of, 41–44
Lubricant testing, 137–164
    for AN, 138
    for aluminum, 139
    for barium, 139
    for base number, 138
    for boron, 139
    for calcium, 139
    for chromium, 139
    for copper, 139
    of engine oil, 141–142
    FTIR for, 137, 144, 146
    of gear oils, 142
    for glycol, 137, 138, 146
    for grease, 159–164
    of hydraulic fluid, 140–141
    ICPMS for, 138–139
    for iron, 139
    for magnesium, 139
    for molybdenum, 139
    for nickel, 139
    for nitration, 157
    for oxidation, 138, 155–157
    for particle count, 137, 138
    particle count counters for, 148–153

Lubricant testing (*continued*)
    for phosphorus, 139
    for potassium, 137, 139, 146
    for PQ index, 138, 158–159
    for silicon, 139
    for silver, 139
    for sodium, 137
    for soot, 138, 145
    for sulfation, 157–158
    TAN and, 137, 154–155
    TBN and, 137, 153–154
    for tin, 139
    for titanium, 139
    of tractor fluid, 143
    for vanadium, 139
    VI and, 140
    for viscosity, 137, 138, 140–143
    for water, 137, 138, 143–145
    for wear metals, 137, 138–139
    for zinc, 139
Lubricity
    AW and, 47
    of diesel fuel, 396–397

Machine fuel
    sampling, 379–380
    tank maintenance, 383
Magnesium
    in DEF, 414, 418
    in gear oil additives, 56
    lubricant testing for, 139
    in oil analysis, 52
Mean, standard deviation and, 203
Median, standard deviation and, 203
Metals. *See* Wear metals; specific metals
Methanol, 362
    energetic values of, 371
Methyl tert-butyl ether (MTBE), 378
Micelles, 47
Microdieseling
    from air, 97
    hydraulic fluid analysis for, 234–235
Middle no. 2 diesel, 361
Molybdenum
    in additives, in oil analysis, 52
    in coolant, 327

    in gear oil additives, 56
    in grease, 82
    lubricant testing for, 139
MTBE. *See* Methyl tert-butyl ether
Multigrade oils
    breakdown of, 41
    HTHS for, 34
    viscosity of, 41
Multipass testing, for filtration beta ratios,
    93
Multi-viscosity fluids (MV), 55

Naphthenic crude oil, 28, 358
Naphthenic grease, 70
Naphthenic oils, 30, 31
National Aerospace Standard (NAS), for
    particle cleanliness, 152
National Lubricating Grease Institute
    (NLGI), 71
    on cone-penetration test, 161
Natural gas, 370–371
    Group IV API base oil group from, 30
Nickel
    as contaminant, 104
    in DEF, 414
    lubricant testing for, 139
    oil analysis for, 178
    wear tables for, 208
Nitration
    lubricant testing for, 157
    oil analysis for, 179
Nitrite-organic acid technology (NOAT)
    coolant, 327
    SCAs for, 333
    testing of, 339
Nitrites, coolant and, 332, 335
Nitrogen
    in light crude, 358
    in petroleum, 28
Nitrogen oxides
    from biodiesel, 363
    in diesel fuel exhaust, 411, 412
    emission control of, 371–372
NLGI. *See* National Lubricating Grease
    Institute
NOAT. *See* Nitrite-organic acid technology

Nonparametric curve, standard deviation and, 202

OAT. *See* Organic acid technology
Octane boosters, for gasoline, 378
Octane rating
    for ethanol, 370
    for gasoline, 368
    for LPG, 370
OEM. *See* Original equipment manufacturer
Oil analysis, 165–195. *See also* Engine oil analysis; Hydraulic fluid analysis; Powertrain oil analysis
    for additives, 52, 178
    for aluminum, 177
    AW additives in, 52
    for axles, 186–190
    in CBM, 6, 7–8, 129, 165
    for chromium, 177
    for contaminants, 112, 176
    for copper, 7, 176, 177, 193–194
    for diesel fuel viscosity, 192–193
    for dirt, 171–173, 181, 184
    for final drive, 186–190
    for fluid mixtures, 181
    for foam inhibitor, 7
    formats for, 167–168
    for gearbox, 183–186
    for glycol, 173–175
    horizontal versus vertical display for, 168–169
    for hydraulic system, 7–8, 123–124, 179–182, 193–194
    interpretation of, 192
    lab comments for, 179
    maintenance applications from, 194–195
    measurement in, 166
    need for, 166
    for nickel, 178
    for nitration, 179
    for oxidation, 184
    for particle count, 181, 185
    for powershift transmission, 183–186
    for PQ index, 185
    required information in, 169
    for silicon, 7

    for silver, 178
    for soot, 175
    for sulfation, 179
    for TAN, 176, 182, 184
    for TBN, 176
    for temperature, 182
    for tin, 178
    for vanadium, 178
    for viscosity, 175, 184
    for viscosity shear-down, 182
    for water, 7, 181, 184
    for wear metals, 176–177, 181, 184
Oil breakdown, 40–41
Oil hours, for axles and final drive, 190
Oil sampling, 129–136
    bottle types for, 133
    cleaning for, 130
    methods of, 130–131
    sampling points in, 134–135
    sampling valve method for, 131–132
    submission of samples, 133–134
    vacuum pump method for, 132–133
    valve dead ends in, 135–136
    valve installation for, 135
Oil-separation test, for grease testing, 161
Operating costs, CBM and, 15–16
Operator, in CBM, 17
Order, in RCA, 23
Organic acid technology (OAT) coolant, 327
    on-site testing of, 338
    pH of, 344–345
Organic clay grease, 79
Original equipment manufacturer (OEM)
    for coolant, 331
    hydraulic fluid for, 54
    oil analysis and, 179
Oxidation
    in axles, 188
    engine oil analysis for, 259–260
    by filtration catalysts, 98
    in final drive, 188
    of gear lube, 188
    of gearbox, 184
    in hydraulic fluid analysis, 223–224
    in hydraulic system, 122
    of lubricant, 59–60

Oxidation (*continued*)
    lubricant testing for, 138, 155–157
    oil analysis for, 179, 184
    oil breakdown and, 40
    of powershift transmission, 184
    powertrain oil analysis for, 277
    R&O hydraulic fluid, 36, 140
    of sulfur, 374
    ZDDP for, 47
Oxygen
    in biodiesel, 361
    in petroleum, 28

PAGs. *See* Polyalkylene glycols
PAOs. *See* Polyalphaolefins
Paraffinic crude oil, 28, 358
Paraffins
    grease, 70
    in Group II API base oils, 29
Parametric bell curve, for standard deviation, 200–202
Particle count
    cleanliness and, 102, 105, 106, 150–153
    of contaminants, 97, 102, 105–107
    for engine oil, 108
    for gearbox, 185
    in hydraulic fluid, 123–124
    in hydraulic fluid analysis, 213–214, 218–219, 228–229
    in hydraulic system, 123–124, 181
    ISO, 106, 111
    lubricant testing for, 137, 138
    oil analysis for, 181, 185
    for powershift transmission, 185
    wear tables for, 210
Particle quantification index (PQ index)
    for axles, 190
    for final drive, 190
    for gearbox, 185
    lubricant testing for, 138, 158–159
    oil analysis for, 185
    for powershift transmission, 185
    wear tables for, 210
Particulate matter (PM)
    from biodiesel, 363
    in diesel fuel, 388
    in diesel fuel exhaust, 411, 412
    filter plugging from, 406
Performance-failure curve (P-F curve), in CBM, 14–15
Petroleum, 18–19
    diesel fuel from, 358–360
    Group IV API base oil group from, 30
P-F curve. *See* Performance-failure curve
pH
    of coolant, 327, 332–333, 338–339, 344
    coolant analysis of, 8
    of OAT coolant, 344–345
Phosphate esters, in Group V API base oil group, 30
Phosphates
    in coolant, 325, 327
    in DEF, 414
Phosphorus
    additive decline and, 59
    in additives in oil analysis, 52
    EP additives for, 49
    in gasoline engine oil, 39
    in gear oils, 57–58, 187
    lubricant testing for, 139
PM. *See* Particulate matter; Preventive maintenance
Polyalkylene glycols (PAGs)
    in Group V API base oil group, 30–31
    as hydraulic fluid, 54
Polyalphaolefins (PAOs)
    grease, 70
    in Group IV API base oil group, 30
Polyol esters, in Group V API base oil group, 30
Polyurea grease, 79
Population standard deviation, 198
Pore blockage counter, for particle count, 150
Portable oil labs, for contaminants, 110–111
Potassium
    as contaminant, 89
    in coolant, 327
    in coolant leak, 174
    in DEF, 414
    lubricant testing for, 137, 139, 146
    wear tables for, 209
Potassium hydroxide, 321

Pour point
  for biodiesel, 376
  for diesel fuel, 394
  for lubricants, 53
  for MV, 55
  for SAE J300, 32
Powershift transmission
  aluminum in, 278–279, 282–284, 290–291
  ATFs for, 53
  chromium in, 281, 284, 287, 288, 290, 291
  copper in, 281–282, 284–285
  dirt in, 287–288
  fluid mixtures in, 285–287
  iron in, 279, 281, 282, 284–285
  oil analysis for, 183–186
  powertrain oil analysis for, 272, 277–291
  PQ index for, 159
  silicon in, 279–281, 290–291
  viscosity in, 184, 288–291
  water in, 287–288
  wear tables for, 278, 279, 280–281, 282,
    284, 285, 287, 288, 290, 291
Powertrain oil analysis, 271–320
  for air, 276
  for aluminum, 278–279, 282–284,
    290–291, 293, 294, 296, 297, 299, 301
  for axles, 273–274, 291–308
  for bogie, 275, 316–317
  for brakes, 273–274, 291–308
  for breather filters, 276
  for chromium, 316–317
  for contaminants, 276
  for copper, 281–282, 284–285, 291, 293,
    294–296, 297, 299, 301
  for dirt, 276, 279, 287–288, 311–312
  for failure modes, 277
  for filtration, 318–319
  for final drive, 274–275, 298–302
  for fluid mixtures, 285–287
  for gear oils, 306–308
  for gearbox, 275, 313–314, 317–319
  for heat, 277
  for iron, 284–285, 291–293, 296–298,
    309–311, 313–314, 316–317
  for lead, 291, 293, 294–296, 297, 299,
    301, 302, 304, 306, 308

  for LS additives, 277
  for oxidation, 277
  for powershift transmission, 272, 277–291
  for pump drives, 275, 311–313
  for silicon, 279–281, 282, 284, 287, 288,
    290–291
  for splitter drive, 275, 311–313
  for TAN, 277, 305–306
  for tandem drive, 275, 314–315
  for torque converters, 275, 277
  for viscosity, 288–291
  for water, 276, 277, 287–288, 293–294,
    314–315
  for wear metals, 276–277, 317–318
  wear tables for, 278, 279, 280–281, 282,
    285, 287, 288, 290, 291, 293, 294, 296,
    297, 299, 301, 302, 304, 306, 308, 313
PQ index. See Particle quantification index
Preignition, with gasoline, 367–368
Preventive maintenance (PM, scheduled
    maintenance)
  CBM and, 1, 5–6, 15
  flaws of, 5–6
Projectiles, for cleaning hydraulic system, 126
Propane. See Liquefied petroleum gas
Propylene glycol, 322
Pump drives
  dirt in, 311–312
  powertrain oil analysis for, 275, 311–313
  wear tables for, 313

Quality check
  for DEF, 420–421
  for fuel, 409–410

RA. See Reserve alkalinity
Rabinowicz, Ernest, 2
RAF. See Repair-after-failure
Rare earth magnets, 96
RCA. See Root-cause analyses
RDE spectrometer. See Rotating disk
    emission spectrometer
Reactive-after-failure. See Repair-after-failure
Refining, of crude oil, 358–360
Remaining Useful Life Evaluation Routine
    (RULER), 163, 164

Repair-after-failure (RAF)
  flaws of, 5–6
  P-F curve and, 14–15
Reserve alkalinity (RA), in coolant, 327,
    334–335, 341
Rheometer test, for grease testing, 164
R&O. *See* Rust and oxidation inhibited
Root-cause analyses (RCA)
  in CBM, 6, 22–25
  determinism in, 23
  discovery in, 23
  failure modes and, 22–25
  Ishikawa cause-and-effect diagram for,
    24–25
  logic tree for, 24
  order in, 23
Rossin-Rammler particle distribution, 90–92
Rotating disk emission spectrometer (RDE
    spectrometer), for lubricant testing,
    138–139
RULER. *See* Remaining Useful Life
    Evaluation Routine
Rust and oxidation inhibited (R&O)
    hydraulic fluid, 36, 140
Rust inhibitors, 49

SAE. *See* Society of Automotive Engineers
SAE 0W-16, 39
SAE J300 standard, 32–36
Saltwater corrosion resistance (CR), grease, 77
Sample information form (SIF), for fuel
    sampling, 383–384
Sample standard deviation, 198–199
Sampling points, in oil sampling, 134–135
Sampling valve method, for oil sampling,
    131–132
Scale inhibitor, coolant and, 325
SCAs. *See* Supplemental coolant additives
Scheduled maintenance. *See* Preventive
    maintenance
SCR. *See* Selective catalytic reducer
Seal-swell agents, for hydraulic fluid, 31
Sediment, 109, 276
  on brakes, 292
  in coolant, 338
  in diesel fuel, 386

Selective catalytic reducer (SCR), 411–412
  failure of, 418–419
Separation processes, in crude oil refining,
    359
Setaflash, 147
Shimadzu ATR method, 145
Shukhov cracking process, for gasoline, 364
SIF. *See* Sample information form
Silicate
  in coolant, 325, 327
  coolant analysis of, 8
Silicate esters, in Group V API base oil
    group, 30
Silicon. *See also* Dirt
  in additives in oil analysis, 52
  as contaminant, 90–92
  in coolant, 327
  in coolant leak, 174
  in gear oil additives, 56
  lubricant testing for, 139
  oil analysis for, 7
  in powershift transmission, 279–281, 282,
    284, 287, 288, 290–291
  powertrain oil analysis for, 279–281, 282,
    284, 287, 288, 290–291
Silt, hydraulic fluid analysis and, 215–216
Silver
  as contaminant, 104
  lubricant testing for, 139
  oil analysis for, 178
  wear tables for, 208
Society of Automotive Engineers (SAE),
    32–36
  ATF grades, 55–56
  gasoline standards of, 365
  gear oil grades, 56
Sodium. *See also* Dirt
  as contaminant, 89
  in coolant, 327
  in coolant leak, 173–174
  in DEF, 414
  grease, 79
  in grease thickeners, 71
  in lubricant testing, 137
  in oil analysis, 52
  wear tables for, 209

Solvency
  of API base oil groups, 32
  of biodiesel, 362
Solvents, for gasoline, 378
Soot, 109–110
  ATR for, 145
  from detergents, 110
  from dispersants, 110
  dispersants for, 50
  engine oil analysis for, 250–252, 262
  idle time and, 19
  lubricant testing for, 138, 145
  oil analysis for, 175
  oil sampling for, 130
Specific gravity, of diesel fuel, 358, 390–391
Specific heat, of coolant, 324
Splitter drive
  dirt in, 311–312
  powertrain oil analysis for, 275, 311–313
  wear tables for, 313
Standard deviation, 197–210
  cleaning and, 205–207
  critical values for, 208
  cumulative curve and, 201–202
  example for, 201
  Excel for, 205–210
  mean and, 203
  measurement and, 203
  median and, 203
  nonparametric curve and, 202
  parametric bell curve for, 200–202
Static current, filtration with, 99
Stoichiometric engines, 371–372
Sugar alcohol, in biodiesel, 362
Sulfate, in coolant, 337–338, 340
Sulfation
  engine oil analysis for, 259–260
  lubricant testing for, 157–158
  oil analysis for, 179
Sulfonates, in gear oils, 56
Sulfur
  in additives in oil analysis, 52
  in API base oil groups, 29
  in diesel fuel, 392
  engine oil analysis for, 260–261
  EP additives for, 49

  in fuel, 374
  injection pump failure and, 408
  in light crude, 358
  in petroleum, 28
  TAN and, 155
  TBN and, 60
  ULSD and, 158, 176, 373, 374, 392
  ZDDP and, 47
Supplemental coolant additives (SCAs), 322
  high concentrations of, 332, 335, 343,
    347–348
  low concentration of, 348–349
  for NOAT, 333
Surge tanks, coolant sampling and, 326
Synthetic blends (Group II API base oil
    group), 29, 30
  for gasoline engine oil, 39
Synthetic esters, in Group V API base oil
    group, 30
Synthetic oils. See Group IV API base oil
    group

TAN. See Total acid number
Tandem drive
  powertrain oil analysis for, 275, 314–315
  PQ index for, 159
  water in, 314–315
  wear tables for, 315
TBN. See Total base number
TCP. See Tricresyl phosphate
TDS. See Total dissolved solids
TEL. See Tetraethyl lead
Telematics
  CAN for, 20
  in CBM, 6, 17–21
Temperature. See also Heat; High-
    temperature, high-shear test; High-
    temperature, high-shear test (HTHS)
  additives and, 48
  coolant sampling and, 325
  foam inhibitor and, 46
  gasoline engine oil and, 39
  for gasoline production, 364–365
  grease and, 71–72
  high-temperature stability, of diesel fuel,
    385, 397–398

Temperature (*continued*)
  LT grease, 78
  oil analysis for, 182
  oil breakdown and, 40
  SAE J300 standard for, 35
Tetraethyl lead (TEL), 365
TGDI. *See* Turbocharged gasoline direct
    injection
Thickeners, for grease, 71, 80–81
Tin
  as contaminant, 104
  lubricant testing for, 139
  oil analysis for, 178
  wear tables for, 208
Tire pressure, telematics for, 21
Titanium
  as contaminant, 104
  hydraulic fluid analysis for, 235–236
  lubricant testing for, 139
  wear tables for, 208
Top tier gasoline, 377
Torque converters, 139
  lubricants for, 27
  powertrain oil analysis for, 275, 277
  viscosity shear-down, 58
  water in, 143
Total acid number (TAN), 59–60
  air filtration and, 98
  in axles, 187, 305–306
  contaminants and, 88
  for diesel fuel, 392
  engine oil analysis for, 256–257
  in final drive, 187
  gearbox and, 184
  in hydraulic system, 122, 182
  injection pump failure and, 408
  lubricant testing for, 137, 154–155
  oil analysis for, 176, 182, 184
  powershift transmission and, 184
  powertrain oil analysis for, 277, 305–306
  standard deviation of, 202
  weak acids and, 262–263
  wear tables for, 210
Total base number (TBN), 59–60
  for diesel fuel, 60, 154
  of diesel fuel, 392

engine oil analysis for, 256
lubricant testing for, 137, 153–154
oil analysis for, 176
standard deviation of, 202
weak acids and, 262–263
wear tables for, 210
Total dissolved solids (TDSs), in coolant,
    336, 339–340, 352–353
Tractor fluid
  hydraulic fluid analysis of, 219–220,
      226–227, 239–241
  lubricant testing of, 143
  particle count for, 106
  signature of, 55–56
  TAN for, 155
  viscosity of, 143
Trans fats, 29
Transmission fluids. *See also* Automatic
    transmission fluids
  optimization of, 65, 66
Triboelectric charging, 99
Tribology, for lubrication, 27
Tricresyl phosphate (TCP), 47
  in additives in oil analysis, 52
  in hydraulic fluid, 54
  in oil analysis, 52
Tropsch, Hans, 360–361
Turbocharged gasoline direct injection
    (TGDI), 368
Turbocharger. *See also* Variable-geometry
    turbochargers
  failure, 405–406

Ultra-low-sulfur diesel fuel (ULSD), 158,
    176, 373, 374, 392
  TBN for, 256
Ultrapressure liquid chromatography
    (UPLC), 336
Upgrading processes, in crude oil refining,
    359–360
UPLC. *See* Ultrapressure liquid
    chromatography
Urea, in DEF, 413, 417

Vacuum pump method, for oil sampling,
    132–133

Vanadium
    lubricant testing for, 139
    oil analysis for, 178
    in petroleum, 28
    wear tables for, 208
Variable-geometry turbochargers (VGTs), 158
    for diesel fuel exhaust, 411
Varnish
    bypass filter for, 127
    from filtration catalyst, 98
    in gasoline engines, 110
    in hydraulic fluid analysis, 232–233
    in hydraulic system, 126–127
    in oil breakdown, 40
VG. See Variable-geometry
VGTs. See Variable-geometry turbochargers
VI. See Viscosity index
Viscosity
    of API base oil groups, 29, 32
    of API Group I base oil group, 31
    in axles, 188
    of diesel fuel, 192–193, 361, 394
    of engine oil, 141–142, 148
    engine oil analysis for, 252–253
    in final drive, 188
    fluid sensors for, 19–20
    in gear oils, 142
    of gearbox, 184
    grease and, 70–71
    of hydraulic fluid, 53–54, 140–141
    hydraulic fluid analysis for, 236–237
    ISO grades of, 36–37
    lubricant testing for, 137, 138, 140–143
    lubrication film and, 113
    of multigrade oils, 41
    oil analysis for, 175, 184
    oil breakdown and, 40–41
    in powershift transmission, 184, 288–291
    powertrain oil analysis for, 288–291
    of SAE J300, 32
    solvency and, 31
    of tractor fluids, 143
    wear tables for, 210
Viscosity index (VI)
    of API base oil groups, 32
    ISO viscosity grades and, 36–37

    lubricant testing and, 140
    of MV hydraulic fluid, 55
    oil breakdown and, 41
Viscosity shear-down
    of gear oils, 59
    of hydraulic fluid, 58
    oil analysis for, 182
Visibility, in CBM, 12
Volatility, of gasoline engine oil, 39

Water
    in axles, 187, 293–294
    bearings and, 113–114
    in brakes, 293–294
    bypass filter for, 95
    as contaminant, 89, 113–114, 121–122
    in coolant, 331, 337–338, 340
    for copper etching, 51
    DEF and, 412–413
    in diesel fuel, 386–387
    engine oil analysis for, 250
    filter plugging from, 406
    filtration of, 99–102
    in final drive, 187
    fluid sensors for, 19
    in fuel sampling, 381–383
    in gearbox, 184
    in hydraulic fluid analysis, 216–217,
        242–244
    in hydraulic system, 121–122, 181, 211
    lubricant testing for, 137, 138, 143–145
    oil analysis for, 7, 181, 184
    oil breakdown and, 40
    particle count and, 106
    in powershift transmission, 184, 287–288
    powertrain oil analysis for, 276, 277,
        287–288, 293–294, 314–315
    specific heat of, 324
    in tandem drive, 314–315
Water inhibitors, 49
Water resistance (WR), grease, 77
Water washout test, for grease testing, 162
Wax, in Group II API base oils, 29
Wax antisettling agents, for diesel fuel, 376,
        377
Weak acids, engine oil analysis for, 262–263

Wear
  CBM and, 2–4, 14–15
  from contaminants, 86
  control of, 4
  detectability of, 14–15
  example of, 209
  lubrication for, 27
  size of, 3
  sources of, 2–4
  technology and, 3
Wear metals. *See also specific metals*
  in axles, 188–189
  contaminants and, 103–104
  in coolant, 337
  in final drive, 103, 188–189
  in gearbox, 184, 317–318
  in hydraulic fluid analysis, 214, 217, 221, 222
  in hydraulic system, 181
  lubricant testing for, 137, 138–139
  oil analysis for, 176–177, 181, 184
  of powershift transmission, 184
  powertrain oil analysis for, 276–277,
    317–318
Wear tables, 7, 197–210
  for aluminum, 208
  for axles, 291, 293, 294, 296, 297, 299,
    301, 302, 304, 306, 308
  for brakes, 291, 293, 294, 296, 297, 299,
    301
  CBM and, 204
  for final drive, 302, 304
  for gearbox, 314, 320
  for hydraulic fluid analysis, 213, 226, 228,
    236, 241
  information collection for, 204
  need for, 204

  for nickel, 208
  for particle count, 210
  for potassium, 209
  for powershift transmission, 278, 279,
    280–281, 282, 284, 285, 287, 288, 290,
    291
  for powertrain oil analysis, 278, 279,
    280–281, 282, 284, 285, 287, 288, 290,
    291, 293, 294, 296, 297, 299, 301, 302,
    304, 306, 308, 313
  for PQ index, 210
  for pump drives, 313
  for silver, 208
  for sodium, 209
  for splitter drive, 313
  for TAN, 210
  for tandem drive, 315
  for TBN, 210
  for tin, 208
  for titanium, 208
  for vanadium, 208
  for viscosity, 210
WR. *See* Water resistance
Wurtz, Charles Adolphe, 321

ZDDP. *See* Zinc dialykl diathiophosphate
Zinc
  in additives in oil analysis, 52
  in DEF, 414, 418
  in gasoline engine oil, 39
  in gear oil additives, 56
  lubricant testing for, 139
Zinc dialykl diathiophosphate (ZDDP),
    46–47
  in additives in oil analysis, 52
  in hydraulic fluid, 54

ABOUT THE AUTHORS

DIEGO NAVARRO

In his 45 years of experience in mobile equipment, which includes working with an equipment dealer, a manufacturer, and a construction company, Diego Navarro understands the demands of a fleet of equipment and the requirements for a successful operation. Diego undertook the development of the condition-based maintenance culture for John Deere Construction and Forestry and produced important maintenance initiatives, including launch of the first intelligent portable filter caddy on the market. Under his command, his team developed Fleet Care, a computerized system capable of processing oil analysis results combined with automated telematics and inspections. Diego has been a regular presenter at the Association of Equipment Management Professionals and Construction Equipment Management Program symposia, as well as at Conexpo and Noria Corporation conferences. Diego retired from John Deere in 2014 after 38 years of service. His last position was global service marketing manager. Diego is an aeronautical technician and has the level 1 certified oil monitoring analyst certification from the Society of Tribologists and Lubrication Engineers.

BLAINE BALLENTINE

Blaine Ballentine is a formulator for heavy-duty lubricant manufacturer Central Petroleum Company with more than 40 years of experience. His career started right out of college as the company's youngest lubricant sales representative, and he became the company's youngest district sales manager two years later. After returning to Iowa State University for an MBA, Blaine was called into the office to help with marketing, but he became more interested in the products themselves. His research allowed him to improve the quality of his company's products as new

technology or information became available, and he created many specialty products along the way for those with special lubrication needs. Blaine was also instrumental in helping his employer achieve two U.S. patents for fuel additive technology. Although Central Petroleum is primarily a heavy-duty lubricant manufacturer, the company has gained much wider recognition from one of its specialty product lines that Blaine formulated—racing oils. In the sport of truck and tractor pulling, over many years and classes, CenPeCo lubricant customers have won more than 230 national points championships.